C0-GMF-017
NORTH DAKOTA
STATE UNIVERSITY

SEP 3 0 1998

SERIALS DEPT.
LIBRARY

WITHDRAWN

STABILITY BY LIAPUNOV'S MATRIX FUNCTION METHOD WITH APPLICATIONS

PURE AND APPLIED MATHEMATICS

A Program of Monographs, Textbooks, and Lecture Notes

EXECUTIVE EDITORS

Earl J. Taft
Rutgers University
New Brunswick, New Jersey

Zuhair Nashed
University of Delaware
Newark, Delaware

EDITORIAL BOARD

M. S. Baouendi
University of California,
San Diego

Jane Cronin
Rutgers University

Jack K. Hale
Georgia Institute of Technology

S. Kobayashi
University of California,
Berkeley

Marvin Marcus
University of California,
Santa Barbara

W. S. Massey
Yale University

Anil Nerode
Cornell University

Donald Passman
University of Wisconsin,
Madison

Fred S. Roberts
Rutgers University

Gian-Carlo Rota
Massachusetts Institute of
Technology

David L. Russell
Virginia Polytechnic Institute
and State University

Walter Schempp
Universität Siegen

Mark Teply
University of Wisconsin,
Milwaukee

MONOGRAPHS AND TEXTBOOKS IN PURE AND APPLIED MATHEMATICS

1. *K. Yano,* Integral Formulas in Riemannian Geometry (1970)
2. *S. Kobayashi,* Hyperbolic Manifolds and Holomorphic Mappings (1970)
3. *V. S. Vladimirov,* Equations of Mathematical Physics (A. Jeffrey, ed.; A. Littlewood, trans.) (1970)
4. *B. N. Pshenichnyi,* Necessary Conditions for an Extremum (L. Neustadt, translation ed.; K. Makowski, trans.) (1971)
5. *L. Narici et al.,* Functional Analysis and Valuation Theory (1971)
6. *S. S. Passman,* Infinite Group Rings (1971)
7. *L. Dornhoff,* Group Representation Theory. Part A: Ordinary Representation Theory. Part B: Modular Representation Theory (1971, 1972)
8. *W. Boothby and G. L. Weiss, eds.,* Symmetric Spaces (1972)
9. *Y. Matsushima,* Differentiable Manifolds (E. T. Kobayashi, trans.) (1972)
10. *L. E. Ward, Jr.,* Topology (1972)
11. *A. Babakhanian,* Cohomological Methods in Group Theory (1972)
12. *R. Gilmer,* Multiplicative Ideal Theory (1972)
13. *J. Yeh,* Stochastic Processes and the Wiener Integral (1973)
14. *J. Barros-Neto,* Introduction to the Theory of Distributions (1973)
15. *R. Larsen,* Functional Analysis (1973)
16. *K. Yano and S. Ishihara,* Tangent and Cotangent Bundles (1973)
17. *C. Procesi,* Rings with Polynomial Identities (1973)
18. *R. Hermann,* Geometry, Physics, and Systems (1973)
19. *N. R. Wallach,* Harmonic Analysis on Homogeneous Spaces (1973)
20. *J. Dieudonné,* Introduction to the Theory of Formal Groups (1973)
21. *I. Vaisman,* Cohomology and Differential Forms (1973)
22. *B.-Y. Chen,* Geometry of Submanifolds (1973)
23. *M. Marcus,* Finite Dimensional Multilinear Algebra (in two parts) (1973, 1975)
24. *R. Larsen,* Banach Algebras (1973)
25. *R. O. Kujala and A. L. Vitter, eds.,* Value Distribution Theory: Part A; Part B: Deficit and Bezout Estimates by Wilhelm Stoll (1973)
26. *K. B. Stolarsky,* Algebraic Numbers and Diophantine Approximation (1974)
27. *A. R. Magid,* The Separable Galois Theory of Commutative Rings (1974)
28. *B. R. McDonald,* Finite Rings with Identity (1974)
29. *J. Satake,* Linear Algebra (S. Koh et al., trans.) (1975)
30. *J. S. Golan,* Localization of Noncommutative Rings (1975)
31. *G. Klambauer,* Mathematical Analysis (1975)
32. *M. K. Agoston,* Algebraic Topology (1976)
33. *K. R. Goodearl,* Ring Theory (1976)
34. *L. E. Mansfield,* Linear Algebra with Geometric Applications (1976)
35. *N. J. Pullman,* Matrix Theory and Its Applications (1976)
36. *B. R. McDonald,* Geometric Algebra Over Local Rings (1976)
37. *C. W. Groetsch,* Generalized Inverses of Linear Operators (1977)
38. *J. E. Kuczkowski and J. L. Gersting,* Abstract Algebra (1977)
39. *C. O. Christenson and W. L. Voxman,* Aspects of Topology (1977)
40. *M. Nagata,* Field Theory (1977)
41. *R. L. Long,* Algebraic Number Theory (1977)
42. *W. F. Pfeffer,* Integrals and Measures (1977)
43. *R. L. Wheeden and A. Zygmund,* Measure and Integral (1977)
44. *J. H. Curtiss,* Introduction to Functions of a Complex Variable (1978)
45. *K. Hrbacek and T. Jech,* Introduction to Set Theory (1978)
46. *W. S. Massey,* Homology and Cohomology Theory (1978)
47. *M. Marcus,* Introduction to Modern Algebra (1978)
48. *E. C. Young,* Vector and Tensor Analysis (1978)
49. *S. B. Nadler, Jr.,* Hyperspaces of Sets (1978)
50. *S. K. Segal,* Topics in Group Kings (1978)
51. *A. C. M. van Rooij,* Non-Archimedean Functional Analysis (1978)
52. *L. Corwin and R. Szczarba,* Calculus in Vector Spaces (1979)
53. *C. Sadosky,* Interpolation of Operators and Singular Integrals (1979)
54. *J. Cronin,* Differential Equations (1980)

55. *C. W. Groetsch,* Elements of Applicable Functional Analysis (1980)
56. *I. Vaisman,* Foundations of Three-Dimensional Euclidean Geometry (1980)
57. *H. I. Freedan,* Deterministic Mathematical Models in Population Ecology (1980)
58. *S. B. Chae,* Lebesgue Integration (1980)
59. *C. S. Rees et al.,* Theory and Applications of Fourier Analysis (1981)
60. *L. Nachbin,* Introduction to Functional Analysis (R. M. Aron, trans.) (1981)
61. *G. Orzech and M. Orzech,* Plane Algebraic Curves (1981)
62. *R. Johnsonbaugh and W. E. Pfaffenberger,* Foundations of Mathematical Analysis (1981)
63. *W. L. Voxman and R. H. Goetschel,* Advanced Calculus (1981)
64. *L. J. Corwin and R. H. Szczarba,* Multivariable Calculus (1982)
65. *V. I. Istrățescu,* Introduction to Linear Operator Theory (1981)
66. *R. D. Järvinen,* Finite and Infinite Dimensional Linear Spaces (1981)
67. *J. K. Beem and P. E. Ehrlich,* Global Lorentzian Geometry (1981)
68. *D. L. Armacost,* The Structure of Locally Compact Abelian Groups (1981)
69. *J. W. Brewer and M. K. Smith, eds.,* Emmy Noether: A Tribute (1981)
70. *K. H. Kim,* Boolean Matrix Theory and Applications (1982)
71. *T. W. Wieting,* The Mathematical Theory of Chromatic Plane Ornaments (1982)
72. *D. B.Gauld,* Differential Topology (1982)
73. *R. L. Faber,* Foundations of Euclidean and Non-Euclidean Geometry (1983)
74. *M. Carmeli,* Statistical Theory and Random Matrices (1983)
75. *J. H. Carruth et al.,* The Theory of Topological Semigroups (1983)
76. *R. L. Faber,* Differential Geometry and Relativity Theory (1983)
77. *S. Barnett,* Polynomials and Linear Control Systems (1983)
78. *G. Karpilovsky,* Commutative Group Algebras (1983)
79. *F. Van Oystaeyen and A. Verschoren,* Relative Invariants of Rings (1983)
80. *I. Vaisman,* A First Course in Differential Geometry (1984)
81. *G. W. Swan,* Applications of Optimal Control Theory in Biomedicine (1984)
82. *T. Petrie and J. D. Randall,* Transformation Groups on Manifolds (1984)
83. *K. Goebel and S. Reich,* Uniform Convexity, Hyperbolic Geometry, and Nonexpansive Mappings (1984)
84. *T. Albu and C. Năstăsescu,* Relative Finiteness in Module Theory (1984)
85. *K. Hrbacek and T. Jech,* Introduction to Set Theory: Second Edition (1984)
86. *F. Van Oystaeyen and A. Verschoren,* Relative Invariants of Rings (1984)
87. *B. R. McDonald,* Linear Algebra Over Commutative Rings (1984)
88. *M. Namba,* Geometry of Projective Algebraic Curves (1984)
89. *G. F. Webb,* Theory of Nonlinear Age-Dependent Population Dynamics (1985)
90. *M. R. Bremner et al.,* Tables of Dominant Weight Multiplicities for Representations of Simple Lie Algebras (1985)
91. *A. E. Fekete,* Real Linear Algebra (1985)
92. *S. B. Chae,* Holomorphy and Calculus in Normed Spaces (1985)
93. *A. J. Jerri,* Introduction to Integral Equations with Applications (1985)
94. *G. Karpilovsky,* Projective Representations of Finite Groups (1985)
95. *L. Narici and E. Beckenstein,* Topological Vector Spaces (1985)
96. *J. Weeks,* The Shape of Space (1985)
97. *P. R. Gribik and K. O. Kortanek,* Extremal Methods of Operations Research (1985)
98. *J.-A. Chao and W. A. Woyczynski, eds.,* Probability Theory and Harmonic Analysis (1986)
99. *G. D. Crown et al.,* Abstract Algebra (1986)
100. *J. H. Carruth et al.,* The Theory of Topological Semigroups, Volume 2 (1986)
101. *R. S. Doran and V. A. Belfi,* Characterizations of C*-Algebras (1986)
102. *M. W. Jeter,* Mathematical Programming (1986)
103. *M. Altman,* A Unified Theory of Nonlinear Operator and Evolution Equations with Applications (1986)
104. *A. Verschoren,* Relative Invariants of Sheaves (1987)
105. *R. A. Usmani,* Applied Linear Algebra (1987)
106. *P. Blass and J. Lang,* Zariski Surfaces and Differential Equations in Characteristic $p > 0$ (1987)
107. *J. A. Reneke et al.,* Structured Hereditary Systems (1987)
108. *H. Busemann and B. B. Phadke,* Spaces with Distinguished Geodesics (1987)
109. *R. Harte,* Invertibility and Singularity for Bounded Linear Operators (1988)
110. *G. S. Ladde et al.,* Oscillation Theory of Differential Equations with Deviating Arguments (1987)

111. *L. Dudkin et al.*, Iterative Aggregation Theory (1987)
112. *T. Okubo*, Differential Geometry (1987)
113. *D. L. Stancl and M. L. Stancl,* Real Analysis with Point-Set Topology (1987)
114. *T. C. Gard,* Introduction to Stochastic Differential Equations (1988)
115. *S. S. Abhyankar,* Enumerative Combinatorics of Young Tableaux (1988)
116. *H. Strade and R. Farnsteiner,* Modular Lie Algebras and Their Representations (1988)
117. *J. A. Huckaba,* Commutative Rings with Zero Divisors (1988)
118. *W. D. Wallis,* Combinatorial Designs (1988)
119. *W. Więsław,* Topological Fields (1988)
120. *G. Karpilovsky,* Field Theory (1988)
121. *S. Caenepeel and F. Van Oystaeyen,* Brauer Groups and the Cohomology of Graded Rings (1989)
122. *W. Kozlowski,* Modular Function Spaces (1988)
123. *E. Lowen-Colebunders,* Function Classes of Cauchy Continuous Maps (1989)
124. *M. Pavel,* Fundamentals of Pattern Recognition (1989)
125. *V. Lakshmikantham et al.,* Stability Analysis of Nonlinear Systems (1989)
126. *R. Sivaramakrishnan,* The Classical Theory of Arithmetic Functions (1989)
127. *N. A. Watson,* Parabolic Equations on an Infinite Strip (1989)
128. *K. J. Hastings,* Introduction to the Mathematics of Operations Research (1989)
129. *B. Fine,* Algebraic Theory of the Bianchi Groups (1989)
130. *D. N. Dikranjan et al.,* Topological Groups (1989)
131. *J. C. Morgan II,* Point Set Theory (1990)
132. *P. Biler and A. Witkowski,* Problems in Mathematical Analysis (1990)
133. *H. J. Sussmann,* Nonlinear Controllability and Optimal Control (1990)
134. *J.-P. Florens et al.,* Elements of Bayesian Statistics (1990)
135. *N. Shell,* Topological Fields and Near Valuations (1990)
136. *B. F. Doolin and C. F. Martin,* Introduction to Differential Geometry for Engineers (1990)
137. *S. S. Holland, Jr.,* Applied Analysis by the Hilbert Space Method (1990)
138. *J. Okniński,* Semigroup Algebras (1990)
139. *K. Zhu,* Operator Theory in Function Spaces (1990)
140. *G. B. Price,* An Introduction to Multicomplex Spaces and Functions (1991)
141. *R. B. Darst,* Introduction to Linear Programming (1991)
142. *P. L. Sachdev,* Nonlinear Ordinary Differential Equations and Their Applications (1991)
143. *T. Husain,* Orthogonal Schauder Bases (1991)
144. *J. Foran,* Fundamentals of Real Analysis (1991)
145. *W. C. Brown,* Matrices and Vector Spaces (1991)
146. *M. M. Rao and Z. D. Ren,* Theory of Orlicz Spaces (1991)
147. *J. S. Golan and T. Head,* Modules and the Structures of Rings (1991)
148. *C. Small,* Arithmetic of Finite Fields (1991)
149. *K. Yang,* Complex Algebraic Geometry (1991)
150. *D. G. Hoffman et al.,* Coding Theory (1991)
151. *M. O. González,* Classical Complex Analysis (1992)
152. *M. O. González,* Complex Analysis (1992)
153. *L. W. Baggett,* Functional Analysis (1992)
154. *M. Sniedovich,* Dynamic Programming (1992)
155. *R. P. Agarwal,* Difference Equations and Inequalities (1992)
156. *C. Brezinski,* Biorthogonality and Its Applications to Numerical Analysis (1992)
157. *C. Swartz,* An Introduction to Functional Analysis (1992)
158. *S. B. Nadler, Jr.,* Continuum Theory (1992)
159. *M. A. Al-Gwaiz,* Theory of Distributions (1992)
160. *E. Perry,* Geometry: Axiomatic Developments with Problem Solving (1992)
161. *E. Castillo and M. R. Ruiz-Cobo,* Functional Equations and Modelling in Science and Engineering (1992)
162. *A. J. Jerri,* Integral and Discrete Transforms with Applications and Error Analysis (1992)
163. *A. Charlier et al.,* Tensors and the Clifford Algebra (1992)
164. *P. Biler and T. Nadzieja,* Problems and Examples in Differential Equations (1992)
165. *E. Hansen,* Global Optimization Using Interval Analysis (1992)
166. *S. Guerre-Delabrière,* Classical Sequences in Banach Spaces (1992)
167. *Y. C. Wong,* Introductory Theory of Topological Vector Spaces (1992)

168. *S. H. Kulkarni and B. V. Limaye*, Real Function Algebras (1992)
169. *W. C. Brown*, Matrices Over Commutative Rings (1993)
170. *J. Loustau and M. Dillon*, Linear Geometry with Computer Graphics (1993)
171. *W. V. Petryshyn*, Approximation-Solvability of Nonlinear Functional and Differential Equations (1993)
172. *E. C. Young*, Vector and Tensor Analysis: Second Edition (1993)
173. *T. A. Bick*, Elementary Boundary Value Problems (1993)
174. *M. Pavel*, Fundamentals of Pattern Recognition: Second Edition (1993)
175. *S. A. Albeverio et al.*, Noncommutative Distributions (1993)
176. *W. Fulks*, Complex Variables (1993)
177. *M. M. Rao*, Conditional Measures and Applications (1993)
178. *A. Janicki and A. Weron*, Simulation and Chaotic Behavior of α-Stable Stochastic Processes (1994)
179. *P. Neittaanmäki and D. Tiba*, Optimal Control of Nonlinear Parabolic Systems (1994)
180. *J. Cronin*, Differential Equations: Introduction and Qualitative Theory, Second Edition (1994)
181. *S. Heikkilä and V. Lakshmikantham*, Monotone Iterative Techniques for Discontinuous Nonlinear Differential Equations (1994)
182. *X. Mao*, Exponential Stability of Stochastic Differential Equations (1994)
183. *B. S. Thomson*, Symmetric Properties of Real Functions (1994)
184. *J. E. Rubio*, Optimization and Nonstandard Analysis (1994)
185. *J. L. Bueso et al.*, Compatibility, Stability, and Sheaves (1995)
186. *A. N. Michel and K. Wang*, Qualitative Theory of Dynamical Systems (1995)
187. *M. R. Darnel*, Theory of Lattice-Ordered Groups (1995)
188. *Z. Naniewicz and P. D. Panagiotopoulos*, Mathematical Theory of Hemivariational Inequalities and Applications (1995)
189. *L. J. Corwin and R. H. Szczarba*, Calculus in Vector Spaces: Second Edition (1995)
190. *L. H. Erbe et al.*, Oscillation Theory for Functional Differential Equations (1995)
191. *S. Agaian et al.*, Binary Polynomial Transforms and Nonlinear Digital Filters (1995)
192. *M. I. Gil'*, Norm Estimations for Operation-Valued Functions and Applications (1995)
193. *P. A. Grillet*, Semigroups: An Introduction to the Structure Theory (1995)
194. *S. Kichenassamy*, Nonlinear Wave Equations (1996)
195. *V. F. Krotov*, Global Methods in Optimal Control Theory (1996)
196. *K. I. Beidar et al.*, Rings with Generalized Identities (1996)
197. *V. I. Arnautov et al.*, Introduction to the Theory of Topological Rings and Modules (1996)
198. *G. Sierksma*, Linear and Integer Programming (1996)
199. *R. Lasser*, Introduction to Fourier Series (1996)
200. *V. Sima*, Algorithms for Linear-Quadratic Optimization (1996)
201. *D. Redmond*, Number Theory (1996)
202. *J. K. Beem et al.*, Global Lorentzian Geometry: Second Edition (1996)
203. *M. Fontana et al.*, Prüfer Domains (1997)
204. *H. Tanabe*, Functional Analytic Methods for Partial Differential Equations (1997)
205. *C. Q. Zhang*, Integer Flows and Cycle Covers of Graphs (1997)
206. *E. Spiegel and C. J. O'Donnell*, Incidence Algebras (1997)
207. *B. Jakubczyk and W. Respondek*, Geometry of Feedback and Optimal Control (1998)
208. *T. W. Haynes et al.*, Fundamentals of Domination in Graphs (1998)
209. *T. W. Haynes et al.*, Domination in Graphs: Advanced Topics (1998)
210. *L. A. D'Alotto et al.*, A Unified Signal Algebra Approach to Two-Dimensional Parallel Digital Signal Processing (1998)
211. *F. Halter-Koch*, Ideal Systems (1998)
212. *N. K. Govil et al.*, Approximation Theory (1998)
213. *R. Cross*, Multivalued Linear Operators (1998)
214. *A. A. Martynyuk*, Stability by Liapunov's Matrix Function Method with Applications (1998)

Additional Volumes in Preparation

STABILITY BY LIAPUNOV'S MATRIX FUNCTION METHOD WITH APPLICATIONS

A. A. Martynyuk
Institute of Mechanics
National Academy of Sciences of Ukraine
Kyïv, Ukraine

MARCEL DEKKER, INC. NEW YORK • BASEL • HONG KONG

Library of Congress Cataloging-in-Publication Data

Martynyuk, A. A. (Anatolil Andreevich)
Stability by Liapunov's matrix function method with applications / A. A. Martynyuk.
p. cm.-(Monographs and textbooks in pure and applied mathematics; 214)
Includes bibliographical references and indexes.
ISBN 0-8247-0191-7 (alk. paper)
1. Lyapunov stability. I. Title. II. Series.
QA871.M327 1998
003'.75-dc21 98-28045
CIP

This book is printed on acid-free paper.

QA
871
.M327
1998

Headquarters
Marcel Dekker, Inc.
270 Madison Avenue, New York, NY 10016
tel: 212-696-9000; fax: 212-685-4540

Eastern Hemisphere Distribution
Marcel Dekker AG
Hutgasse 4, Postfach 812, CH-4001 Basel, Switzerland
tel: 44-61-261-8482; fax: 44-61-261-8896

World Wide Web
http://www.dekker.com

The publisher offers discounts on this book when ordered in bulk quantities. For more information, write to Special Sales/Professional Marketing at the headquarters address above.

Copyright © 1998 by Marcel Dekker, Inc. All Rights Reserved.

Neither this book nor any part may be reproduced or transmitted in any form or by any means, electronic or mechanical, including photocopying, microfilming, and recording, or by any information storage and retrieval system, without permission in writing from the publisher.

Current printing (last digit):
10 9 8 7 6 5 4 3 2 1

PRINTED IN THE UNITED STATES OF AMERICA

DEDICATION

To the memory of my mother
Tat'jana Fomovna Martynyuk
(1901 - 1991)
and my father
Andrej Gerasimovich Martynyuk
(1900 - 1996)

PREFACE

One can hardly name a branch of natural science or technology in which the problems of stability do not claim the attention of scholars, engineers, and experts who investigate natural phenomena or operate designed machines or systems. If, for a process or a phenomenon, for example, atom oscillations or a supernova explosion, a mathematical model is constructed in the form of a system of differential equations, the investigation of the latter is possible either by a direct (numerical as a rule) integration of the equations or by its analysis by qualitative methods.

The direct Liapunov method based on scalar auxiliary function proves to be a powerful technique of qualitative analysis of the real world phenomena.

This volume examines new generalizations of the matrix-valued auxiliary function. Moreover the matrix-valued function is a structure the elements of which compose both scalar and vector Liapunov functions applied in the stability analysis of nonlinear systems.

Due to the concept of matrix-valued function developed in the book, the direct Liapunov method becomes yet more versatile in performing the analysis of nonlinear systems dynamics.

The possibilities of the generalized direct Liapunov method are opened up to stability analysis of solutions to ordinary differential equations, singularly perturbed systems, and systems with random parameters.

The reader with an understanding of fundamentals of differential equations theory, elements of motion stability theory, mathematical analysis, and linear algebra should not be confused by the many formulas in the book. Each of these subjects is a part of the mathematics curriculum of any university.

In view of the fact that beginners in motion stability theory usually face some difficulties in its practical application, the sets of problems taken from various branches of natural sciences and technology are solved at the end of each chapter. The problems of independent value are integrated in Chapter 5.

A certain contribution to the development of the Liapunov matrix function method has been made by the scientists and experts of Belgrade University, Technical University in Zurich, and Stability of Processes Department of Institute of Mechanics National Academy of Sciences of Ukraine.

The useful remarks by the reviewers of Marcel Dekker, Inc., have been taken into account in the final version of the book. Great assistance in preparing the manuscript for publication has been rendered by S.N. Rasshivalova, L.N. Chernetzkaya, A.N. Chernienko, and V.I. Goncharenko. The author expresses his sincere gratitude to all these persons.

A. A. Martynyuk

CONTENTS

NOTATION

R the set of all real numbers

$R_+ = [0, +\infty) \subset R$ the set of all nonnegative numbers

R^k k-th dimensional real vector space

$R \times R^n$ the Cartesian product of R and R^n

$\mathcal{T} = [-\infty, +\infty] = \{t\colon -\infty \leq t \leq +\infty\}$ the largest time interval

$\mathcal{T}_\tau = [\tau, +\infty) = \{t\colon \tau \leq t < +\infty\}$ the right semi-open unbounded interval associated with τ

$\mathcal{T}_i \subseteq R$ a time interval of all initial moments to under consideration (or, all admissible t_0)

$\mathcal{T}_0 = [t_0, +\infty) = \{t\colon t_0 \leq t < +\infty\}$ the right semi-open unbounded interval associated with t_0

$\|x\|$ the Euclidean norm of x in R^n

$\chi(t; t_0, x_0)$ a motion of a system at $t \in R$ iff $x(t_0) = x_0$, $\chi(t_0; t_0, x_0) \equiv x_0$

$B_\varepsilon = \{x \in R^n\colon \|x\| < \varepsilon\}$ open ball with center at the origin and radius $\varepsilon > 0$

$\delta_M(t_0, \varepsilon) = \max\{\delta\colon \delta = \delta(t_0, \varepsilon) \ni x_0 \in B_\delta(t_0, \varepsilon) \Rightarrow \chi(t; t_0, x_0) \in B_\varepsilon, \forall t \in \mathcal{T}_0\}$ the maximal δ obeying the definition of stability

$\Delta_M(t_0) = \max\{\Delta\colon \Delta = \Delta(t_0),\ \forall \rho > 0,\ \forall x_0 \in B_\Delta,\ \exists \tau(t_0, x_0, \rho) \in (0, +\infty) \ni \chi(t; t_0, x_0) \in B_\rho,\ \forall t \in \mathcal{T}_\tau\}$ the maximal Δ obeying the definition of attractivity

$\tau_m(t_0, x_0, \rho) = \min\{\tau\colon \tau = \tau(t_0, x_0, \rho) \ni \chi(t; t_0, x_0) \in B_\rho,\ \forall t \in \mathcal{T}_\tau\}$ the minimal τ satisfying the definition of attractivity

$\mathcal{N}$ a time-invariant neighborhood of original of R^n

$f\colon R \times \mathcal{N} \to R^n$ a vector function mapping $R \times \mathcal{N}$ into R^n

$C(\mathcal{T}_\tau \times \mathcal{N})$ the family of all functions continuous on $\mathcal{T}_\tau \times \mathcal{N}$

$C^{(i,j)}(\mathcal{T}_\tau \times \mathcal{N})$ the family of all functions i-times differentiable on $\mathcal{T}_\tau$ and j-times differentiable on $\mathcal{N}$

$D^+v(t, x)\,(D^-v(t, x))$ the upper right (left) Dini derivative of v along $\chi(t; t_0, x_0)$ at (t, x)

$D_+v(t,x)\,(D_-v(t,x))$ the lower right (left) Dini derivative of v along $\chi(t;t_0,x_0)$ at (t,x)

$D^*v(t,x)$ denotes that both $D^+v(t,x)$ and $D_+v(t,x)$ can be used

$Dv(t,x)$ the Eulerian derivative of v along $\chi(t;t_0,x_0)$ at (t,x)

$\lambda_i(\cdot)$ the i-th eigenvalue of a matrix $(\cdot)$

$\lambda_M(\cdot)$ the maximal eigenvalue of a matrix $(\cdot)$

$\lambda_m(\cdot)$ the minimal eigenvalue of a matrix $(\cdot)$

1

PRELIMINARIES

1.1 Introduction

Nonlinear dynamics of systems is a branch of science that studies actual equilibriums and motions of natural or artificial real objects. However it is known that hardly every state of a really functioning system is observed in practice that corresponds to a mathematically strict solution of either equilibrium or differential motion equations. It has been found out that only those equilibriums and motions of real systems are evident that possess certain "resistivity" to the outer perturbations. The equilibrium states and motions of this kind are referred to as stable while the others are called unstable.

The notion of stability had been clearly intuited but difficult to formulate and only Liapunov (see Liapunov [101]) managed to give accurate definitions (for the historical aspect see Moiseev [146]).

Section 1.2 presents recent strict definitions of stability of nonautonomous systems and other general information necessary for proper understanding of the monograph. Presently there is a series of monographs and textbooks that expose the direct Liapunov method of motion stability investigation based on auxiliary scalar function and provide a lot of many illustrative examples of its application. The books by Chetaev [19], Malkin [107], Lur'e [104], Duboshin [32], Demidovich [24], Krasovskii [89], Barbashin [10], Zubov [177], Letov [99], Bellman [15], Hahn [66], Harris and Milles [68], Yoshizawa [174], LaSalle and Lefschetz [98], Coppel [23], Lakshmikantham, Leela and Martynyuk [94] and others show the modern level of Liapunov method development in qualitative theory of equations.

Section 1.3 (subsection 1.3.1) gives a brief account of results obtained in this direction.

In 1962 it was proposed by Bellman [16], Martosov [132], and Melnikov [139] to apply Liapunov functions consisting of more than one component. Such functions were referred to as vector Liapunov functions. A quick development of investigations in the field has been summarized in a series of monographs such as in Grujić [55], Michel and Miller [143], Šiljak [167], Rouche, Habets and Laloy [159], LaSalle [97], Grujić, Martynyuk and Ribbens-Pavella [57], Lakshmikantham, Matrosov and Sivasundaram [96], Abdullin, Anapolskii et al. [1].

Section 1.3 (subsection 1.3.2) provides a short survey of the direct Liapunov method development in terms of vector function.

The preliminary information and the survey of the direct Liapunov method development in terms of both scalar and vector auxiliary functions are cited here with the aim to prepare the reader to the study of a new method in qualitative theory of equations called the method of matrix Liapunov functions.

1.2 On Definition of Stability

1.2.1 Liapunov's original definition

Liapunov started his investigations with the following (see Liapunov [101], p.11):

> Let us consider any material system with k degrees of freedom. Let $q_1, q_2, \dots, q_k$ be k independent variables, which we use to determine its position.
>
> We shall assume that quantities taking real values for all real system positions are taken for such variables.
>
> Considering the mentioned variables as functions in time t we shall denote their first time derivatives by $q_1', q_2', \dots, q_k'$.
>
> In every dynamic problem, in which forces are prespecified in a certain way, such functions will satisfy some k second order differential equations.
>
> Let any particular solution for such equations be found
>
> $$q_1 = f_1(t),\ q_2 = f_2(t),\ \dots,\ q_k = f_k(t),$$
>
> in which the quantities q_j are expressed as real functions in t, which at every t give only possible values to them.[1]

[1] It can happen that the quantities q_j by their choice do not take all real values but only those not greater than – and not less than certain bounds.

To that particular solution will correspond a definite motion of our system. Comparing it in a known sense with others, which are possible under the same forces, we shall call that motion *unperturbed*, and all others, with which it is compared, *perturbed*.

For t_0 understood a given instant, let us denote the values corresponding to it of quantities q_j, q'_j along any motion with q_{j0}, q'_{j0}.

Let

$$q_{10} = f_1(t_0) + \varepsilon_1, q_{20} = f_2(t_0) + \varepsilon_2, \ldots, q_{ko} = f_k(t_0) + \varepsilon_k,$$
$$q'_{10} = f'_1(t_0) + \varepsilon'_1, q'_{20} = f'_2(t_0) + \varepsilon'_2, \ldots, q'_{k0} = f'_2(t_k) + \varepsilon'_k,$$

where ε_j, ε'_j are real-valued constants.

Prespecifying the constants, which will be called *perturbations*, a perturbed motion is determined. We shall assume that we may prescribe them every number sufficiently small.

By speaking about perturbed motions, *close* to the unperturbed one, we shall comprehend motions, for which the perturbations are numerically small.

Let $Q_1, Q_2, \ldots, Q_n$ be any given continuous real-valued functions of quantities

$$q_1, q_2, \ldots, q_k, \quad q'_1, q'_2, \ldots, q'_k.$$

Along the unperturbed motion they become known functions of t, which will be denoted by $F_1, F_2, \ldots, F_n$. Along a perturbed motion they will be functions of quantities

$$t, \ \varepsilon_1, \ \varepsilon_2, \ \ldots, \ \varepsilon_k, \ \varepsilon'_1, \ \varepsilon'_2, \ \ldots, \ \varepsilon'_k.$$

When all ε_j, ε'_j are equal to zero, then quantities

$$Q_1 - F_1, \ Q_2 - F_2, \ \ldots, \ Q_n - F_n$$

will be equal to zero for every t. However, if the constants ε_j, ε'_j are not zero, but all are infinitely small, then a question rises: is it possible to specify such the latter never become grater than their values?

A solution of the question, which is the topic of our investigations, depends on both a character of the considered unperturbed motion and a choice of the functions $Q_1, Q_2, \ldots, Q_n$ and the instant

t_0. Under a specific choice of the latter, the reply to the question, respectively, will characterize in some sense the unperturbed motion, by determining a feature of the latter, which will be called *stability*, or that contrary to it, will be called *instability*.

We shall be exclusively interested in those cases in which the solution of the considered question does not depend on a choice of the instant t_0, when perturbations are acting. Thus we accept the following definition.

Let $L_1, L_2, \ldots, L_n$ be arbitrary given positive numbers. If all L_s, regardless of how small they are, can be selected positive numbers $E_1, E_2, \ldots, E_k$, $E'_1, E'_2, \ldots, E'_k$ so that for all real ε_j, ε'_j, satisfying the conditions[1]

$$|\varepsilon_j| \leq E_j, \ |\varepsilon'_j| \leq E'_j \qquad (j = 1, 2, \ldots, k),$$

and for all t, greater than t_0, the inequalities

$$|Q_1 - F_1| < L_1, \ |Q_2 - F_2| < L_2, \ \ldots, \ |Q_n - F_n| < L_n,$$

are satisfied, then the unperturbed motion is stable with respect to the quantities $Q_1, Q_2, \ldots, Q_n$; otherwise it is unstable with respect to the same quantities.

1.2.2 Comments on Liapunov's original definition

COMMENT 1.2.1. The inequalities on $|\varepsilon_j|$ and $|\varepsilon'_j|$ are weak and those on $|Q_j - F_j|$ are strong. This asymmetry is usually avoided imposing the same type of inequalities on all $|\varepsilon_j|, |\varepsilon'_j|$ and $|Q_j - F_j|$, which yields stability definitions equivalent to Liapunov's original definition. This equivalence can be easily proved.

COMMENT 1.2.2. Stability of the reference motion was defined by Liapunov with respect to arbitrary functions Q_j that are continuous in all q_i, q'_i. This has been very thoughtful and physically important because Q_j can represent energy or material flow. In this connection Liapunov introduced new variables x_i,

$$x_i = Q_i - F_i, \qquad i = 1, 2, \ldots, n,$$

[1] In general $|x|$ means the absolute value of a real-, or modulus of a complex quantity x.

and accepted the following (Liapunov [101], p.15):

> We shall assume that the number n and the functions Q_s, are such, that the order of the system is n and that it is reducible to the normal form
>
> $$\frac{dx_1}{dt} = X_1, \quad \frac{dx_2}{dt} = X_2, \quad \ldots, \quad \frac{dx_n}{dt} = X_n, \tag{1}$$
>
> and everywhere in the sequel we shall consider these last equations, calling them the differential equations of a perturbed motion.
>
> All X_s in the equations (1) are known functions of quantities
>
> $$x_1, x_2, \ldots, x_n, \ t,$$
>
> vanishing for
>
> $$x_1 = x_2 = \cdots = x_n = 0.$$

COMMENT 1.2.3. Stability of the reference motion requires arbitrary closeness of the perturbed motions to the reference motion provided their sufficient closeness is assured at the initial instant t_0.

COMMENT 1.2.4. The closeness of the perturbed motions to the reference motion is to be realized over unbounded time interval $\mathcal{T}_0^* = (t_0, +\infty]$, i.e. for *all* t *greater* than t_0. This point has been commonly neglected in the literature. Namely, the closeness has been commonly required either on $\overline{\mathcal{T}}_0 = [t_0, +\infty]$ or on $\mathcal{T}_0 = [t_0, +\infty)$, i.e. for *all* t *not less* then t_0. This difference can be crucial in cases when system motions are discontinuous at $t = t_0$.

COMMENT 1.2.5 A. M. Liapunov defined stability of the reference motion for cases when it is not influenced by t_0. However, the initial moment can essentially influence stability of the reference motion in cases when system motions are not continuous in t. Besides, t_0 can essentially influence the maximal admissible values of all E_j and E_j' even when all system motions are continuous in t.

COMMENT 1.2.6. The stability of the reference motion was defined by A.M.Liapunov with respect to initial perturbations of the general coordinates q_j, q_j', rather than with respect to persistent external disturbances.

COMMENT 1.2.7. The stability definition does not care about the values E_j and E'_j except that they must be positive. Hence, for large values of all L_j, the maximal admissible E_j and E'_j can be so small that they are not useful for engineering needs.

1.2.3 Relationship between the reference motion and the zero solution

Let $2k$ be the order of the system and y_i, $i = 1, 2, \ldots, 2k$, be its i-th state variable. Using basic physical laws (e.g. the law of the energy conservation and the law of the material conservation) we can for a large class of systems get state differential equations in the following scalar form

$$\frac{dy_i}{dt} = Y_i(t, y_1, \ldots, y_{2k}), \qquad i = 1, 2, \ldots, 2k, \tag{1.2.1}$$

or in the equivalent vector form

$$\frac{dy}{dt} = Y(t, y), \tag{1.2.2}$$

where* $y = (y_1, y_2, \ldots, y_{2k})^{\mathrm{T}} \in R^{2k}$ and $Y = (Y_1, Y_2, \ldots, Y_{2k})^{\mathrm{T}}$, $Y\colon \mathcal{T} \times R^{2k} \to R^{2k}$. A motion of (1.2.2) is denoted by $\eta(t; t_0, y_0)$, $\eta(t_0; t_0, y_0) \equiv y_0$, and the reference motion $\eta_r(t; t_0, y_{r0})$. From the physical point of view the reference motion should be realizable by the system. From the mathematical point of view this means that the reference motion is a solution of (1.2.2),

$$\frac{d\eta_r(t; t_0, y_{r0})}{dt} \equiv Y[t, \eta_r(t; t_0, y_{r0})]. \tag{1.2.3}$$

Let the Liapunov transformation of coordinates be used,

$$x = y - y_r, \tag{1.2.4}$$

where $y_r(t) \equiv \eta_r(t; t_0, y_{r0})$. Let $f\colon \mathcal{T} \times R^{2k} \to R^{2k}$ be defined by

$$f(t, x) = Y[t, y_r(t) + x] - Y[t, y_r]. \tag{1.2.5}$$

*In Liapunov's notation $y = (q_1, q_2, \ldots, q_k, q'_1, q'_2, \ldots, q'_k)^{\mathrm{T}}$.

It is evident that

$$f(t,0) \equiv 0. \tag{1.2.6}$$

Now (1.2.2) – (1.2.5) yield

$$\frac{dx}{dt} = f(t,x). \tag{1.2.7}$$

In this way, the behavior of perturbed motions related to the reference motion (in total coordinates) is represented by the behavior of the state deviation x with respect to the zero state deviation. The reference motion in the total coordinates y_i is represented by the zero deviation $x = 0$ in state deviation coordinates x_i. With this in mind, the following result emphasizes complete generality of both Liapunov's second method and results represented in Liapunov [101] for the system (1.2.7). Let $Q\colon R^{2k} \to R^n$, $n = 2k$ is admissible but not required.

THEOREM 1.2.1. *Stability of $x = 0$ of the system (1.2.7) with respect to $Q = x$ is necessary and sufficient for stability of the reference motion η_r of the system (1.2.2) with respect to every vector function Q that is continuous in y.*

PROOF. *Necessity.* This part is true because $Q(y) = y$ is contionuous in y and evidently stability of $x = 0$ with respect to x is implied by stability of η_r with respect to $Q(y) = y$.

Sufficiency. Let $L_i > 0$, $i = 1,2,\dots,n$, be arbitrarily chosen. Continuity of Q in y implies existence of $l_i > 0$, $l_i = l_i(L, y_r)$, $L = (L_1, L_2, \dots, L_n)^{\mathrm{T}}$, $i = 1,2,\dots,n$, such that $|y_i - y_{ri}| < l_i$, $\forall i = 1,2,\dots,2k$, implies $|Q_i(y) - Q_i(y_r)| < L_i$, $i = 1,2,\dots,n$. Stability of $x = 0$ of (1.2.7) (with respect to x) guarantees existence of $\delta_i > 0$, $\delta_i = \delta_i(l)$, $l = (l_1, l_2, \dots, l_{2k})^{\mathrm{T}}$, such that $|x_{i0}| < \delta_i$, $i = 1,2,\dots,2k$, where $\chi(t; t_0, x_0)$, $\chi(t_0; t_0, x_0) \equiv x_0$, is the solution of (1.2.7), $\chi = (\chi_1, \chi_2, \dots, \chi_{2k})^{\mathrm{T}}$. Finally, for every $L_i > 0$, $i = 1,2,\dots,n$, there is $\delta_j^* > 0$, $\delta_j^* = \frac{1}{2}\delta_j$, $j = 1,2,\dots,n$, such that $|y_{j0} - y_{rj0}| \le \delta_j^*$, $j = 1,2,\dots,n$, implies

$$|Q_i[\eta(t; t_0, y_0)] - Q_i[\eta_r(t; t_0, y_{r0})]| < L_i, \qquad \forall t \ge t_0,\ i = 1,2,\dots,n.$$

This theorem reduced the problem of the stability of the reference motion of (1.2.2) with respect to Q to the stability problem of $x = 0$ of (1.2.7) with respect to x; it is stated and proved herein for the first time.

1.2.4 Accepted definitions of stability

By the very definition, stationary (time-invariant) systems are those whose motions are not effected by (the choice of) the initial instant $t_0 \in R$. However, such property is not characteristic for nonstationary (time-varying) systems. It is therefore natural to consider the influence of t_0 on stability properties of nonstationary systems, which is motivation for accepting the next definitions.

DEFINITION 1.2.1. The state $x = 0$ of the system (1.2.7) is:

(i) *stable with respect to* $\mathcal{T}_i$ iff for every $t_0 \in \mathcal{T}_i$ and every $\varepsilon > 0$ there exists $\delta(t_0, \varepsilon) > 0$, such that $\|x_0\| < \delta(t_0, \varepsilon)$ implies

$$\|\chi(t; t_0, x_0)\| < \varepsilon, \quad \forall t \in \mathcal{T}_0;$$

(ii) *uniformly stable with respect to* $\mathcal{T}_0$ iff both (i) holds and for every $\varepsilon > 0$ the corresponding maximal δ_M obeying (i) satisfies

$$\inf[\delta_M(t, \varepsilon) \colon t \in \mathcal{T}_i] > 0;$$

(iii) *stable in the whole with respect to* $\mathcal{T}_i$ iff both (i) holds and

$$\delta_M(t, \varepsilon) \to +\infty \quad \text{as} \quad \varepsilon \to +\infty, \ \forall t \in \mathcal{T}_i;$$

(iv) *uniformly stable in the whole with respect to* $\mathcal{T}_i$ iff both (ii) and (iii) hold;

(v) *unstable with respect to* $\mathcal{T}_i$ iff there are $t_0 \in \mathcal{T}_i$, $\varepsilon \in (0, +\infty)$ and $\tau \in \mathcal{T}_0$, $\tau > t_0$, such that for every $\delta \in (0, +\infty)$ there is x_0, $\|x_0\| < \delta$, for which

$$\|\chi(\tau; t_0, x_0)\| \geq \varepsilon.$$

The expression "*with respect to* $\mathcal{T}_i$" is omitted from (i)–(v) iff $\mathcal{T}_i = R$. These stability properties hold as $t \to +\infty$ but not for $t = +\infty$.

EXAMPLE 1.2.1. (see Grujić [45]). Let $x \in R$ and $\dot{x} = (1 - t)^{-1} x$. Then,

$$\chi(t; t_0, x_0) = (t - 1)^{-1}(t_0 - 1)x_0 \qquad \text{for} \quad t_0 \neq 1 \quad \text{and} \quad t \neq 1.$$

For $t_0 = 1$ the motion is not defined and

$$\|\chi(t; t_0, x_0)\| \to +\infty \quad \text{as} \quad t \to (1-0), \quad \forall\, t_0 \in (-\infty, 1), \quad \forall\, (x_0 \neq 0) \in R.$$

Hence,

$$\delta_M(t, \varepsilon) = 0, \quad \forall \varepsilon > 0, \quad \forall\, t \in (-\infty, 1].$$

However,

$$\delta_M(t, \varepsilon) = \varepsilon, \quad \forall\, t \in (1, +\infty).$$

The state $x = 0$ is uniformly stable in the whole with respect to every $\mathcal{T}_i \subseteq (-1, +\infty)$, but it is not stable.

EXAMPLE 1.2.2. (see Grujić [45]). The first order nonstationary system is defined by

$$\frac{dx}{dt} = \frac{(1 + t\sin t + t^2 \cos t)x \cdot \exp\{-\frac{1}{2}\pi\}}{\frac{1}{2}\pi \cdot \exp\{-t \sin t\} + t \cdot \exp\{-\frac{1}{2}\pi\}}.$$

Solutions are found in the form

$$\chi(t; t_0, x_0) = \frac{\frac{1}{2}\pi + t_0 \exp\{-\frac{1}{2}\pi + t_0 \sin t_0\}}{\frac{1}{2}\pi + t \exp\{-\frac{1}{2}\pi + t \sin t\}}\, x_0, \qquad t_0 \neq -\frac{\pi}{2},\ t \neq -\frac{\pi}{2},$$

so that

$$|\chi(t; t_0, x_0)| \to +\infty \quad \text{as} \quad t \to \left(\frac{\pi}{2}, -0\right),$$
$$\forall\, t_0 \in \left(-\infty, -\frac{\pi}{2}\right),\ \forall\, (x_0 \neq 0) \in R.$$

This result and analysis of $\chi(t; t_0, x_0)$ yield

$$\delta_M(t, \varepsilon) = \begin{cases} 0, & t \in \left(-\infty, -\frac{\pi}{2}\right]; \\ \varepsilon, & t \in \left(-\frac{\pi}{2}, 0\right]; \\ \varepsilon\pi \left[\pi + 2t \cdot \exp\{-\frac{\pi}{2} + t \sin t\}\right]^{-1}, & t \in [0, +\infty). \end{cases}$$

The state $x = 0$ is stable in the whole with respect to $\left(-\frac{\pi}{2}, +\infty\right)$ and uniformly stable in the whole with respect to every bounded $\mathcal{T}_i \subset \left(-\frac{\pi}{2}, +\infty\right)$, but it is not stable.

In these examples, the motions χ are not continuous in all $t \in R$.

PROPOSITION 1.2.1. *If there is a time-invariant neighborhood* $\mathcal{N} \subseteq R^n$ *of* $x = 0$ *such that* $\chi(t; t_0, x_0)$ *is continuous in* $(t; t_0, x_0) \in \mathcal{T}_0 \times R \times \mathcal{N}$, *then stability of* $x = 0$ *of the system (1.2.7) with respect to some non-empty* $\mathcal{T}_i$ *implies its stability.*

This result can be easily proved as well as the following:

PROPOSITION 1.2.2. *If* $x = 0$ *of (1.2.7) is stable (in the whole) then, respectively, it is uniformly stable (in the whole) with respect to every bounded* $\mathcal{T}_i \subset R$.

EXAMPLE 1.2.3. (see Grujić [45]). Solutions of the first order non-stationary system

$$\frac{dx}{dt} = -\frac{\beta + 2\gamma t}{\alpha + \beta t + \gamma t^2}x, \qquad \alpha > 0, \quad \beta^2 < 4\alpha\gamma, \quad \gamma > 0$$

are given by

$$\chi(t; t_0, x_0) = (\alpha + \beta t_0 + \gamma t_0^2)(\alpha + \beta t + \gamma t^2)^{-1} x_0.$$

In this case

$$\delta_M(t, \varepsilon) = \frac{(4\alpha\gamma - \beta^2)\varepsilon}{8\gamma(\alpha + \beta t + \gamma t^2)} \left[1 - \operatorname{sign}\left(t + \frac{\beta}{2\gamma}\right)\right] + \frac{\varepsilon}{2}\left[1 + \operatorname{sign}\left(t + \frac{\beta}{2\gamma}\right)\right].$$

Hence,

$$\inf\,[\delta_M(t, \varepsilon) \colon t \in R] = 0, \qquad \forall \varepsilon \in (0, +\infty),$$

and

$$\delta_M(t, \varepsilon) \to +\infty \quad \text{as} \quad \varepsilon \to +\infty, \ \forall t \in R.$$

The state $x = 0$ is stable in the whole but not uniformly.

However, it is uniformly stable in the whole with respect to $\mathcal{T}_i = [\zeta, +\infty)$ for any $\zeta \in (-\infty, +\infty)$.

DEFINITION 1.2.2. The state $x = 0$ of the system (1.2.7) is:

(i) *attractive with respect to* $\mathcal{T}_i$ iff for every $t_0 \in \mathcal{T}_i$ there exists $\Delta(t_0) > 0$ and for every $\zeta > 0$ there exists $\tau(t_0; x_0, \zeta) \in [0, +\infty)$ such that $\|x_0\| < \Delta(t_0)$ implies $\|\chi(t; t_0, x_0)\| < \zeta$, $\forall t \in (t_0 + \tau(t_0; x_0, \zeta), +\infty)$;

(ii) x_0 – *uniformly attractive with respect to* $\mathcal{T}_i$ iff both (i) is true and for every $t_0 \in \mathcal{T}_i$ there exists $\Delta(t_0) > 0$ and for every $\zeta \in (0, +\infty)$ there exists $\tau_u[t_0, \Delta(t_0), \zeta] \in [0, +\infty)$ such that

$$\sup\,[\tau_m(t_0; x_0, \zeta) \colon x_0 \in \mathcal{T}_i] = \tau_u(\mathcal{T}_i, x_0, \zeta);$$

(iii) t_0 – *uniformly attractive with respect to* $\mathcal{T}_i$ iff (i) is true, there is $\Delta > 0$ and for every $(x_0, \zeta) \in B_\Delta \times (0, +\infty)$ there exists $\tau_u(\mathcal{T}_i, x_0, \zeta) \in [0, +\infty)$ such that

$$\sup\left[\tau_m(t_0); x_0, \zeta\right): t_0 \in \mathcal{T}_i] = \tau_u(\mathcal{T}_i, x_0, \zeta);$$

(iv) *uniformly attractive with respect to* $\mathcal{T}_i$ iff both (ii) and (iii) hold, that is, that (i) is true, there exists $\Delta > 0$ and for every $\zeta \in (0, +\infty)$ there is $\tau_u(\mathcal{T}_i, \Delta, \zeta) \in [0, +\infty)$ such that

$$\sup\left[\tau_m(t_0; x_0, \zeta) : (t_0, x_0) \in \mathcal{T}_i \times B_\Delta\right] = \tau(\mathcal{T}_i, \Delta, \zeta);$$

(v) The properties (i) – (iv) hold *"in the whole"* iff (i) true for every $\Delta(t_0) \in (0, +\infty)$ and every $t_0 \in \mathcal{T}_i$.

The expression *"with respect to* $\mathcal{T}_i$*"* is omitted iff $\mathcal{T}_i = R$.

EXAMPLE 1.2.4. For the system of Example 1 the following are found:

$$\Delta_M(t) = \begin{cases} 0, & t \in (-\infty, 1) \\ +\infty, & t \in (1, +\infty) \end{cases};$$

$$\tau_m(t, x, \zeta) = \begin{cases} +\infty, & t \in (-\infty, 1) \\ \dfrac{t-1}{\zeta}|x| + 1, & t \in (1, +\infty) \end{cases}.$$

The state $x = 0$ is:

(a) attractive in the whole with respect to $\mathcal{T}_i = (1, +\infty)$,
(b) t_0 – uniformly attractive in the whole with respect to any bounded $\mathcal{T}_i \subset (1, +\infty)$,
(c) x_0 – uniformly attractive with respect to $\mathcal{T}_i = (1, +\infty)$,
(d) uniformly attractive with respect to any bounded $\mathcal{T}_i \subset (1, +\infty)$,
(e) not attractive.

The next results can be easily verified.

PROPOSITION 1.2.3. *If there is a time-invariant neighborhood* $\mathcal{N} \subseteq R^n$ *of* $x = 0$ *such that* $\chi(t; t_0, x_0)$ *is continuous in* $(t; t_0, x_0) \in \mathcal{T}_0 \times R \times \mathcal{N}$, *then attraction of* $x = 0$ *of the system (1.2.7) with respect to some nonempty* $\mathcal{T}_i$ *implies its attraction.*

EXAMPLE 1.2.5. We consider the system of Example 1.2.3 once again and find:

$$\inf [\Delta_M(t) \colon t \in R] = +\infty,$$

$$\tau_m(t,\Delta,\zeta) = \begin{cases} \min [0,\ (2\gamma)^{-1}\{[\beta^2 - 4\alpha\gamma + 4\gamma\zeta^{-1}\,\Delta(\alpha+\beta t+\gamma t^2)]^{\frac{1}{2}} - \beta\} \\ \qquad \text{for } \Delta \geq (4\alpha\gamma-\beta^2)\zeta[4\gamma(\alpha+\beta t+\gamma t^2)]^{-1}, \\ 0, \quad \text{for } \Delta < (4\alpha\gamma-\beta^2)\zeta[4\gamma(\alpha+\beta t+\gamma t^2)]^{-1}. \end{cases}$$

Hence,

$$\sup [\tau_m(t,\Delta,\zeta) \colon t \in R] = +\infty \quad \text{for} \quad \Delta \geq (4\alpha\gamma-\beta^2)\zeta[4\gamma(\alpha+\beta t+\gamma t^2)]^{-1}.$$

The state $x = 0$ is:

(a) attractive in the whole,
(b) x_0 – uniformly attractive in the whole,
(c) t_0 – uniformly attractive in the whole with respect to any bounded $\mathcal{T}_i \subset R$,
(d) uniformly attractive in the whole with respect to any bounded $\mathcal{T}_i \subset R$,
(e) not uniformly attractive.

DEFINITION 1.2.3. The state $x = 0$ of the system (1.2.7) is:

(i) *asymptotically stable with respect to* $\mathcal{T}_i$ iff it is both stable with respect to $\mathcal{T}_i$ and attractive with respect to $\mathcal{T}_i$;
(ii) *equi-asymptotically stable with respect to* $\mathcal{T}_i$ iff it is both stable with respect to $\mathcal{T}_i$ and x_0-uniformly attractive with respect to $\mathcal{T}_i$;
(iii) *quasi-uniformly asymptotically stable with respect to* $\mathcal{T}_i$ iff it is both uniformly stable with respect to $\mathcal{T}_i$ and t_0-uniformly attractive with respect to $\mathcal{T}_i$;
(iv) *uniformly asymptotically stable with respect to* $\mathcal{T}_i$ iff it is both uniformly stable with respect to $\mathcal{T}_i$ and uniformly attractive with respect to $\mathcal{T}_i$;
(v) the properties (i) – (iv) hold *"in the whole"* iff both the corresponding stability of $x = 0$ and the corresponding attraction of $x = 0$ hold in the whole;
(vi) *exponentially stable with respect to* $\mathcal{T}_i$ iff there are $\Delta > 0$ and real numbers $\alpha \geq 1$ and $\beta > 0$ such that $||x_0|| < \Delta$ implies

$$||\chi(t;t_0,x_0)|| \leq \alpha ||x_0|| \exp[-\beta(t-t_0)], \quad \forall t \in \mathcal{T}_0,\ \forall t_0 \in \mathcal{T}_i.$$

This holds *in the whole* iff it is true for $\Delta = +\infty$.

The expression *"with respect to $\mathcal{T}_i$"* is omitted iff $\mathcal{T}_i = R$.

EXAMPLE 1.2.6. (see Grujić [45]). The second order system is described by

$$\frac{dx}{dt} = A(t)x, \qquad A(t) = \frac{1}{1+t^2}\begin{bmatrix} -t, & 1 \\ -1, & -t \end{bmatrix},$$

and its solutions are found in the form

$$\chi(t; t_0, x_0) = \frac{1}{1+t^2}\begin{bmatrix} 1+t_0t, & t-t_0 \\ t_0-t, & 1+t_0t \end{bmatrix} x_0.$$

Hence,

$$\delta_M(t,\varepsilon) = \frac{\varepsilon}{2}\left[1 + (1+t^2)^{-1}(1-\operatorname{sign} t) + \operatorname{sign} t\right],$$

which implies

$$\inf\left[\delta_M(t,\varepsilon) \colon t \in R\right] = 0, \qquad \forall \varepsilon \in (0, +\infty),$$

and

$$\tau_m(t, \|x\|, \zeta) = \begin{cases} \left[(1+t^2)^{\frac{1}{2}}\,\|x\|\zeta^{-1} - 1\right]^{\frac{1}{2}}, & \text{for } \|x\| \geq \zeta(1+t^2)^{-\frac{1}{2}}, \\ 0, & \text{for } 0 < \|x\| \leq \zeta(1+t^2)^{-\frac{1}{2}}, \end{cases}$$

which yields

$$\sup\left[\tau_m(t, \Delta, \zeta) \colon t \in R\right] = +\infty \quad \text{for} \quad 0 < \zeta \leq \Delta(1+t^2)^{\frac{1}{2}}, \ \forall \Delta \in (0, +\infty).$$

Therefore, the state $x = 0$ is:

(a) asymptotically stable in the whole,
(b) equi-asymptotically stable,
(c) uniformly asymptotically stable with respect to any bounded $\mathcal{T}_i \subset R$,
(d) not equi-asymptotically stable in the whole,
(c) not uniformly asymptotically stable in the whole with respect to any bounded $\mathcal{T}_i \subset R$.

Notice that the system is linear.

The next results are straightforward corollaries to Propositions 1.2.1 – 1.2.4.

PROPOSITION 1.2.5. *If there is a time-invariant neighborhood $\mathcal{N} \subset R^n$ of $x = 0$ such that $\chi(t; t_0, x_0)$ is continuous in $(t; t_0, x_0) \in \mathcal{T}_0 \times R \times \mathcal{N}$ then asymptotic stability of $x = 0$ of the system (1.2.7) with respect to some nonempty $\mathcal{T}_i$ implies its asymptotic stability.*

PROPOSITION 1.2.6. *If $x = 0$ of (1.2.7) is asymptotically stable then it is uniformly asymptotically stable with respect to every bounded $\mathcal{T}_i \subset R$.*

1.2.5 Equilibrium states

For the sake of clarity we state

DEFINITION 1.2.4. State x^* of the system (1.2.7) is its *equilibrium state over $\mathcal{T}_i$* iff

$$\chi(t; t_0, x^*) = x^*, \qquad \forall\, t \in \mathcal{T}_0, \quad \forall\, t_0 \in \mathcal{T}_i. \tag{1.2.8}$$

The expression *"over $\mathcal{T}_i$"* is omitted iff $\mathcal{T}_i = R$.

PROPOSITION 1.2.7. *For $x^* \in R^n$ to be an equilibrium state of the system (1.2.7) over $\mathcal{T}_i$ it is necessary and sufficient that both*

(1) *for every $t_0 \in \mathcal{T}_i$ there is the unique solution $\chi(t; t_0, x^*)$ of (1.2.7), which is defined for all $t_0 \in \mathcal{T}_0$*

and

(2) $f(t, x^*) = 0, \quad \forall\, t \in \mathcal{T}_0, \ \forall\, t_0 \in \mathcal{T}_i.$

PROOF. *Necessity.* Necessity of (i) and (ii) for x^* to be an equilibrium state of (1.2.7) is evidently implied by (1.2.8).

Sufficiency. If x^* satisfies the condition (ii) then $x(t) = x(t; t_0, x^*) = x^*$, $\forall\, t \in \mathcal{T}_0$ and $\forall\, t_0 \in \mathcal{T}_i$, obeys

$$\frac{dx(t)}{dt} = 0 = f(t, x^*) = f[t, x(t)], \qquad \forall\, t \in \mathcal{T}_0, \quad \forall\, t_0 \in \mathcal{T}_i.$$

Hence, $\chi(t; t_0, x^*) = x^*$ is a solution of (1.2.7) at (t_0, x^*) for all $t_0 \in \mathcal{T}_i$, which is unique due to the condition (i).

Hence (1.2.8) holds.

The conditions for existence and uniqueness of the solutions can be found in the books by Bellman [15], Hartman [69], Halanay [67] and Pontriagin [154] (see also Kalman and Bertram [80]).

PROPOSITION 1.2.8. *If $x = 0$ of the system (1.2.7) is stable with respect to $\mathcal{T}_i$ then it is then it is an equilibrium state of the system over $\mathcal{T}_i$.*

PROOF. Let $x = 0$ of (1.2.7) be stable with respect to $\mathcal{T}_i$ and $\varepsilon > 0$ be arbitrarily small. Then $||\chi(t; t_0, 0)|| < \varepsilon$ for all $t \in \mathcal{T}_0$ and every $t_0 \in \mathcal{T}_i$ because $x_0 = 0$ and $||x_0|| = 0 < \delta_M(t_0, \varepsilon)$. Let χ_1 and χ_2 be two solutions of (1.2.7) through $(t_0, 0)$, $t_0 \in \mathcal{T}_i$. Then,

$$(1.2.9) \qquad ||\chi_1(t; t_0, 0) - \chi_2(t; t_0, 0)|| \leq ||\chi_1(t; t_0, 0)|| + ||\chi_2(t; t_0, 0)|| < \varepsilon_n$$

for all $t \in \mathcal{T}_0$ and every $t_0 \in \mathcal{T}_i$ because

$$||x_0|| = 0 < \delta_M\left(t_0, \frac{\varepsilon_n}{2}\right).$$

Let $\varepsilon_n \to 0$ as $n \to +\infty$. It now follows from (1.2.9) that $||\chi_1(t; t_0, 0) - \chi_2(t; t_0, 0)||$ is less than ε_n no matter how large integer n is taken. Hence,

$$\chi_1(t; t_0, 0) \equiv \chi_2(t; t_0, 0)$$

and

$$||\chi_i(t; t_0, 0)|| < \varepsilon_n, \qquad i = 1,\ 2,$$

for arbitrarily large integer n. It follows that $\chi(t; t_0, 0) \equiv 0$ is the unique solution of (1.2.7) on $\mathcal{T}_0$ for all $t_0 \in \mathcal{T}_i$, which proves that $x = 0$ is an equilibrium state of (1.2.7) over $\mathcal{T}_i$.

Let $g\colon R^n \to R^n$ define an autonomous system

$$(1.2.10) \qquad \frac{dx}{dt} = g(x).$$

Every stability property of $x = 0$ of (1.2.10) is uniform in $t_0 \in R$. Besides, Proposition 1.2.8 yield the following.

COROLLARY 1.2.1. If $x = 0$ of the system (1.2.10) is its equilibrium state over some nonempty interval $\mathcal{T}_i \subset R$ then it is an equilibrium state of the system.

1.3 Brief Outline of Trends in Liapunov's Stability Theory

1.3.1 Of Liapunov's original results

Liapunov ([101], p.25) defined two essentially different approaches to solving stability problems as follows:

> All ways, which we can present for solving the question we are interested in, we can divide in two categories.
>
> With one we associate all those, which lead to a direct investigation of a perturbed motion and in the basis of which there is a determination of general and particular solutions of the differential equation (1.2.1).
>
> In general the solutions should be searched in the form of infinite series, the simplest type of which can be considered from those in the preceding paragraph. They are series ordered in terms of integer powers of fixed variables. However we shall meet series of another character in the sequel.
>
> The collection of all ways for the stability investigation, which are in this category, we call *the first method.*
>
> With another one we associate all those, which are based on principles independent of a determination of any solution of the differential equations of a perturbed motion.
>
> One such example is the well-known way for an investigation of equilibrium stability in the case that there is a force function.
>
> All these ways can be reduced to a determination and an investigation of integrals of the equations (1.2.1), and in general in the basis of all of them, which we shall meet in the sequel, there will be always a determination of functions of variables $x_1, x_2, \ldots, x_n, t$ according to given conditions, which should be satisfied by their total derivatives in t, taken under an assumption that $x_1, x_2, \ldots, x_n$ are functions of t satisfying the equations (1.2.1).
>
> The collection of all ways of such a category we shall call *the second method.*

In order to effectively develop the second method Liapunov introduced the concept of semi-definite and definite functions and the notion of decreasing functions as follows (Liapunov [101], p.59):

We shall consider herein real-valued functions of real variables

$$x_1, x_2, \ldots, x_n, t, \tag{39}$$

obeying conditions of the norm

$$t \geq T, \quad |x_s| \leq H \quad (s = 1, 2, \ldots, n), \tag{40}$$

where T and H are constants, the former of which can be arbitrarily large and the latter may be arbitrarily small (but different than zero).

Then we shall consider only functions which are continuous and one-to-one under the conditions (40) and vanish at

$$x_1 = x_2 = \cdots = x_n = 0.$$

Such properties will possess all functions considered by us (even if it were not mentioned). But, besides that, they can possess special features; for definitions we shall introduce several terms.

Consider a function V such that under the conditions T sufficiently large and H sufficiently small, it can take, apart from those equal to zero, only values of one arbitrary sign.

Such a function we shall call *signconstant.* When we wish to underline its sign, then we shall say that it is a *positive* or *negative function.*

In addition to that, if the function V does not depend on t, and the constant H can be chosen sufficiently small so that, under the conditions (40) the equation $V = 0$ can hold only for one set of values of the variables

$$x_1 = x_2 = \cdots = x_n = 0,$$

then we shall call the function V *signdefinite* one, and wishing to underline its sign – *positive-definite* or *negative-definite.*

We shall use the last notions also with respect to functions depending on t. However, in such a case the function V will be called *signdefinite* only under the condition, if for it is possible to find such a t-independent positive-definite function W, for which one of two expressions

$$V - W \qquad -V - W$$

would represent a positive function.

Hence, each of functions

$$x_1^2 + x_2^2 - 2x_1x_2 \cos t, \qquad t(x_1^2 + x_2^2) - 2x_1x_2 \cos t$$

is signconstant. However, the former is only signconstant, and the latter, if $n = 2$, is simultaneously signdefinite.

Every function V, for which the constant H can be chosen so small that for numerical values of that function under the conditions (40) there is an upper bound, will be called *bounded.*

In view of the properties which, under our assumption, possess all functions considered by us, will be such, for example, every function independent of t.

A bounded function can be such that for every positive ε, regardless how small, there is such nonzero number h, for which for all values of variables, satisfying conditions

$$t \geq T, \quad |x_s| \leq h \quad (s = 1, 2, \ldots, n),$$

will hold the following:

$$|V| \leq \varepsilon.$$

This condition will satisfy, for example, every function independent of t. However functions depending on t, even bounded, can violate it. Such a case represents, for example, a function

$$\sin\left[(x_1 + x_2 + \cdots + x_n)t\right].$$

When the function V fulfills the preceding requirement, then we shall say that it *admits infinitely small upper bound.*

Such an example is the function

$$(x_1 + x_2 + \cdots + x_n) \sin t.$$

Let V be a function admitting infinitely small upper bound. Then, if we know that the variables satisfy a condition

$$t \geq T, \quad |V| \geq \ell,$$

where ℓ is a positive number, hence we conclude that there is another positive number λ, less than which cannot be the greatest quantity among $|x_1|, |x_2|, \ldots, |x_n|$.

In order to examine behavior of the values of a definite function V along system motions without using the motions themselves Liapunov ([101], p.61) proposed the following:

Simultaneously with the function V we shall often consider an expression

$$V' = \frac{\partial V}{\partial x_1} X_1 + \frac{\partial V}{\partial x_2} X_2 + \cdots + \frac{\partial V}{\partial x_n} X_n + \frac{\partial V}{\partial t},$$

representing its total time derivative, taken under the assumption that $x_1, x_2, \ldots, x_n$ are functions of t, which satisfy differential equations of a perturbed motion.

In such cases we shall always assume that the function V is such that V' as a function of the variables (39) would be continuous and one-to-one under the conditions (40).

Speaking further about the derivative of the function V, we shall mean that it is the total derivative.

These concepts have been the keystone of the second Liapunov method and for a solution of (uniform) stability of $x = 0$ (Liapunov [101], p.61):

THEOREM 1. If the differential equations of a perturbed motion are such that it is possible to find a signdefinite function V, the derivative V' of which in view of these equations would be either a signconstant function with the opposite sign to that of V, or identically equal to zero, then the unperturbed motion is stable.

In addition to this result Liapunov [101] made the *"Remark 2"* that has become the foundation of the asymptotic stability concept and for a solution of (uniform) asymptotic stability of $x = 0$.

In order to illustrate deepness, generality and importance of Liapunov's results once again, let following his results be cited (Liapunov [101], p.79-80):

THEOREM 1. When the roots $k_1, k_2, \ldots, k_n$ of the characteristic equation are such that for a given natural number m it is impossible to have any relationship of the form

$$m_1k_1 + m_2k_2 + \cdots + m_nk_n = 0,$$

in which all m_s are nonnegative integers, giving their sum equal to m, then it is always possible to find just one whole homogenous function V of the power m of the quantities k_s satisfying the equation

$$\sum_{s=1}^{n} (p_{s1}x_1 + p_{s2}x_2 + \cdots + p_{sn}x_n) \frac{\partial V}{\partial x_s} = U \tag{9}$$

for arbitrarily given whole homogenous function U of the quantities x_s of the same power m.

THEOREM 2. When the real parts of all roots k_s are negative and when in the equation (9) there is the function U being sign-definite form of any even power m, then the form V of the power m satisfying that equation is also sign definite with the opposite sign to that of U.

Gantmakher [38] recognized the fundamental potential of these results and deduced the Liapunov matrix theorem (see Barnett and Storey [14]). This theorem is a fundamental theorem for stability theory. For its presentation the following is needed.

DEFINITION 1.3.1. A matrix $H = (h_{ij}) \in R^{n\times n}$ is:

(i) *positive (negative) semi-definite* iff its quadratic form $V(x) = x^{\mathrm{T}}Hx$ is positive (negative) semi-definite, respectively;

(ii) *positive (negative) definite* iff its quadratic form $V(x) = x^{\mathrm{T}}Hx$ is positive (negative) definite, respectively.

Let a k-th order principal minor of the matrix H be denoted by

$$H\begin{bmatrix} i_1 & i_2 & \ldots & i_k \\ i_1 & i_2 & \ldots & i_k \end{bmatrix} = \begin{bmatrix} h_{i_1i_1} & h_{i_1i_2} & \ldots & h_{i_1i_k} \\ h_{i_2i_1} & h_{i_2i_2} & \ldots & h_{i_2i_k} \\ \ldots & \ldots & \ldots & \ldots \\ h_{i_ki_1} & h_{i_ki_2} & \ldots & h_{i_ki_k} \end{bmatrix},$$

where

$$i_j \in \{1, 2, \ldots, n\}, \quad i_j < i_{j+1}, \quad j = 1, 2, \ldots, k, \quad k = 1, 2, \ldots, n.$$

The leading principal minor of the k-th order of H is

$$H\begin{bmatrix} 1 & 2 & \ldots & k \\ 1 & 2 & \ldots & k \end{bmatrix} = \begin{bmatrix} h_{11} & h_{12} & \ldots & h_{1k} \\ h_{21} & h_{22} & \ldots & h_{2k} \\ \ldots & \ldots & \ldots & \ldots \\ h_{k1} & h_{k2} & \ldots & h_{kk}, \end{bmatrix}, \quad k = 1, 2, \ldots, n.$$

The following criteria are well known (see Gantmacher [38]).

THEOREM 1.3.1. *Necessary and sufficient for a symmetric* $n \times n$ *matrix* H *to be:*

(1) *positive semi-definite is that all its principal minors are non-negative*

$$H\begin{bmatrix} i_1 & i_2 & \ldots & i_k \\ i_1 & i_2 & \ldots & i_k \end{bmatrix} \geq 0, \quad 1 \leq i_1 < i_2 < \cdots < i_k \leq n, \quad k = 1, 2, \ldots, n;$$

(2) *negative semi-definite is that both all its even order principal minors are non-negative and all its odd order principal minors are non-positive*

$$H\begin{bmatrix} i_1 & i_2 & \ldots & i_k \\ i_1 & i_2 & \ldots & i_k \end{bmatrix} \begin{cases} \geq 0, & k = 2, 4, \ldots \\ \leq 0, & k = 1, 3, \ldots \end{cases};$$

(3) *positive definite is that all its leading principal minors are positive*

$$H\begin{bmatrix} 1 & 2 & \ldots & k \\ 1 & 2 & \ldots & k \end{bmatrix} > 0, \quad k = 1, 2, \ldots, n;$$

(4) *negative definite is that both its first order leading principal minor is negative and all its leading principal minors are alternatively negative and positive*

$$(-1)^k H\begin{bmatrix} 1 & 2 & \ldots & k \\ 1 & 2 & \ldots & k \end{bmatrix} > 0, \quad k = 1, 2, \ldots, n.$$

Notice that a square matrix A with all real valued elements is (semi-) definite iff its symmetric part $A_s = \frac{1}{2}(A + A^{\mathrm{T}})$ is (semi-) definite, and a square matrix A with complex valued elements is (semi-) definite iff its Hermitian part $A_H = \frac{1}{2}(A + A^*)$ is (semi-) definite, where A^* is the transpose conjugate matrix of the matrix A.

Now, the fundamental theorem of the stability theorem – the Liapunov matrix theorem – can be stated as a corollary to the preceding Theorems 1 and 2 by Liapunov.

THEOREM 1.3.2. *In order that real parts of all eigenvalues of a matrix* A, $A \in R^{n \times n}$, *be negative it is necessary and sufficient that for any positive definite symmetric matrix* G, $G \in R^{n \times n}$, *there exists the unique solution* H, $H \in R^{n \times n}$, *of the (Liapunov) matrix equation*

$$A^{\mathrm{T}}H + HA = -G,$$

which is also positive definite symmetric matrix.

For solving the Liapunov matrix equation, see for example Barnett and Storey [14], and Barbashin [10].

1.3.2 Classical and novel developments of the scalar Liapunov functions method

Following Liapunov [101], the classical development of his second method consists of a number of stability theorems providing stability conditions are imposed on appropriate *scalar function* V and its total time derivative along system motions over a *time-invariant* neighborhood of $x = 0$. Adequate expositions of the classic development of the Liapunov second method can be found in the books by Yoshizawa [174, 175] and Rouche, Habets and Laloy [159].

1.3.2.1 Comparison functions. Comparison functions are used as upper or lower estimates of the function V and its total time derivative. They are usually denoted by φ, $\varphi\colon R_+ \to R_+$. The main contributor to the investigation of properties of and use of the comparison functions is Hahn [66]. What follows is mainly based on his definitions and results.

DEFINITION 1.3.2. A function φ, $\varphi\colon R_+ \to R_+$, belongs to

(i) the *class* $K_{[0,\alpha)}$, $0 < \alpha \leq +\infty$, iff both it is defined, continuous and strictly increasing on $[0, \alpha)$ and $\varphi(0) = 0$;
(ii) the *class* K iff (i) holds for $\alpha = +\infty$, $K = K_{[0,+\infty)}$;
(iii) the *class* KR iff both it belongs to the class K and $\varphi(\zeta) \to +\infty$ as $\zeta \to +\infty$;
(iv) the *class* $L_{[0,\alpha)}$ iff both it is defined, continuous and strictly decreasing on $[0, \alpha)$ and $\lim\,[\varphi(\zeta)\colon \zeta \to +\infty] = 0$;
(v) the *class* L iff (iv) holds for $\alpha = +\infty$, $L = L_{[0,+\infty)}$.

Let φ^I denote the *inverse function* of φ, $\varphi^I[\varphi(\zeta)] \equiv \zeta$.
The next result was established by Hahn [66].

PROPOSITION 1.3.1.

(1) *If* $\varphi \in K$ *and* $\psi \in K$ *then* $\varphi(\psi) \in K$;
(2) *If* $\varphi \in K$ *and* $\sigma \in L$ *then* $\varphi(\sigma) \in L$;
(3) *If* $\varphi \in K_{[0,\alpha)}$ *and* $\varphi(\alpha) = \xi$ *then* $\varphi^I \in K_{[0,\xi)}$;
(4) *If* $\varphi \in K$ *and* $\lim[\varphi(\zeta)\colon \zeta \to +\infty] = \xi$ *then* φ^I *is not defined on* $(\xi, +\infty]$;
(5) *If* $\varphi \in K_{[0,\alpha)}$, $\psi \in K_{[0,\alpha)}$ *and* $\varphi(\zeta) > \psi(\zeta)$ *on* $[0,\alpha)$ *then* $\varphi^I(\zeta) < \psi^I(\zeta)$ *on* $[0,\beta]$, *where* $\beta = \psi(\alpha)$.

DEFINITION 1.3.3. A function φ, $\varphi\colon R_+ \times R_+ \to R_+$, belongs to:

(i) the *class* $KK_{[0;\alpha,\beta)}$ iff both $\varphi(0,\zeta) \in K_{[0,\alpha)}$ for every $\zeta \in [0,\beta)$ and $\varphi(\zeta,0) \in K_{[0,\beta)}$ for every $\zeta \in [0,\alpha)$;
(ii) the *class* KK iff (i) holds for $\alpha = \beta = +\infty$;
(iii) the *class* $KL_{[0;\alpha,\beta)}$ iff both $\varphi(0,\zeta) \in K_{[0,\alpha)}$ for every $\zeta \in [0,\beta)$ and $\varphi(\zeta,0) \in L_{[0,\beta)}$ for every $\zeta \in [0,\alpha)$;
(iv) the *class* KL iff (iii) holds for $\alpha = \beta = +\infty$.

DEFINITION 1.3.4. Two *functions* $\varphi_1, \varphi_2 \in K$ or $\varphi_1, \varphi_2 \in KR$ are said to be *of the same order of magnitude* if there exist positive constants α_i, β_i, $i = 1, 2$, such that

$$\alpha_i\varphi_i(\zeta) \le \varphi_j(\zeta) \le \beta_i\varphi_i(\zeta), \quad i \ne j;\ i,j \in [1,2].$$

1.3.2.2 Some generalizations of the theory by Liapunov. We shall set out some generalizations of Liapunov theorems with regard to the results obtained by Zubov [178, 179].

DEFINITION 1.3.5. A function $v\colon R \times R^n \to R$ is *positive definite on* $\mathcal{T}_\tau$, $\tau \in R$, if and only if there is a time-invariant connected neighborhood $\mathcal{N}$ of $x = 0$, $\mathcal{N} \subseteq R^n$ and $a \in K_{[0,\alpha)}$, where $\alpha = \sup\{\|x\|\colon x \in \mathcal{N}\}$ such that $v(t,0) = 0$, $\forall t \in \mathcal{T}_\tau$, and $a(\|x\|) \le v(t,x)$ $\forall (t,x) \in \mathcal{T}_\tau \times \mathcal{N}$.

THEOREM 1.3.3. *Let the vector function* f *in system (1.2.7) be continuous on* $R \times \mathcal{N}$ *(on* $\mathcal{T}_\tau \times \mathcal{N}$*). If there exist*

(1) *an open connected time-invariant neighborhood* $\mathcal{G}$ *of point* $x = 0$;
(2) *a positive definite function* v *on* $\mathcal{G}$ *(on* $\mathcal{T}_\tau \times \mathcal{N}$*) such that:*
    (a) $v(t,0) = 0$ *and for a fixed* $t \in R$ $(t \in \mathcal{T}_\tau)$ *the function* $v(t,x)$ *is continuous at the point* $x = 0$;

(b) *$v(t,x)$ is definite on any integral curve $x = x(t;t_0,x_0)$ of the system (1.2.7) unless the curve leaves the definition domain of function $v(t,x)$ and on every such curve the function*

$$v(t) = v\,(t,\ x(t;t_0,x_0))$$

does not increase when $t \in R$ (for all $t \in \mathcal{T}_\tau$), then and only then the state $x = 0$ of the system (1.2.7) is stable (on $\mathcal{T}_\tau$).

The proof of sufficiency of the theorem conditions is a routine of the Liapunov functions method (see e.g. Liapunov [101], Demidovich [23], etc.).

In the proof of necessity of the Theorem 1.3.3 conditions one employes the function

$$v(t_0,x_0) = \begin{cases} \sup\limits_{t\geq t_0} \|x(t;t_0,x_0)\|, & \text{if } \sup\limits_{t\geq t_0} \|x(t;t_0,x_0)\| \leq 1, \\ 1, & \text{if } \sup\limits_{t\geq t_0} \|x(t;t_0,x_0)\| > 1. \end{cases}$$

It is easy to verify that these functions satisfy all conditions of the Theorem 1.3.3.

DEFINITION 1.3.6. A function $v\colon R\times R^n \to R$ is *decreasing on $\mathcal{T}_\tau$*, $\tau \in R$, if and only if there is a time-invariant neighborhood $\mathcal{N}$ of $x = 0$ and a function $b \in K_{[0,\alpha)}$, such that

$$v(t,x) \leq b(\|x\|) \quad \forall\, t \in \mathcal{T}_\tau \times \mathcal{N}.$$

THEOREM 1.3.4. *In order that the solution $x = 0$ of the system (1.2.7) is t_0-uniformly stable (on $\mathcal{T}_\tau$), it is necessary and sufficient that the function $v(t,x)$ mentioned in Theorem 1.3.3 be decreasing on $\mathcal{G}$ (on $\mathcal{T}_\tau \times \mathcal{G}$) and all conditions of Theorem 1.3.1 be satisfied.*

THEOREM 1.3.5. *For the solution $x = 0$ of the system (1.2.7) to be asymptotically stable (on $\mathcal{T}_\tau$), it is necessary and sufficient that the conditions of Theorem 1.3.3 be satisfied and along any integral curve $x(t;t_0,x_0)$ the function $v(t,x)$ tend to zero as $t \to +\infty$, i.e.*

$$v(t) = v\,(t,\ x(t;t_0,x_0)) \to 0 \quad \text{for } t \to +\infty,$$
$$\|x_0\| < \gamma(t_0), \quad t > t_0,\ t_0 \in \mathcal{T}_i.$$

Theorems 1.3.3 – 1.3.5 have the condition associated with the function $v(t,x)$ nonincreasing or decreasing along the integral curves of the system

(1.2.7). As the explicit representation of the integral curves $x(t; t_0, x_0)$ of the system (1.2.7) is not known, it is impossible to test this condition. Therefore, when these theorems actually are employed, various sufficient conditions of function $v(t, x)$ nonincreasing (decreasing) that are easier to check become of great importance.

THEOREM 1.3.6. *Let a vector function f in (1.2.7) be continuous on $R \times \mathcal{N}$ (on $\mathcal{T}_\tau \times \mathcal{N}$). If there exist*

(1) *an open connected time-invariant neighborhood $\mathcal{G}$ of point $x = 0$;*
(2) *the function $v(t, x)$ satisfying condition (2)(a) of the Theorem 1.3.3;*
(3) *the nonpositive function $w(t, x)$ that is a total derivative of the function $v(t, x)$ along the solutions of the system (1.2.7) such that*

$$w(t, x) \leq \varphi_\alpha(t) \leq 0 \quad \text{for } \|x\| \geq \alpha^2$$

and

(4)
$$\int_I \varphi(s)\, ds = -\infty,$$

where I is any infinite system of closed nonintersecting segments on the interval $[t_0, \infty)$, $t_0 \in \mathcal{T}_i$, such that the lengh of each one is not less than a fixed positive constant, then the state $x = 0$ of the system (1.2.7) is asymptotically stable (on $\mathcal{T}_\tau$).

THEOREM 1.3.7. *For the solution $x = 0$ of the system (1.2.7) to be t_0-uniformly asymptotically stable (on $\mathcal{T}_\tau$), it is necessary and sufficient that the function $v(t, x)$ satisfy all conditions of Theorem 1.3.5 and be decreasing on $\mathcal{G}$ (on $\mathcal{T}_\tau \times \mathcal{G}$).*

This theorem is an immediate corollary of Theorems 1.3.4 and 1.3.5.

THEOREM 1.3.8. *Let a vector function f in system (1.2.7) be continuous on $R \times \mathcal{N}$ (on $\mathcal{T}_\tau \times \mathcal{N}$). If there exist*

(1) *an open connected time-invariant neighborhood $\mathcal{G}$ of point $x = 0$;*
(2) *the function $v(t, x)$ being positive definite on $\mathcal{G}$ (on $\mathcal{T}_\tau \times \mathcal{G}$) and decreasing on $\mathcal{G}$ (on $\mathcal{T}_\tau \times \mathcal{G}$);*
(3) *the function $w(t, x)$ that is negative definite on $\mathcal{G}$ (on $\mathcal{T}_\tau \times \mathcal{G}$) and decreasing on $\mathcal{G}$ (on $\mathcal{T}_\tau \times \mathcal{G}$);*
(4) *the correlation*

$$Dv(t, x) = w(t, x) \quad \text{for} \quad (t, x) \in R \times \mathcal{G} \ (\forall\, (t, x) \in \mathcal{T}_\tau \times \mathcal{G}),$$

then and only then the state $x = 0$ of the system (1.2.7) is uniformly asymptotically stable (on $\mathcal{T}_\tau$) and uniformly attractive (on $\mathcal{T}_\tau$).

COROLLARY 1.3.1. If the state $x = 0$ of the system (1.2.7) is asymptotically stable (on $\mathcal{T}_\tau$), then an independent variable t can be transformed so that the zero solution of the newly obtained system is uniformly attractive (on $\mathcal{T}_\tau$).

THEOREM 1.3.9. *Let the vector function f in system (1.2.7) be continuous on $R \times \mathcal{N}$ (on $\mathcal{T}_\tau \times \mathcal{N}$). If there exist*

(1) *the function $v(t,x)$ decreasing on $\mathcal{G}$ and taking negative values in the arbitrarily small semiaxis neighborhood (for any fixed $t > T$), $x_1 = x_2 = \cdots = x_n = 0,\ t \geq 0$;*
(2) *an integrable function $\varphi_\alpha(t)$ such that*

$$Dv(t,x) = w(t,x)$$

and for $\|x\|^2 \geq \alpha^2,\ w(t,x) \leq \varphi_\alpha(t)$,

$$\int_0^t \varphi_\alpha(t)\, dt \to -\infty \quad \text{for } t \to +\infty,$$

then the equilibrium state $x = 0$ of the system (1.2.7) is unstable.

COROLLARY 1.3.2. If the function $v(t,x)$ satisfies conditions (1) – (2) of Theorem 1.3.9, and the function $w(t,x)$ is negative definite, then the equlibrium state $x = 0$ of the system (1.2.7) is unstable.

This corollary is the first Liapunov theorem on instability (see Liapunov [101], p.65).

Following Krasovskii [89] it is easy to prove.

THEOREM 1.3.10. *If χ is continuous on $\mathcal{T}_0 \times R \times \mathcal{N}$ (on $\mathcal{T}_0 \times \mathcal{T}_\tau \times \mathcal{N}$) then existence of a time-invariant neighborhood $\mathcal{S}$ of $x = 0$, a function v, positive real numbers η_1, η_2 and η_3 and a positive integer p such that $v(t,x) \in C(\mathcal{T}_0 \times \mathcal{N})$ and both, respectively,*

$$(1) \qquad \eta_1 \|x\|^p \leq v(t,x) \leq \eta_2 \|x\|^p, \quad \forall\,(t,x) \in R \times \mathcal{S}\ (\forall\,(t,x) \in \mathcal{T}_\tau \times \mathcal{S}),$$

and

$$(2) \qquad D^* v(t,x) \leq -\eta_3 \|x\|^p, \quad \forall\,(t,x) \in R \times \mathcal{S}\ (\forall\,(t,x) \in \mathcal{T}_\tau \times \mathcal{S}),$$

is necessary and sufficient for exponential stability (on $\mathcal{T}_\tau$) of $x = 0$ of the system (2.1.7).

THEOREM 1.3.11. *If χ is continuous on $\mathcal{T}_0 \times R \times R^n$ (on $\mathcal{T}_0 \times \mathcal{T}_\tau \times R^n$) then existence of a function v, positive real numbers η_1, η_2 and η_3 and a positive integer p such that $v(t,x) \in C(\mathcal{T}_0 \times R^n)$ and both, respectively,*

$$(1) \quad \eta_1\|x\|^p \leq v(t,x) \leq \eta_2\|x\|^p, \quad \forall\,(t,x) \in R \times R^n \;\; (\forall\,(t,x) \in \mathcal{T}_\tau \times R^n),$$

and

$$(2) \quad D^*v(t,x) \leq -\eta_3\|x\|^p, \quad \forall\,(t,x) \in R \times R^n \;\; (\forall\,(t,x) \in \mathcal{T}_\tau \times R^n),$$

is necessary and sufficient for exponential stability in the whole (on $\mathcal{T}_\tau$) of $x = 0$ of the system (2.1.7).

1.3.2.3 Partial stability. We return back to the system (1.2.7) and represent the vector x of the system state as

$$x^{\mathrm{T}} = \left(x_1^{\mathrm{T}}, x_2^{\mathrm{T}}\right)^{\mathrm{T}},$$

where $x_1 \in R^{n_1}$, $x_2 \in R^{n_2}$, $n_1 + n_2 = n$. Then we assume on system of the equations (1.2.7) that:

(H_1). In domain $t \in R$, $\mathcal{T}_\tau \times \Omega(H) \times D$ the right-hand parts of the system (1.2.7) are continuous and locally Lipschitzian in x, i.e. $f \in C(R \times \Omega(H) \times D,\ R^n)$, where $\Omega(H) = \{x_1 \in R^{n_1} : \|x_1\| < H,\ H = \mathit{const} > 0\}$, $D = \{x_2 \in R^{n_2} : 0 < \|x_2\| < +\infty\}$.

(H_2). The solution of the system (1.2.7) are x_2-continuable, i.e. any solution $x(t;t_0,x_0)$ of the system (1.2.7) is definite for all $t \geq 0$ ($t \in \mathcal{T}_\tau$) such that $\|x_1(t)\| \leq H$.

It was noted by Liapunov [102] that a more general problem on motion stability with respect to a part of variables may be studied.

The theory of motion stability with respect to a part of variables is exposed by Rumyantzev and Oziraner [161]. In this presentation we restrict ourselves to a few results obtained in this direction.

DEFINITION 1.3.7. The state $x = 0$ of the system (1.2.7) is x_1-*stable with respect to* $\mathcal{T}_\tau$, iff for every $t_0 \in \mathcal{T}_i$ and every $\varepsilon > 0$ there exists a $\delta(t_0,\varepsilon) > 0$ such that $\|x_0\| < \delta(t_0,\varepsilon)$ implies $\|x_1(t;t_0,x_0)\| < \varepsilon \ \forall\, t \in \mathcal{T}_0$.

The other types of x_1-stability are defined in the same way as Definition 1.3.7 taking into account Definitions 1.2.1 – 1.2.3.

Following the results by Rumyantzev [160] we shall set out the following result.

THEOREM 1.3.12. *Let the vector function f in system (1.2.7) satisfy conditions H_1 and H_2. If there exist the function $v(t,x)$ and comparison functions a, b and c of class K such that*

(1) $a(\|x_1\|) \leq v(t,x) \leq b(\|x_1\|) \quad \forall\,(t,x) \in R \times \Omega(H) \times D \; (\forall\,(t,x) \in \mathcal{T}_\tau \times \Omega(H) \times D))$;

(2) $D^+v(t,x) \leq -c(\|x_1\|) \quad \forall\,(t,x) \in R \times \Omega(H) \times D \; (\forall\,(t,x) \in \mathcal{T}_\tau \times \Omega(H) \times D))$.

Then

(a) *any* $\alpha > 0$ *and any* $(t_0, x_0) \in R \times (B_\alpha \cap \Omega(H)) \times D \; ((t_0, x_0) \in \mathcal{T}_\tau \times (B_\alpha \cap \Omega(H)) \times D)$, *the solution* $x_1(t; t_0, x_0) \to 0$ *uniformly relatively* (t_0, x_0) *as* $t \to +\infty$;

(b) *the state* $x = 0$ *of the system (1.2.7) is uniformly asymptotically* x_1*-stable (on* $\mathcal{T}_\tau$*).*

1.3.2.4 The development of Marachkov's idea. One of the trends in generalization of Liapunov's theorems is the establishment of conditions that could replace the condition of function v decreasing in the theorems on asymptotic stabiliy. The Marachkov's theorem [108] is the first result in this direction.

THEOREM 1.3.13. *Let the vector function f in the system (1.2.7) be bounded on $R \times \mathcal{N}$ (on $\mathcal{T}_\tau \times \mathcal{N}$). If there exist*

(1) *a positive definite function* $v \in C^1(R \times \mathcal{N}, R_+)\, (v \in C^1(\mathcal{T}_\tau \times \mathcal{N}, R_+))$, $v(t,0) = 0, \quad \forall\, t \in R\, (\forall\, t \in \mathcal{T}_\tau)$;

(2) *a function c of class K such that*

$$Dv(t,x) \leq -c(\|x\|)$$
$$\forall\,(t,x) \in R \times N \;\; (\forall\,(t,x) \in \mathcal{T}_\tau \times N),$$

then the equilibrium state $x = 0$ of the system (1.2.7) is asymptotically stable (on $\mathcal{T}_\tau$).

The Marachkov's theorem was generalized by Salvadori [162] via the application of two auxiliary functions. We shall formulate this result in the following way.

THEOREM 1.3.14. *Let the vector function f in the system (1.2.7) be continuous on $R \times \mathcal{N}$ (on $\mathcal{T}_\tau \times \mathcal{N}$). If there exist*

(1) *a positive definite function* $v \in C^1(R \times \mathcal{N}, R_+)\; (v \in C^1(\mathcal{T}_\tau \times \mathcal{N}, R_+))$, $v(t,0) = 0, \quad \forall\, t \in R\, (\forall\, t \in \mathcal{T}_\tau)$;

(2) *the function* $w \in C^1(R \times \mathcal{N},\, R_+)$ $(w \in C^1(\mathcal{T}_\tau \times \mathcal{N},\, R_+))$, $w(t,x)$ *is positive definite and* $Dw(t,x)|_{(1.2.7)}$ *is bounded from above or below on* $R \times \mathcal{N}$ *(on* $\mathcal{T}_\tau \times \mathcal{N}$*);*

(3) *a function* c *of class* K *such that*

$$Dv(t,x) \le -c(w(t,x)) \quad on\ R \times \mathcal{N}\ (on\, \mathcal{T}_\tau \times \mathcal{N}).$$

Then the equilibrium state $x = 0$ *of the system (1.2.7) is asymptotically stable (on* $\mathcal{T}_\tau$*).*

Below we shall cite a result showing that the positive definiteness condition in Theorem 1.3.13 may be replaced by the condition of positive semidefiniteness.

THEOREM 1.3.15. *Let in condition (1) of Theorem 1.3.13 the function* $v \in C^1(R \times \mathcal{N})$, $v \in C^1(\mathcal{T}_\tau \times \mathcal{N})$, $v(t,x) \ge 0$ *and* $v(t,0) = 0$ $\forall t \in R$ $(\forall t \in \mathcal{T}_\tau)$ *and condition (2) be satisfied.*

Then the equilibrium state $x = 0$ *of the system (1.2.7) is asymptotically stable (on* $\mathcal{T}_\tau$*).*

1.3.2.5 Generalized comparison principle. Further alongside the system (1.2.7) the equation

$$\frac{du}{dt} = g(t,u,x) \tag{1.3.1}$$

is considered, where $u \in R_+$, $g \in C(\mathcal{T}_\tau \times R_+ \times R^n,\, R)$, $g(t,0,0) = 0$ for all $t \in \mathcal{T}_\tau$.

We recall that equation (1.3.1) emerges as a result of estimation of the total derivative $D^+v(t,x)$ along a solution of the system (1.2.7) in terms of the inequality

$$\begin{gathered} D^+v(t,x) \le g(t,\, v(t,x),\, x) \\ (t,x) \in R \times \mathcal{N}\ (\forall\, (t,x) \in \mathcal{T}_\tau \times \mathcal{N}). \end{gathered} \tag{1.3.2}$$

Sometimes an obvious dependence of function g on vector x widens the possibility to apply the principle of comparison with scalar Liapunov function (cf. Corduneanu [20]).

THEOREM 1.3.16. *Let the vector function f in the system (1.2.7) be continuous on $R \times \mathcal{N}$ (on $\mathcal{T}_\tau \times \mathcal{N}$). If there exist scalar function $v(t,x)$ and $g(t,u,x)$ and comparison functions a and b of class K such that*

(1) $a(\|x\|) \leq v(t,x) \leq b(\|x\|) \quad \forall\,(t,x) \in R \times \mathcal{N}\ (\forall\,(t,x) \in \mathcal{T}_\tau \times \mathcal{N})$;

(2) $D^+v(t,x) \leq g(t,\, v(t,x),\, x) \quad \forall\,(t,x) \in R \times \mathcal{N}\ (\forall\,(t,x) \in \mathcal{T}_\tau \times \mathcal{N})$,

then the property of u-stability of the extended system

$$
\begin{aligned}
\frac{dx}{dt} &= f(t,x), \quad x(t_0) = x_0;\\
\frac{du}{dt} &= g(t,u,x), \quad u(t_0) = u_0 \geq 0,
\end{aligned}
$$

implies the corresponding property of stability of solution $x = 0$ to the system (1.2.7).

For the proof of this theorem when $\mathcal{T}_\tau = R$ see Hatvani [71] and for its generalization see Martynyuk [110].

We note that for the case when estimate (1.3.2) holds with an inverse inequality and the function $g(t,u,x) = g(t,u)$ the theorems on instability of solution $x = 0$ to system (1.2.7) are known (see Rouche, Habets, Laloy [159]) that are based on the principle of comparison with scalar Liapunov function.

1.3.3 A survey of development of the method of vector Liapunov functions

With the purpose to weaken the requirements to the Liapunov functions used in the theory of motion stability it was proposed by Duhem [33] in 1902 to apply several Liapunov functions instead of one.

In modern terms he discovered a multicomponent Liapunov function. After 60 years this idea of multicomponent function was developed by Bellman [16], Matrosov [132] and Melnikov [139]. The papers by Corduneanu [20, 21] where the scalar Liapunov function were aplied together with differential inequalities and the works by Kamke [81] and Ważewski [171] have become a background for a series of important results in motion stability theory obtained via the principle of comparison with vector Liapunov function. This section reviews basic ideas and results developed lately while working out the method of vector Liapunov functions.

1.3.3.1 Scalar approach. We return back to the system (1.2.7) and consider also a *vector function*

$$V(t,x) = (v_1(t,x), v_2(t,x), \dots, v_m(t,x))^{\mathrm{T}}, \tag{1.3.3}$$

where $v_s \in C(\mathcal{T}_0 \times R^n, R_+)$, $s = 1, 2, \dots, m$ and its total derivative along solutions of the system (1.2.7)

$$D^+V(t,x) = \lim\sup\{[V(t+\theta, x+\theta f(t,x)) - V(t,x)]\,\theta^{-1} : \theta \to \theta^+\} \tag{1.3.4}$$

for $(t,x) \in \mathcal{T}_0 \times R^n$.

The notion of the property of having a fixed sign of function (1.3.3) is introduced as follows. By means of a real vector $\alpha \in R^m$ one constructs a scalar function

$$v(t,x,\alpha) = \alpha^{\mathrm{T}}V(t,x) \qquad (t,x) \in \mathcal{T}_0 \times R^n. \tag{1.3.5}$$

DEFINITION 1.3.8. A vector function $V : \mathcal{T}_0 \times R^n \to R^m$ is

(i) *positive semi-definite on* $\mathcal{T}_\tau = [\tau, +\infty)$, $\tau \in R$ iff there exist a connected time-invariant neighborhood $\mathcal{N}$ of point $x = 0$, $\mathcal{N} \subseteq R^n$ and a real vector $\alpha \in R^n$ such that
  (a) $v(t,x,\alpha)$ is continuous in $(t,x) \in \mathcal{T}_\tau \times \mathcal{N}$;
  (b) $v(t,x,\alpha)$ is nonnegative on $\mathcal{N}$; $v(t,x,\alpha) \geq 0$ $\forall\,(t,x,\alpha \neq 0) \in \mathcal{T}_\tau \times \mathcal{N} \times R^m$;
  (c) $v(t,x,\alpha)$ vanishes whenever $x = 0$ for any $(t, \alpha \neq 0) \in \mathcal{T}_\tau \times R^m$.

REMARK 1.3.1. Taking Definition 1.3.8 for the sample the other definitions for function (1.3.3) are introduced in a similar way.

The state vector x of system (1.2.7) is divided into m subvectors, i.e. $x = \left(x_1^{\mathrm{T}}, \dots, x_m^{\mathrm{T}}\right)^{\mathrm{T}}$, where $x_s \in R^{n_s}$ and $n_1 + n_2 + \dots + n_m = n$.

Assume that

$$a_{i1}\psi_{i1}^{\frac{1}{2}}(\|x_i\|) \leq v_i(t,x) \leq a_{i2}\psi_{i2}^{\frac{1}{2}}(\|x_i\|), \quad i = 1, 2, \dots, m, \tag{1.3.6}$$

where a_{i1} and a_{i2} are some positive constants and ψ_{i1} and ψ_{i2} are of class K (KR).

Actually the condition (1.3.6) means that the components $v_i(t,x)$ of the vector function (1.3.3) are positive definite and decreasing with respect to a part of variables.

Let us introduce designations

$$\begin{aligned} A_1 &= \operatorname{diag}[a_{11}, a_{12}, \dots, a_{1m}], \\ A_2 &= \operatorname{diag}[a_{21}, a_{22}, \dots, a_{2m}]. \end{aligned} \tag{1.3.7}$$

PROPOSITION 1.3.2. *For the vector function (1.3.3) to be positive definite and decreasing, it is necessary and sufficient that the bilateral inequalities*

$$u_1^{\mathrm{T}} A_1 u_1 \le v(t,x,\alpha) \le u_2^{\mathrm{T}} A_2 u_2 \tag{1.3.8}$$

be satisfied, where

$$u_1^{\mathrm{T}} = \left(\psi_{11}^{\frac{1}{2}}(\|x_1\|), \dots, \psi_{1m}^{\frac{1}{2}}(\|x_m\|)\right)^{\mathrm{T}},$$
$$u_2^{\mathrm{T}} = \left(\psi_{21}^{\frac{1}{2}}(\|x_1\|), \dots, \psi_{1m}^{\frac{1}{2}}(\|x_m\|)\right)^{\mathrm{T}}.$$

REMARK 1.3.2. If $\psi_{i1} = \psi_{i2} = \|x_i\|$, then the estimates (1.3.8) are known (see Krasovskii [89]) as the estimates characteristics of the quadratic forms.

Taking into account (1.3.4) we get for the function (1.3.5)

$$D^+V(t,x,\alpha) = \alpha^{\mathrm{T}} D^+V(t,x). \tag{1.3.9}$$

Let for $(t,x) \in \mathcal{T}_0 \times R^n$ there exist an $m \times m$ matrix $S(t,x)$, for which

$$D^+V(t,x,\alpha) \le \psi_3^{\mathrm{T}} S(t,x)\, \psi_3, \tag{1.3.10}$$

where $\psi_3 = \left(\psi_{13}^{\frac{1}{2}}(\|x_1\|), \psi_{23}^{\frac{1}{2}}(\|x_2\|), \dots, \psi_{m3}^{\frac{1}{2}}(\|x_m\|)\right)^{\mathrm{T}}$.

Estimates (1.3.8) – (1.3.10) allows us to establish stability conditions for the state $x = 0$ of system (1.2.7) as follows.

THEOREM 1.3.17. *Let the vector function f in system (1.2.7) be continuous on $R \times \mathcal{N}$ (on $\mathcal{T}_\tau \times \mathcal{N}$). If there exist*

(1) *an open connected time-invariant neighborhood $\mathcal{G}$ of point $x = 0$;*
(2) *the decreasing positive definite vector function V on $\mathcal{G}$ (on $\mathcal{T}_\tau \times \mathcal{G}$);*
(3) *the $m \times m$-matrix $S(t,x)$ on $\mathcal{G}$ (on $\mathcal{T}_\tau \times \mathcal{G}$) such that inequality (1.3.10) is satisfied.*

Then

(a) *the state $x = 0$ of the system (1.2.7) is uniformly stable if the matrix $S(t,x)$ is negative semidefinite on $\mathcal{G}$ (on $\mathcal{T}_\tau \times \mathcal{G}$);*
(b) *the state $x = 0$ of the system (1.2.7) is uniformly asymptotically stable (on $\mathcal{T}_\tau$) providing the matrix $S(t,x)$ is negative definite on $\mathcal{G}$ (on $\mathcal{T}_\tau \times \mathcal{G}$).*

PROOF. Formula (1.3.5) and estimates (1.3.8) and (1.3.10) allow us to repeat all points of the proof of Theorems 8.1 and 8.3 by Yoshizawa [174] on uniform (asymptotic) stability. The theorem is proved.

REMARK 1.3.3. New points of the theorem resulting from the application of vector function (1.3.3) are

(a) a possibility to apply the components $v_i(t,x)$, $i = 1, 2, \dots, m$ being of a fixed sign with respect to a part of variables;
(b) a possibility to check the property of having a fixed sign of the matrix $S(t,x)$ via the algebraic method.

A specific way of constructing $m \times m$-matrix $S(t,x)$ enables us to derive from Theorem 1.3.17 the assertions found in the monographs by Michel and Miller [143], Šiljak [167] and Grujić, Martynyuk, Ribbens-Pavella [57]. Thus, Theorem 1.3.17 proves to be quite universal in the framework of the scalar approach of the vector Liapunov function application.

Also, within the scalar approach the application of the vector Liapunov function together with the comparison principle is developed.

THEOREM 1.3.18. *Let the vector function f in system (1.2.7) be continuous on $R \times \mathcal{N}$ (on $\mathcal{T}_\tau \times \mathcal{N}$). If there exist*

(1) an open connected time-invariant neighborhood $\mathcal{G}$ of point $x = 0$;
(2) the vector function $V(t,x)$ and a vector $\alpha \in R^m$ for which inequalities (1.3.8) are satisfied;
(3) the function $w \in C(\mathcal{T}_\tau \times R_+, R)$, $w(t,0) = 0$ such that

$$D^+ v(t,x,\alpha) \le w\,(t,\, v(t,x,\alpha)) \quad \forall\ (t,x) \in \mathcal{T}_\tau \times \mathcal{N};$$

(4) the solution $r^* = 0$ of the comparison equation

$$\frac{dr}{dt} = w(t,r), \quad r(t_0) = r_0 \ge 0 \tag{1.3.11}$$

existing for $t \ge t_0$.

Then

(a) the stability of state $r = 0$ of the equation (1.3.11) implies the stability of state $x = 0$ of the system (1.2.7);
(b) the asymptotic stability of state $r = 0$ of (1.3.11) implies the asymptotic stability of state $x = 0$ of the system (1.2.7);
(c) if, moreover, $v(t,x,\alpha) \to 0$ as $\|x\| \to 0$ uniformly on $\mathcal{T}_\tau$, then the uniform stability or uniform asymptotic stability of state $r = 0$ of system (1.3.11) implies the corresponding stability of state $x = 0$ of system (1.2.7).

For the analysis of various partial cases of inequality (1.3.11) or the same inequality in the integral form see Grujić, Martynyuk and Ribbens-Pavella [57].

One of the Theorem 1.3.18 generalizations is based on the application of a majorizing function $w \in C(\mathcal{T}_\tau \times R^n \times R_+, R)$, $w(t,r,x) = 0$ when $r = 0$ and $x = 0$.

Besides an extended system

$$(1.3.12) \qquad \begin{aligned} \frac{dx}{dt} &= f(t,x), \quad x(t_0) = x_0, \\ \frac{dr}{dt} &= w(t,r,x), \quad r(t_0) = r_0 \geq 0 \end{aligned}$$

is treated for which certain type of r-stability of the zero solution $\left(x^{\mathrm{T}}, r\right)^{\mathrm{T}} = 0$ yields an appropriate type of stability of the state $x = 0$ of (1.2.7).

The theorem has been developed and applied for the cases when the function $w(t,r) = w(r)$, i.e. it is independent of $t \in \mathcal{T}_\tau$. These and other results obtained in this direction are set out by Grujić, Martynyuk and Ribbens-Pavella [57].

1.3.3.2 Vector approach. The combination of vector function (1.3.3) with the comparison system

$$(1.3.13) \qquad \frac{du}{dt} = \Omega(t,u), \quad u(t_0) = u_0 \geq 0,$$

where $u \in R^m_+$, $\Omega \in C(\mathcal{T}_\tau \times R^m_+, R^m)$, $\Omega(t,0) = 0$ for all $t \in \mathcal{T}_\tau$, leads to the following general result of the method of vector Liapunov functions.

THEOREM 1.3.19. *Let the vector function f in system (1.2.7) be continuous on $R \times \mathcal{N}$ (on $\mathcal{T}_\tau \times \mathcal{N}$). If there exist*

(1) *an open connected time-invariant neighborhood $\mathcal{G}$ of point $x = 0$;*
(2) *the vector function $V \in C(\mathcal{T}_\tau \times \mathcal{N}, R^m_+)$, $V(t,x)$ is locally Lipschitzian in x and a real vector $\alpha \in R^m$ such that function (1.3.5) satisfies bilateral inequality (1.3.8);*
(3) *the function $\Omega \in C(\mathcal{T}_\tau \times R^m_+, R^m)$, $\Omega(t,0) = 0$ and $\Omega(t,u)$ is quasimonotone nondecreasing in u when all $t \in \mathcal{T}_\tau$, so that*

$$D^+V(t,x) \leq \Omega\left(t, V(t,x)\right), \quad (t,x) \in \mathcal{T}_\tau \times \mathcal{N}.$$

Then certain stability properties of the state $u = 0$ of the system (1.3.13) imply appropriate stability properties of the state $x = 0$ of the system (1.2.7).

PROOF. We shall cite first an assertion that establishes a relationship between the vector function variation and maximal solution to comparison system (1.3.13).

PROPOSITION 1.3.3. *Let $V \in C(\mathcal{T}_\tau \times \mathcal{N},\, R^m_+,)$ and $V(t,x)$ be locally Lipschitzian in x. Let the vector function $D^+V(t,x)$ specified by (1.3.4) satisfy the inequality*

$$D^+V(t,x) \le \Omega\,(t,\, V(t,x))\,, \quad \forall\,(t,x) \in \mathcal{T}_\tau \times \mathcal{N},$$

where $\Omega \in C(\mathcal{T}_\tau \times R^m_+,\, R^m_+)$ and the function $\Omega(t,u)$ be quasimonotone increasing in u.

Assume that the maximal solution $u_M(t;t_0,r_0)$ of the comparison system

$$\frac{du}{dt} = \Omega(t,u)$$

exists on the interval $\mathcal{T}_\tau$ and passes through the point $(t_0,r_0) \in \mathcal{T}_\tau \times R^m_+$. If $x(t;t_0,x_0)$ is any solution to system (1.2.7) defined on $[t_0,\, t_0+\delta)$, $t_0 \in \mathcal{T}_\tau$ and passing through the point $(t_0,x_0) \in \mathcal{T}_\tau \times \mathcal{N}$, then the condition

$$V(t_0,x_0) \le r_0 \tag{1.3.14}$$

yields the estimate

$$V\,(t,\, x(t;t_0,x_0)) \le u_M(t;t_0,r_0) \quad \forall\, t \in [t_0,\, t_0+\delta). \tag{1.3.15}$$

Further the fact that function (1.3.5) satisfies bilateral inequality (1.3.8) implies that the vector function $V(t,x)$ is positive definite and decreasing.

Estimate (1.3.15) and the fact that the solution $u = 0$ of the system (1.3.15) possesses a certain type of stability allow the conclusion that the solution $x = 0$ of the system (1.2.8) has a corresponding type of stability (for further details see Lakshmikantham, Leela and Martynyuk [94], etc.).

In the case when system (1.3.15) is autonomous

$$\frac{du}{dt} = \Omega(u), \quad u \in R^m_+, \tag{1.3.16}$$

where $\Omega \in C(R^m_+,\, R^m)$, $\Omega(u)$ satisfies the quasimonotonicity condition and the solution of the system (1.3.16) is locally unique for any $u_0 \in R^m_+$ we establish a criterion of asymptotic stability of the state $u = 0$ of the system (1.3.16) as follows.

THEOREM 1.3.20. *Let for the system (1.3.16) there exist a neighborhood $\mathcal{U}$ of state $u = 0$ such that for all $u \in \mathcal{U}$, $u \neq 0$, $\Omega(u) \neq 0$ and $\Omega(u) = 0$ when $u = 0$.*

The isolated equlibrium state $u = 0$ of the system (1.3.16) is asymptotically stable iff there exists a positive vector $u^0 = K^0 \cap \mathcal{U}$ such that the system of inequalities

$$\Omega_s(u_1^0, \dots, u_m^0) < 0 \quad \forall\, s \in [1, m]$$

is joint.

Besides, $K^0 = \operatorname{int} K$ and $K = \{u \in R^m \colon u_s \geq,\ s = 1, 2, ..., m\}$.

Under some additional conditions the theorem is proved as well for the case when the comparison system (1.3.16) has a nonisolated singular point (see Martynyuk and Obolenskii [129]).

Further we assume that the vector function $\Omega(t, u)$ has bounded partial derivatives in u.

Designate

$$\left.\frac{\partial \Omega}{\partial u}\right|_{u=0} = P(t), \qquad \Phi(t, u) = \Omega(t, u) - P(t)u.$$

Consider a system comparison equations

$$\frac{du}{dt} = P(t)u + \Phi(t, u), \quad u(t_0) = u_0 \geq 0, \tag{1.3.17}$$

and its linear approximation

$$\frac{d\xi}{dt} = P(t)\xi, \quad \xi(t_0) = \xi_0 \geq 0. \tag{1.3.18}$$

DEFINITION 1.3.9. (Šiljak [167]). Matrix $P(t)$ is called *a nonautonomous M-matrix* iff

$$p_{ij}(t) \begin{cases} < 0 & \text{for all } t \in \mathcal{T}_0,\ i = j; \\ \geq 0 & \text{for all } t \in \mathcal{T}_0,\ i \neq j,\ i, j = 1, 2, \dots, m. \end{cases}$$

DEFINITION 1.3.10. Nonautonomous linear system (1.3.18) is called a *reducible comparison system*, provided that there exists a Liapunov transformation $\xi = Q(t)y$ by means of which it can be reduced to the system

$$\frac{dy}{dt} = By,$$

with a constant M-matrix B. Moreover

$$B = Q^{-1}\left(PQ - \frac{dQ}{dt}\right).$$

Recall that for the *Liapunov transformation*

$$\xi = Q(t)y \tag{1.3.19}$$

there exists $Q^{-1}(t)$ and $Q \in C^1(\mathcal{T}_0,\, R^{m\times m})$.

Besides, the values

$$k = \sup_{t\geq 0} ||Q(t)|| \quad \text{and} \quad l = \sup_{t\geq 0} ||Q^{-1}(t)||$$

are finite.

THEOREM 1.3.21. *Let for the system (1.2.7) the following conditions hold true*

(1) *there exists a positive definite decreascent vector function $V(t,x)$ such that*

$$D^+V(t,x) \leq P(t)V(t,x) + \Phi\left(t,\, V(t,x)\right), \tag{1.3.20}$$

where $P(t)$ is a nonautonomous M-matrix and $\Phi(t,u)$ is quasi-monotone in u and

$$\lim_{||u||\to 0} \frac{||\Phi(t,u)||}{||u||} = 0 \quad \textit{uniformly in } t \geq t_0;$$

(2) *a matrix $P(t)$ reducible in the sense of Liapunov.*

Then the following assertions are valid

(a) *if the matrix B in the system*

$$\frac{dy}{dt} = By + Q^{-1}\Phi(t,\, Qy) \tag{1.3.21}$$

has all eigenvalues with negative real parts, then the zero solution of comparison system (1.3.17) is uniformly asymptotically stable;

(b) *if the matrix B in the system (1.3.21) has all eigenvalues with negative real parts and in addition $V(t,x) \geq \Delta \,||x||^2$ for some $\Delta > 0$, then the zero solution of comparison equation (1.3.17) is exponentially stable;*

(c) *if the inequality (1.3.20) holds with a reversed sign and the matrix B in system (1.3.21) has at least one eigenvalues with positive real parts, then the zero solution of comparison system (1.3.17) is unstable.*

PROOF. We apply to system (1.3.17) the Liapunov transformation $u = Q(t)y$ and get system (1.3.21). By condition (1) of the Theorem 1.3.21

$$||\Phi(t,\, Qy)|| \leq \varepsilon ||Qy||$$

for some $\varepsilon > 0$ and hence, the fact that $||y|| \leq \dfrac{\eta}{k}$ yields

$$||Q^{-1}\Phi(t,\, Qy)|| \leq \varepsilon lk ||y||.$$

So, it is clear that if all eigenvalues of the matrix B in the system (1.3.21) have negative real parts, then the solutions of the system

$$\frac{dy}{dt} = By$$

vanish and furthermore the solutions of the systems (1.3.21) and (1.3.17) respectively possess the same property.

Assertions (b) and (c) are proved in the same manner.

If in Theorem 1.3.21 inequality (1.3.20) is satisfied with a constant matrix P being an M-matrix, then all assertions of the Theorem 1.3.21 remain valid without the transformation of the system (1.3.17) to (1.3.21).

1.4 Notes

1.2. The work by Liapunov [101] was published more than 100 years ago; nevertheless its ideas still inspire many investigations today. Therefore in Sections 1.2 and 1.3 are included not to repeat the contents of this paper but to cite the basic statements of the second Liapunov method according to the original (see Liapunov [101]).

Comments 1.2.1 – 1.2.7, Theorem 1.2.1 and Definitions 1.2.1 – 1.2.3 are set out according to Grujić, Martynyuk and Ribbens-Pavella [57], where a huge bibliography on stability theory is available as well.

1.3. A short survey of main directions of the method of Liapunov functions begins with a review of its original results (see Liapunov [101]). The survey of classical and new trends of the method of scalar Liapunov functions is based on the results by Zubov [178, 179] (Theorems 1.3.3 - 1.3.9), Hahn [66] and Krasovskii [89] (Theorems 1.3.10 and 1.3.11). Theorem 1.3.12 is due to Rumyantzev [160] and Theorem 1.3.13 is due to Marachkov [108]. Theorem 1.3.14 is based on the results by Salvadori [162], while Theorem 1.3.16 is due to Hatvani [71]. For recent development in the method of scalar Liapunov functions see Lakshmikantham and Martynyuk [92].

The survey of the development of the method of the vector Liapunov function takes into account the results by Bellman [16], Matrosov [132], Melnikov [139], Corduneanu [20, 21], Kamke [81], etc. Theorem 1.3.17 is due to Michel and Miller [143]. Theorem 1.3.18 is a generalization of results by Corduneanu [20, 21] and is related to the results by Gruijć, Martynyuk and Ribbens-Pavella [57]. Theorem 1.3.19 is a development of Theorem 1.6.1 by Matrosov, Lakshmikantham and Sivasundaram [96]. Theorem 1.3.20 is due to Martynyuk and Obolenskii [129]. Theorem 1.3.21 is new.

2

MATRIX LIAPUNOV FUNCTION METHOD IN GENERAL

2.1 Introduction

The short survey of the direct Liapunov method development cited in Chapter 1 shows that the generalizations of this method in terms of multicomponent functions make this method more versatile in applications. On the other hand unsolved still is the problem of construction of appropriate functions or systems of functions in terms of which the further development of this fruitful technique is possible. In this regard a two indices system of functions (a matrix-valued function) is proposed in this chapter as a basis for construction of both scalar or vector Liapunov functions.

This chapter gives an account of the foundations of the method of matrix Liapunov functions that is a new method of qualitative analysis of nonlinear systems.

The Chapter is organized as follows.

In Section 2.2 all necessary notions of the direct Liapunov method based on matrix-valued function are introduced.

In Section 2.3 the theorems of direct Liapunov method on motion stability are set out where a scalar function constructed on the set of the two-indices system of functions is applied.

In Section 2.4 a scalar function constructed in terms of a matrix-valued function is incorporated together with the principle of comparison.

The basic theorems of the method of matrix Liapunov functions are presented in Section 2.5. Also the aggregation forms are developed for autonomous large scale systems in terms of matrix-valued functions and the estimates of asymptotic stability domains are discussed.

Section 2.6 deals with a new direction in stability theory refered to as a "multistability of motion". For the analysis of multistability of large scale systems consisting of two, three or four subsustems the method of matrix-valued Liapunov functions is employed in combination with the method of comparison with scalar and vector Liapunov functions.

Section 2.7 presents applications of some general results in the problems of mechanics, automatics regulation and mathematical biology.

2.2 Definition of Matrix-Valued Liapunov Functions

2.2.1 The property of having a fixed sign of the matrix-valued function

Together with the system (1.2.10) we shall consider a two-indices system of functions

$$U(x) = [v_{ij}(x)], \qquad i, j = 1, 2, \dots, m, \tag{2.2.1}$$

where $v_{ii} \in C(R^n, R_+)$ and $v_{ij} \in C(R^n, R)$ for all $i \neq j$. Moreover it is assumed that

(i) $v_{ij}(x)$ are locally Lipschitzian in x;
(ii) $v_{ij}(0) = 0$ for all $i, j = 1, 2, \dots, m$;
(iii) $v_{ij}(x) = v_{ji}(x)$ in any open connected neighborhood of point $x = 0$.

REMARK 2.2.1. If $v_{ij} \equiv 0$ for all $i \neq j = 1, 2, \dots, m$ then $U(x) = \operatorname{diag}[v_{11}(x), \dots, v_{mm}(x)]$ and

$$V(x) = U(x)e, \qquad e \in R^m \tag{2.2.2}$$

is a vector function.

REMARK 2.2.2. If $v_{ij} \equiv 0$ for all $i \neq j = 1, 2, \dots, m$ and there exists at least one value of $k \in [1, m]$ such that $v_{ii} \equiv 0$ for all $i = 1, 2, \dots, k-1, k+1, \dots, m$ and $v_{kk}(x) > 0$ satisfies the conditions (i) – (ii), then

$$U(x) = v_{kk}(x) \quad \text{for all} \quad x \in N,\ N \subseteq R^n, \tag{2.2.3}$$

is a positive definite scalar function.

Thus the two-indices system of functions (2.2.1) is a basis for construction of both scalar and vector Liapunov functions.

However, for the matrix-valued function (2.2.1) to solve the stability problem for the equilibrium state $x = 0$ of the system (1.2.10) it should possess the property of having a fixed sign in the sense of Liapunov.

It runs as follows:

(i) the concept of positive definiteness of a matrix-valued function (2.2.1) should be compatible with the well-known concept of positive definiteness of a matrix;

(ii) the concept of positive definiteness of a matrix function (2.2.1) should be compatible with Liapunov's original concept of positive definiteness of scalar functions;

(iii) the concept of positive definiteness of a matrix function (2.2.1) should be directly applicable to stability analysis and adequate to Liaponuv's (second) method.

For the sake of preciseness the following definition will be used throught the book, which is based on the corresponding definition by Liapunov [101] and Hahn [66], Grujić [47] and Martynyuk [116].

DEFINITION 2.2.1. The matrix-valued function $U\colon R^n \to R^{m\times m}$ is:

(i) *positive semi-definite* iff there is a time-invariant neighborhood $\mathcal{N}$ of $x = 0$, $\mathcal{N} \subseteq R^n$, such that
   (a) U is continuous on $\mathcal{N}$: $U(x) \in C(\mathcal{N})$,
   (b) U vanishes at the origin: $U(0) = 0$,
   (c) $v(x, y) = y^{\mathrm{T}}U(x)y \geq 0 \;\; \forall (x \neq 0,\, y \neq 0) \in \mathcal{N} \times R^m$;

(ii) *positive semi-definite on a neighborhood* $\mathcal{S}$ *of* $x = 0$ iff (i) holds for $\mathcal{N} = \mathcal{S}$;

(iii) *positive semi-definite in the whole* iff (i) holds for $\mathcal{N} = R^n$;

(iv) *negative semi-definite (on a neighborhood* $\mathcal{S}$ *of* $x = 0$ *in the whole)* iff $(-U)$ is positive semi-definite (on the neighborhood $\mathcal{S}$ or in the whole, respectively).

REMARK 2.2.3. Stability analysis shows sufficiency of using a fixed vector $\eta \in R^m$ insted of any y in (c), that is $v = R^n \to R$ is defined by

$$v(x) = \eta^{\mathrm{T}}U(x)y, \quad \eta = (\eta_1, \dots, \eta_m)^{\mathrm{T}}, \quad \eta_i \neq 0, \;\; i = 1, 2, \dots, m.$$

Iff all $\eta_i = 1$ in η, then $\eta = I = (1, 1, \dots, 1)^{\mathrm{T}} \in R^s$ and

$$U(x) = \sum_{i,j=1}^{m} u_{ij}(x), \qquad u_{ij}(x) = u_{ji}(x).$$

REMARK 2.2.4. In case $m = 1$, then Definition 2.2.1 reduces to Liapunov's original definition of positive definiteness concept (cf. Liapunov [101]).

REMARK 2.2.5. It is to be noted that matrix-valued function U defined by $U(x) = 0$ for all $x \in R^n$ is both positive and negative semi-definite. This ambiguity can be avoided by introducing the notion of strictly positive (negative) semi-definite and there is $\widetilde{x} \in \mathcal{N}$ such that $U(\widetilde{x}) > 0$ $(U(\widetilde{x}) < 0)$, respectively.

DEFINITION 2.2.2. The matrix-valued function $U\colon R^n \to R^{m\times m}$ is:

(i) *positive definite* iff there is a time-invariant neighborhood $\mathcal{N}$, $\mathcal{N} \subseteq R^n$, of $x = 0$, such that it is both positive semi-definite on $\mathcal{N}$ and $v(x, y) = y^{\mathrm{T}}U(x)y > 0$ $\forall\,(x \neq 0,\, y \neq 0) \in \mathcal{N} \times R^m$;
(ii) *positive definite on a neighborhood* $\mathcal{S}$ of $x = 0$, iff (i) holds for $\mathcal{N} = \mathcal{S}$;
(iii) *positive definite in the whole*, iff (i) holds for $\mathcal{N} = R^n$;
(iv) *negative definite (on a neighborhood* $\mathcal{S}$ *of* $x = 0$ *in the whole)* iff $(-U)$ is positive definite (on the neighborhood $\mathcal{S}$ or in the whole, respectively).

The expression "on $\mathcal{T}_\tau$" is omitted iff all corresponding requirements hold for every $\tau \in R$.

Together with the system (1.2.7) we shall consider a two-indices system of functions

$$U(t, x) = [v_{ij}(t, x)], \qquad i, j = 1, 2, \ldots, m, \tag{2.2.4}$$

where $v_{ii} \in C(\mathcal{T}_\tau \times R^n, R_+)$, $v_{ij} \in C(\mathcal{T}_\tau \times R^n, R)$ for all $i \neq j$. Moreover the next conditions are making

(i) $v_{ij}(t, x)$ are locally Lipschitzian in x;
(ii) $v_{ij}(t, 0) = 0$ for all $t \in R$ $(t \in \mathcal{T}_\tau)$ $i, j = 1, 2, \ldots, m$;
(iii) $v_{ij}(t, x) = v_{ji}(t, x)$ in any open connected neighborhood $\mathcal{N}$ of point $x = 0$ for all $t \in R$ $(t \in \mathcal{T}_\tau)$.

PROPOSITION 2.2.1. *The matrix-valued function* $U\colon R \times R^n \to R^{m\times m}$ *is positive definite on* $\mathcal{T}_\tau$, $\tau \in R$ *iff it can be written as*

$$y^{\mathrm{T}}U(t, x)y = y^{\mathrm{T}}U_+(t, x)y + a(\|x\|),$$

where $U_+(t, x)$ *is a positive semi-definite matrix-valued function and* $a \in K$.

DEFINITION 2.2.3. Set $v_\zeta(t)$ is *the largest connected neighborhood of* $x = 0$ *at* $t \in R$ which can be associated with a function $U\colon R \times R^n \to R^{m\times m}$ so that $x \in v_\zeta(t)$ implies $v(t,x,y) < \zeta$, $y \in R^m$.

REMARK 2.2.6. In order to understand and appreciate deepness and importance of Liapunov's concept of definite functions let scalar functions v and w be considered, $v,\ w\colon R \times R^n \times R^m \to R$. Let them obey the following on $\mathcal{T}_\tau \times \mathcal{N}$, where $\mathcal{N}$ is a connected neighborhood of $x = 0$:

(i) v is *positive definite on* $\mathcal{T}_\tau \times \mathcal{N}$;

(ii) w *is positive semi-definite on* $\mathcal{T}_\tau \times \mathcal{N}$ and $w(t,x,y) > 0\ \ \forall\,(t,\, x \neq 0) \in \mathcal{T}_\tau \times \mathcal{N}$, but it is not positive on $\mathcal{T}_\tau \times \mathcal{N}$.

Let $v_\zeta(t)$ and $w_\zeta(t)$ be associated with v and w in sence of Definition 2.2.3. Then, the following is true:

(a) there is $\xi \in (0, +\infty)$ such that $v_\zeta(t) \subseteq \mathcal{N}$, $\forall\, t \in \mathcal{T}_\tau$, $\forall\,\zeta \in (0,\xi)$;

(b) for any $\xi \in (0,+\infty)$ for which $w_\zeta(\tau) \subseteq \mathcal{N}$ there is $t \in \mathcal{T}_\tau$, $t > \tau$, such that $w_\zeta(t) \setminus \mathcal{N} \neq \emptyset$.

DEFINITION 2.2.4. The matrix-valued function $U\colon R \times R^n \to R^{s\times s}$ is:

(i) *decreasing on* $\mathcal{T}_\tau$, $\tau \in R$, iff there is a time-invariant neighborhood $\mathcal{N}$ of $x = 0$ and a positive definite function w on $\mathcal{N}$, $w\colon R^n \to R$, such that $y^{\mathrm{T}}U(t,x)y \le w(x)$, $\forall\,(t,x) \in \mathcal{T}_\tau \times \mathcal{N}$;

(ii) *decreasing on* $\mathcal{T}_\tau \times \mathcal{S}$ iff (i) holds for $\mathcal{N} = \mathcal{S}$;

(iii) *decreasing in the whole on* $\mathcal{T}_\tau$ iff (i) holds for $\mathcal{N} = R^n$.

The expression "on $\mathcal{T}_\tau$" is omitted iff all corresponding conditions still hold for every $\tau \in R$.

PROPOSITION 2.2.2. *The matrix-valued function* $U\colon R \times R^n \to R^{m\times m}$ *is decreasing on* $\mathcal{T}_\tau$, $\tau \in R$, *iff it can be written as*

$$y^{\mathrm{T}}U(t,x)y = y^{\mathrm{T}}U_-(t,x)y + b(\|x\|), \qquad (y \neq 0) \in R^m,$$

where $U_-(t,x)$ *is a negative semi-definite matrix-valued function and* $b \in K$.

Barbashin and Krasovskii [12, 13] discovered the concept of radially unbounded functions. They showed necessity of it for asymptotic stability in the whole.

DEFINITION 2.2.5. The matrix-valued function $U\colon R \times R^n \to R^{m\times m}$ is:

(i) *radially unbounded on* $\mathcal{T}_\tau$, $\tau \in R$, iff $\|x\| \to \infty$ implies $y^{\mathrm{T}}U(t,x)y \to +\infty$, $\forall\, t \in \mathcal{T}_\tau$, $y \in R^m$;

(ii) *radially unbounded* iff $\|x\| \to \infty$ implies $y^{\mathrm{T}}U(t,x)y \to +\infty$, $\forall\, t \in \mathcal{T}_\tau$, $\forall\, \tau \in R$, $y \in R^m$.

PROPOSITION 2.2.3. *The matrix-valued function $U\colon R \times R^n \to R^{m\times m}$ is radially unbounded in the whole (on $\mathcal{T}_\tau$) iff it can be written as*

$$y^{\mathrm{T}}U(t,x)y = y^{\mathrm{T}}U_+(t,x)y + a(\|x\|) \qquad \forall\, x \in R^n,$$

where $U_+(t,x)$ is a positive semi-definite matrix-valued function in the whole (on $\mathcal{T}_\tau$) and $a \in KR$.

2.2.2 Dini derivative and Eulerian derivative

In this section the notations of upper and lower limit of a function $\psi\colon R \to R$ are needed (see McSchane [138]). In brief (see Demidovich [24]) they can be explained as follows.

Let t_k be a member of a sequence S_τ^- (S_τ^+) obeying

(i) $t_k \in R$ for every integer k, $t_k < \tau$ $(t_k > \tau)$

and

(ii) $t_k \to \tau^-$ $(t_k \to \tau^+)$ as $k \to +\infty$.

DEFINITION 2.2.6.

(i) Number $\alpha \in R$ is *the partial limit of the function ψ over the sequence S_τ^- (S_τ^+)* iff for every $\varepsilon > 0$ there is an integer N such that $k > N$ implies $|\psi(t_k) - \alpha| < \varepsilon$;

(ii) the symbol $\alpha = +\infty$ $(\alpha = -\infty)$ is *the partial limit of the function ψ over the sequence S_τ^- (S_τ^+)* iff for every $\varepsilon \in (0,\infty)$ there is an integer N such that, respectively, $k > N$ implies $\psi(t_k) > 1/\varepsilon$ $(\psi(t_k) < -1/\varepsilon)$;

(iii) *the greatest (smallest) partial limit of the function ψ over the sequence S_τ^-* is its left upper (lower) limit at $t = \tau$, respectively, which is denoted by $\limsup\,[\psi(t)\colon t \to \tau^-]$, $(\liminf\,[\psi(t)\colon t \to \tau^-])$;

(iv) *right upper (lower) limit of ψ* at $t = \tau$ is analogously defined when everywhere in (iii) τ^- and S_τ^- are respectively replaced by τ^+ and S_τ^+.

DEFINITION 2.2.7. Let U be a continuous function $U\colon \mathcal{T}_\tau \times R^n \to R^{m\times m}$, $U \in C(\mathcal{T}_\tau \times \mathcal{N})$ and let solutions χ of the system (2.1.7) exist and be defined on $\mathcal{T}_\tau \times \mathcal{N}$. Then, for $(t,x) \in \mathcal{T}_\tau \times \mathcal{N}$

(i)

$$D^+U(t,x) = \limsup \left\{ \frac{U\,[t+\theta,\ \chi(t+\theta(t,x)] - U(t,x)}{\theta} : \theta \to 0^+ \right\}$$

is *the upper right Dini derivative of* U *along the motion* χ *at* (t,x);

(ii)

$$D_+U(t,x) = \liminf \left\{ \frac{U\,[t+\theta,\ \chi(t+\theta(t,x)] - U(t,x)}{\theta} : \theta \to 0^+ \right\}$$

is *the lower right Dini derivative of* U *along the motion* χ *at* (t,x);

(iii)

$$D^-U(t,x) = \limsup \left\{ \frac{U\,[t+\theta,\ \chi(t+\theta(t,x)] - U(t,x)}{\theta} : \theta \to 0^- \right\}$$

is *the upper left Dini derivative of* U *along the motion* χ *at* (t,x);

(iv)

$$D_-U(t,x) = \liminf \left\{ \frac{U\,[t+\theta,\ \chi(t+\theta(t,x)] - U(t,x)}{\theta} : \theta \to 0^- \right\}$$

is *the lower left Dini derivative of* U *along the motion* χ *at* (t,x).

(v) The function U has Eulerian derivative $\dot U$, $\dot U(t,x) = \frac{d}{dt}U(t,x)$ at (t,x) along the motion χ iff

$$D^+U(t,x) = D_+U(t,x) = D^-U(t,x) = D_-U(t,x) = DU(t,x)$$

and then

$$\dot U(t,x) = DU(t,x).$$

If u_{ij} is differentiable at (t,x) then (see Liapunov [101])

$$\dot u_{ij}(t,x) = \frac{\partial u_{ij}}{\partial t} + (\operatorname{grad} u_{ij})^{\mathrm T} f(t,x)$$

and

$$\operatorname{grad} u_{ij} = \left(\frac{\partial u_{ij}}{\partial x_1}, \frac{\partial u_{ij}}{\partial x_2}, \dots, \frac{\partial u_{ij}}{\partial x_n} \right), \qquad i,\ j = 1,2,\dots,s.$$

Effective application of D^+U in the framework of the second Liapunov method is based on the result by Yoshizawa [174], which enables calculation of D^+U without utilizing system motions themselves.

THEOREM 2.2.1. *Let the matrix-valued function U be continuous and locally Lipschitzian in x over $\mathcal{T}_\tau \times \mathcal{S}$ and $\mathcal{S}$ be an open set. Then,*

$$D^+U(t,x) = \limsup \left\{ [U(t+\theta,\, x+\theta f(t,x)) - U(t,x)]\,\theta^{-1} : \theta \to 0^+ \right\}$$

holds along solutions χ of the system (2.1.7) at $(t,x) \in \mathcal{T}_\tau \times \mathcal{G}$.

D^*U will mean that both D^+U and D_+U can be used.

2.3 Direct Liapunov's Method in Terms of Matrix-Function

The following results are useful in the subsequent sections.

PROPOSITION 2.3.1. *Suppose $m(t)$ is continuous on (a,b). Then $m(t)$ is nondecreasing (nonincreasing) on (a,b) iff*

$$D^+m(t) \ge 0 \ (\le 0) \qquad \textit{for every} \quad t \in (a,b),$$

where

$$D^+m(t) = \limsup \{[m(t+\theta) - m(t)]\,\theta^{-1} :\ \theta \to 0^+\}.$$

Following Liapunov [101], Persidskii [152], Yoshizawa [174] and Grujić, Martynyuk and Ribbens-Pavella [54], the next result is obtained.

THEOREM 2.3.1. *Let the vector function f in system (2.1.7) be continuous on $R \times \mathcal{N}$ (on $\mathcal{T}_\tau$). If there exist*

(1) *an open connected time-invariant neighborhood $\mathcal{S} \subseteq \mathcal{N}$ of point $x = 0$;*
(2) *a positive definite on $\mathcal{G}$ (on $\mathcal{T}_\tau \times \mathcal{G}$) matrix-valued function $U(t,x)$ and vector $y \in R^m$ such that function $v(t,x,y) = y^{\mathrm{T}}U(t,x)y$ is locally Lipschitzian in x and $D^+v(t,x,y) \le 0$.*

Then

(a) *the state $x = 0$ of system (2.1.7) is stable (on $\mathcal{T}_\tau$), provided $U(t,x)$ is weakly decreasing on $\mathcal{G}$ (on $\mathcal{T}_\tau \times \mathcal{G}$);*
(b) *the state $x = 0$ of system (2.1.7) is uniformly stable (on $\mathcal{T}_\tau$), provided $U(t,x)$ is decrescent on $\mathcal{G}$ (on $\mathcal{T}_\tau \times \mathcal{G}$).*

PROOF. We shall prove first assertion (a) of Theorem 2.3.1. The fact that function $U(t,x)$ is weakly decreasing on $\mathcal{G}$ (on $\mathcal{T}_\tau \times \mathcal{G}$) implies that for any $t_0 \in R$ ($t_0 \in \mathcal{T}_\tau$) and $x_0 \in \mathcal{G}$ there exists a constant $\delta_0 = \delta(t_0) > 0$, a vector $y \in R^m$ and a function $b \in CK$ such that

$$y^{\mathrm{T}}U(t_0, x_0)y \le b(t_0, \|x_0\|), \qquad \|x_0\| < \delta_0. \tag{2.3.1}$$

Further, since $U(t,x)$ is positive definite on $\mathcal{G}$ (on $\mathcal{T}_\tau \times \mathcal{G}$) then

$$a(\|x\|) \le y^{\mathrm{T}}U(t,x)y \qquad \forall\,(t,x) \in R \times \mathcal{G} \quad (\forall\,(t,x) \in \mathcal{T}_\tau \times \mathcal{G}). \tag{2.3.2}$$

Let $\varepsilon > 0$ and $t_0 \in R\,(t_0 \in \mathcal{T}_\tau)$ are arbitrary. The properties of functions $a \in K$ and $b \in CK$ yield the existence of a $\delta_1 = \delta_1(t_0, \varepsilon) > 0$ continuously dependent on t_0 and such that

$$b(t_0, \delta_1) < a(\varepsilon). \tag{2.3.3}$$

We define $\delta(t_0) = \min\{\delta_0, \delta_1\}$. It is clear that inequalities (2.3.1) – (2.3.3) are satisfied for $\|x_0\| < \delta$. Therefore,

$$a(\|x_0\|) < y^{\mathrm{T}}U(t_0, x_0)y \le b(t\,\|x_0\|) < a(\varepsilon) \tag{2.3.4}$$

which yield $\|x_0\| < \varepsilon$.

Now we claim that for any solution $\chi(t; t_0, x_0)$ of system (2.1.7) with the initial conditions $x_0 \colon \|x_0\| < \delta$ the inequality $\|\chi(t; t_0, x_0)\| < \varepsilon \;\; \forall\, t \in \mathcal{T}_0$ holds. If not, there exists a $t_1 > t_0$ such that

$$\|\chi(t_1; t_0, x_0)\| = \varepsilon \quad \text{and} \quad \|\chi(t; t_0, x_0)\| < \varepsilon \quad \forall\, t \in [t_0, t_1) \tag{2.3.5}$$

for some solution $\chi(t; t_0, x_0)$ of system (2.7.1). Let

$$m(t) = y^{\mathrm{T}}U(t,\, \chi(t; t_0, x_0))y \quad \text{when } t \in [t_0, t_1].$$

Since $v(t,x,y)$ is locally Lipschitzian in x, then we get by condition (2) $D^+v(t,x,y) = D^+m(t) \le 0$.

Hence, we find in view of Proposition 2.3.1 that $m(t)$ is a nonincreasing function on $[t_0, t_1]$. Thus, we have

$$\begin{aligned} a(\varepsilon) = a\left(\|\chi(t_1; t_0, x_0)\|\right) &\le y^{\mathrm{T}}U\left(t_1,\, \chi(t_1; t_0, x_0)\right)y \\ &\le y^{\mathrm{T}}U(t_0, x_0)y \le a(\varepsilon). \end{aligned}$$

The contradiction obtained shows that the state $x = 0$ of system (2.1.7) is stable (on $\mathcal{T}_\tau$).

To prove assertion (b) of Theorem 2.3.1 it is sufficient to note that by condition (b) of Theorem 2.3.1 function $U(t,x)$ is decreasing and function b in inequality (2.3.1) can be taken independent of $t_0 \in R$. This proves the theorem.

THEOREM 2.3.2. *Let the vector function f in system (2.1.7) be continuous on $R \times R^n$ (on $\mathcal{T}_\tau \times R^n$). If there exist*

(1) *radially unbounded positive definite in the whole matrix-valued function $U \in C(R \times R^n, R^{m\times m})$ (or $U \in C(\mathcal{T}_\tau \times R^n, R^{m\times m})$ (on $\mathcal{T}_\tau$) and vector $y \in R^m$ such that the function $v(t,x,y) = y^{\mathrm{T}}U(t,x)y$ is locally Lipschitzian in x and*

$$D^+v(t,x,y) \le 0 \qquad \forall\,(t,x) \in R \times R^n \quad (\forall\,(t,x) \in \mathcal{T}_\tau \times R^n)\,.$$

Then

(a) *the state $x = 0$ of system (2.1.7) is stable in the whole (on $\mathcal{T}_\tau$), provided $U(t,x)$ is weakly decreasing in the whole (on $\mathcal{T}_\tau$);*
(b) *the state $x = 0$ of system (2.1.7) is uniformly stable in the whole (on $\mathcal{T}_\tau$), provided $U(t,x)$ is decreasing in the whole (on $\mathcal{T}_\tau$).*

REMARK 2.3.1. If f is locally Lipschitzian on $R\times\mathcal{N}$ (on $\mathcal{T}_\tau$) then U in the preceding theorems is also locally Lipschitzian on $R\times\mathcal{N}$ (on $\mathcal{T}_\tau$) which enables effective calculation of D^+U via Theorem 2.2.1.

REMARK 2.3.2. The proceding theorems hold also when D^+U is replaced by D_+U (McShane [138] and LaSalle [97]).

Following Liapunov [101], Massera [130, 131], Yoshizawa [174], Halanay [67], Hahn [66], Grujić, Martynyuk and Ribbens-Pavella [57] the next result is obtained.

THEOREM 2.3.3. *Let the vector function f in system (2.1.7) be continuous on $R \times \mathcal{N}$ (on $\mathcal{T}_\tau \times \mathcal{N}$). If there exist*

(1) *open connected time-invariant neighborhood $\mathcal{G} \subseteq \mathcal{N}$ of the point $x = 0$;*
(2) *positive definite on $\mathcal{G}$ (on $\mathcal{T}_\tau \times \mathcal{G}$) matrix-valued function $U(t,x)$, a vector $y \in R^m$ and positive definite on $\mathcal{G}$ function ψ such that the function $v(t,x,y) = y^{\mathrm{T}}U(t,x)y$ is locally Lipschitzian in x and*

$$D^+v(t,x,y) \le -\psi(x) \qquad \forall\,(t,x,y) \in R \times \mathcal{G} \times R^m \quad (\mathcal{T}_\tau \times \mathcal{G} \times R^m)\,.$$

Then

(a) *iff $U(t,x)$ is weakly decreasing on $\mathcal{G}$ (on $\mathcal{T}_\tau \times \mathcal{G}$), the state $x = 0$ of system (2.1.7) is asymptotically stable (on $\mathcal{T}_\tau$);*
(b) *iff $U(t,x)$ is decreasing on $\mathcal{G}$ (on $\mathcal{T}_\tau \times \mathcal{G}$), the state $x = 0$ of system (2.1.7) is uniformly asymptotically stable (on $\mathcal{T}_\tau$).*

PROOF. *Necessity.* Consider assertion (b) of Theorem 2.3.3. Let $x = 0$ of (2.1.7) be uniformly asymptotically stable (on $\mathcal{T}_\tau$). Let $\varepsilon > 0$ be arbitrarily chosen, ζ be such that $B_\zeta \subseteq \mathcal{N}$, $\Delta \in (0, +\infty)$ and $\xi = \min\{\delta_M(\varepsilon), \Delta, \zeta\}$. Let $\mathcal{G} = B_\xi$, $\varphi \in K_{[0,\varepsilon)}$, $\alpha \in (1, +\infty)$, $y = (1,1, \dots, 1)^{\mathrm{T}} \in R^m$ and

$$y^{\mathrm{T}}U(t,x)y = v(t,x) = \sup\{\varphi\,[\|\chi(t+\sigma;\, t,x)\|]\,(1+\alpha\sigma)(1+\sigma)^{-1} : \sigma \in [0,\, +\infty)\},\ \ \forall\, t \in R.$$

The function v is decreasing and positive definite on $\mathcal{N}$ (on $\mathcal{T}_\tau \times \mathcal{G}$) because χ is continuous in all its arguments, $\varphi \in K_{[0,\varepsilon)}$, $(1+\alpha\sigma)(1+\sigma)^{-1}$ is also continuous, $\chi(t;t_0,0) \equiv 0$, $\varphi(0) = 0$, and $\varphi(\|x\|) \le v(t,x) \le \varphi[\Pi(x)]$ $\forall\, t \in \mathcal{T}_0$, $\forall\, t_0 \in R$ ($\forall\, t_0 \in \mathcal{T}_\tau$), $\forall\, x \in \mathcal{G}$, where $\Pi \in K_{[0,\varepsilon)}$.

Let $x^* = \chi(t+\theta;t,x)$, $x = \chi(t;t_0,x_0)$, $\theta > 0$, so that

$$\begin{aligned}
&v(t+\theta,\, x^*)\\
&= \sup\{\varphi\,[\|\chi(t+\theta+\sigma;\, t+\theta,\, x^*)\|]\,(1+\alpha\sigma)(1+\sigma)^{-1} : \sigma \in [0,+\infty)\}\\
&= \sup\{\varphi\,[\|\chi(t+\theta+\sigma;\, t,x)\|]\,(1+\alpha\sigma)(1+\sigma)^{-1} : \sigma \in [0,+\infty)\}\\
&= \varphi\,[\|\chi(t+\theta+\sigma^*;\, t,x)\|]\,(1+\alpha\sigma^*)(1+\sigma^*)^{-1}, \qquad \forall\, t \in R.
\end{aligned}$$

Let $\Delta = \min\left\{1,\ \dfrac{\Delta}{\alpha}\right\}$. The existence of $\sigma^* \in [0,\, \tau_u(\Delta,\nu)]$ obeying the last equation is guaranteed by continuity of χ, $\varphi \in K_{[0,\varepsilon)}$ continuity of $(1+\alpha\sigma)(1+\sigma)^{-1}$ and uniform attraction of $x = 0$.

Let $\sigma = \theta + \sigma^*$. Then (see Halanay [67]),

$$\frac{1+\alpha\sigma^*}{1+\sigma^*} = \frac{1+\alpha\sigma}{1+\sigma}\left[1 - \frac{(\alpha-1)\theta}{(1+\alpha\sigma^*)(1+\sigma)}\right] > 0$$

so that

$$\begin{aligned}
v(t+\theta,\, x^*) &= \varphi\,[\|\chi(t+\theta;\, t,x)\|]\;\frac{1+\alpha\sigma}{1+\sigma}\left[1 - \frac{(\alpha-1)\theta}{(1+\alpha\sigma^*)(1+\sigma)}\right]\\
&\le v(t,x)\left[1 - \frac{(\alpha-1)\theta}{(1+\alpha\sigma^*)(1+\sigma)}\right],\\
&\forall\, t_0 \in R \quad (\forall\, t_0 \in \mathcal{T}_\tau) \quad \forall\, t \in \mathcal{T}_0,
\end{aligned}$$

or $\forall\, t \in \mathcal{T}_0$

$$\frac{v(t+\theta,\, x^*) - v(t,x)}{\theta} \le -\frac{(\alpha-1)v(t,x)}{(1+\sigma^*)(1+\alpha\sigma^*+\alpha\theta)} \qquad \forall\, t_0 \in R \quad (\forall\, t_0 \in \mathcal{T}_\tau).$$

This inequality in limit as $\theta \to 0^+$ takes the form

$$D^*v(t,x) \leq -\psi(x) \qquad \forall\,(t,x) \in R \times \mathcal{G} \quad (\forall\,(t,x) \in \mathcal{T}_\tau \times \mathcal{G}),$$

where

$$\psi(x) = \frac{(\alpha - 1)\varphi(\|x\|)}{[1 + T(\Delta)][1 + \alpha T(\Delta)]},$$

is the continuous function, $T(\Delta) \in [\tau_u(\Delta, \nu), +\infty)$, because $x = 0$ is uniformly attractive (see Halanay [67]).

Hence, ψ is positive definite on $\mathcal{G}$ (on $\mathcal{T}_\tau \times \mathcal{G}$).

Sufficiency. Under the conditions of Theorem 2.3.3 all conditions of Theorem 2.3.1 are fulfilled. Hence, $x = 0$ of (2.1.7) is uniformly stable (on $\mathcal{T}_\tau$). Its uniform attraction (on $\mathcal{T}_\tau$) is proved as follows.

Let ζ be such that $B_\zeta \subseteq \mathcal{G}$. Let φ_1 and $\varphi_2 \in K_{[0,\zeta]}$ obey

$$\begin{gathered} \varphi_1(\|x\|) \leq v(t,x) \leq \varphi_2(\|x\|) \\ \forall\,(t,x) \in R \times B_\zeta \quad (\forall\,(t,x) \in \mathcal{T}_\tau \times B_\zeta). \end{gathered} \tag{2.3.6}$$

Let

$$\Delta = \varphi_2^I[\varphi_1(\zeta)]. \tag{2.3.7}$$

As shown in the proof of the sufficiency part of Theorem 2.3.1, the conditions (2.3.3) and (2.3.4) guarantee that $\|x_0\| < \Delta$ implies

$$v(t,\, \chi(t; t_0, x_0)) \leq v(t_0, x_0) \qquad \forall\, t \in \mathcal{T}_0, \quad \forall\, t_0 \in R \quad (\forall\, t_0 \in \mathcal{T}_\tau)$$

and that v is decreasing in t along motions χ of (2.1.7).

Let

$$\begin{gathered} \inf\,\{v(t,\, \chi(t; t_0, x_0))\colon\ t \in \mathcal{T}_0\} = \nu, \\ \forall\, t_0 \in R \quad (\forall\, t_0 \in \mathcal{T}_\tau), \quad \|x_0\| < \Delta. \end{gathered} \tag{2.3.8}$$

Obviously $\nu \geq 0$. If $\nu > 0$ then

$$D^*v(t,\, \chi(t; t_0, x_0)) \leq -\gamma,$$

where

$$\gamma = \inf\,\{w(x)\colon\ x \in \partial B_\rho,\ \rho = \varphi_2^I(\nu)\}.$$

Therefore,

$$v(t,\, \chi(t; t_0, x_0)) \leq v(t_0, x_0) - \gamma(t - t_0),$$

so that

$$v(t,\ \chi(t;t_0,x_0)) < \nu \quad \text{for} \quad t \in \left(\frac{v(t_0,x_0)-\nu}{\gamma} + t_0,\ \infty\right),$$

which contradicts (2.3.8). Hence, $\gamma = 0$ which together with (2.3.5), (2.3.8) and positive definiteness of v on $\mathcal{G}$ (on $\mathcal{T}_\tau \times \mathcal{G}$) prove that $\|x_0\| < \Delta$ implies

$$\lim\ [\|\chi(t;t_0,x_0)\|\colon\ t \to +\infty] = 0, \qquad \forall\, t \in R \quad (\forall\, t_0 \in \mathcal{T}_\tau),$$

i.e. that $x = 0$ is attractive. Let now $\rho > 0$ be arbitrarily chosen,

$$\gamma = \lim\{w(x)\colon x \in B_\Delta \setminus B_\rho\}, \qquad \gamma = \gamma(\rho),$$

and

$$\tau_u(\Delta,\rho) = \min\left\{0,\ \frac{\varphi_2(\Delta)-\varphi_1(\rho)}{\gamma(\rho)}\right\}.$$

Then

$$D^* v(t,x) \le -\gamma,$$

$$\forall\,(t,x) \in R \times (B_\Delta \setminus B_\rho) \quad (\forall\,(t,x) \in \mathcal{T}_\tau \times (B_\Delta \setminus B_\rho))$$

and for $t = \tau_u(\Delta,\rho) + t_0$, $\tau_u(\Delta,\rho) > 0$, $\tau_u(\Delta,\rho) = 0$ implies

$$\|\chi(t;t_0,x_0)\| < \rho \qquad \forall\, t \in \mathcal{T}_0,$$

$$v(t,\ \chi(t;t_0,x_0)) \le v(t_0,x_0) - \gamma(t-t_0) \le \varphi_2(\Delta) - \varphi_2(\Delta) + \varphi_1(\rho) = \varphi_1(\rho)$$

so that

$$\varphi_1(\|\chi(t;t_0,x_0)\|) \le \varphi_1(\rho), \qquad \forall\ t_0 \in R \quad (\forall\ t_0 \in \mathcal{T}_\tau)$$

yields

$$\|\chi(t;t_0,x_0)\| \le \rho \quad \text{at} \quad t = \tau_u(\Delta,\rho) + t_0, \quad \forall\ x_0 \in B_\Delta.$$

For $t \in (\tau_u(\Delta,\rho),\ +\infty)$

$$v(t,\ \chi(t;t_0,x_0)) < v(t_0 + \tau_u(\Delta,\rho);$$

$$\chi(t_0 + \tau_u(\Delta,\rho),\ t_0,x_0)) \le \varphi_1(\rho)$$

so that

$$\|\chi(t;t_0,x_0)\| < \rho,$$

$$\forall\ t \in (t_0 + \tau_u(\Delta,\rho),\ +\infty), \quad \forall\, t_0 \in R\ (t_0 \in \mathcal{T}_\tau), \quad \forall\, x_0 \in B_\Delta$$

which proves that attraction of $x = 0$ is uniform (on $\mathcal{T}_\tau$).

Following Barbashin and Krasovskii [12, 13] and Martynyuk [116], and the proceding proof in which we choose $\varphi \in KR$ it is easy to prove.

THEOREM 2.3.4. *Let the vector function f in system (2.1.7) be continuous on $R \times R^n$ (on $\mathcal{T}_\tau \times R^n$). If there exist*

(1) *radially unbounded positive definite in the whole matrix-valued function $U(t,x) \in C(R \times R^n, R^{m\times m})$ (or $U(t,x) \in C(\mathcal{T}_\tau \times R^n, R^{m\times m})$ (on $\mathcal{T}_\tau$), a vector $y \in R^m$ and a positive definite in the whole function θ, such that the function*

$$v(t,x,y) = y^{\mathrm{T}}U(t,x)y$$

is locally Lipschitzian in x and

$$D^+v(t,x,y) \le -\theta(x) \qquad \forall\,(t,x,y) \in R \times R^n \times R^m$$
$$\forall\,(t,x,y) \in \mathcal{T}_\tau \times R^n \times R^m.$$

Then

(a) *iff $U(t,x)$ is weakly decreasing in the whole (on $\mathcal{T}_\tau$), the state $x = 0$ of system (2.1.7) is asymptotically stable in the whole (on $\mathcal{T}_\tau$);*

(b) *iff $U(t,x)$ is decreasing in the whole (on $\mathcal{T}_\tau$), the state $x = 0$ of system (2.1.7) is uniformly asymptotically stable in the whole (on $\mathcal{T}_\tau$).*

Following Krasovskii [89], Grujić, Martynyuk and Ribbens-Pavella [57] and He and Wang [72] and utilizing $\varphi(\zeta) = \zeta^p$ in the proof of Theorem 2.3.3, it is easy to prove the following result.

THEOREM 2.3.5. *Let the vector function f in system (2.1.7) be continuous on $R \times \mathcal{N}$ (on $\mathcal{T}_\tau \times \mathcal{N}$). If there exist*

(1) *an open connected time-invariant neighborhood $\mathcal{G} \subseteq \mathcal{N}$ of the point $x = 0$;*

(2) *a matrix-valued function $U(t,x)$ and a vector $y \in R^m$ such that the function $v(t,x,y) = y^{\mathrm{T}}U(t,x)y$ is locally Lipschitzian in x;*

(3) *functions φ_1, $\varphi_2 \in K$ and a positive real number η_1 and positive integer p such that*

$$\eta_1\|x\|^p \le v(t,x,y) \le \varphi_1(\|x\|) \qquad \forall\,(t,x,\eta \ne 0) \in R \times \mathcal{G} \times R^m$$

and

$$D^+v(t,x,y) \le -\varphi_2(\|x\|) \qquad \forall\,(t,x,\eta \ne 0) \in R \times \mathcal{G} \times R^m$$
$$(\forall\,(t,x,\eta \ne 0) \in \mathcal{T}_\tau \times \mathcal{G} \times R^m).$$

Then, iff the comparison functions φ_1 *and* φ_2 *are of the same magnitude, the state* $x = 0$ *of system (2.1.7) is exponentially stable (on* $\mathcal{T}_\tau$*).*

REMARK 2.3.2. The statement of Theorem 2.3.6 remains valid, if $\varphi_1(\|x\|) = \eta_2\|x\|^p$ and $\varphi_2(\|x\|) = \eta_3\|x\|^p$, $\eta_2, \eta_3 = \text{const} > 0$.

THEOREM 2.3.6. *Let the vector function* f *in system (2.1.7) be continuous on* $R \times R^n$ *(on* $\mathcal{T}_\tau \times R^n$*). If there exist*

(1) *radially unbounded positive definite in the whole matrix-valued function* $U(t,x) \in C(R \times R^n, R^{m\times m})$ *(or* $U(t,x) \in C(\mathcal{T}_\tau \times R^n, R^{m\times m})$*) (on* $\mathcal{T}_\tau$*) and vector* $y \in R^m$ *such that the function*

$$v(t,x,\eta) = y^{\mathrm{T}}U(t,x)y$$

is locally Lipschitzian in x*;*

(2) *functions* $\psi_1, \psi_2 \in KR$ *a positive real number* η_1 *and positive integer* q *such that*

$$\eta_2\|x\|^q \le v(t,x,y) \le \psi_1(\|x\|) \qquad \forall\,(t,x,\,y \ne 0) \in R \times R^n \times R^m$$
$$\forall\,(t,x,\,y \ne 0) \in \mathcal{T}_\tau \times R^n \times R^m$$

and

$$D^+v(t,x,y) \le -\psi_2(\|x\|) \qquad \forall\,(t,x,\,y \ne 0) \in R \times R^n \times R^m$$
$$\forall\,(t,x,\,y \ne 0) \in \mathcal{T}_\tau \times R^n \times R^m.$$

Then, if the comparison functions ψ_1, ψ_2 *, are of the same magnitude, the state* $x = 0$ *of system (2.1.7) is exponentially stable in the whole (on* $\mathcal{T}_\tau$*).*

PROOF. The proof is similar to that of Theorem 2.3.5.

REMARK 2.3.3. The assertion of Theorem 2.3.6 remains valid, if $\varphi_1(\|x\|) = \eta_2\|x\|^q$ and $\varphi_2(\|x\|) = \eta_3\|x\|^q$.

PROPOSITION 2.3.2. *In order that the state* $x = 0$ *of system (2.1.7) be exponentially stable (on* $\mathcal{T}_\tau$*) in the whole, it is necessary and sufficient for it to be exponentially stable (on* $\mathcal{T}_\tau$*) and uniformly asymptotically stable in the whole (on* $\mathcal{T}_\tau$*).*

Following Zubov [178] and taking into account the results by Martynyuk [116] we shall formulate and prove a result on instability.

THEOREM 2.3.7. *Let the vector function f in system (2.1.7) be continuous on $R \times \mathcal{N}$ (on $\mathcal{T}_\tau \times \mathcal{N}$). If there exist*

(1) *an open connected time-invariant neighborhood $\mathcal{G} \subseteq \mathcal{N}$ of the point $x = 0$;*
(2) *a matrix-valued function $U(t,x) \in C^{1,1}(R \times \mathcal{G}, R^{m\times m})$ or $U(t,x) \in C^{1,1}(\mathcal{T}_\tau \times \mathcal{G}, R^{m\times m})$ and a vector $y \in R^m$ such that the function $v(t,x,y) = y^{\mathrm{T}}U(t,x)y$ is strictly positive semi-definite (on $\mathcal{T}_\tau$) and satisfies the relation*

$$\frac{dv}{dt} = \lambda v + \widetilde{\theta}(x), \qquad \lambda = \lambda(t,x),$$

where $\widetilde{\theta}(x)$ is a positive semi-definite function on $\mathcal{G}$;
(3) *a number $\varepsilon > 0$ such that when $\delta > 0$ $(\delta < \varepsilon)$ for continuous on $\mathcal{T}_0 \times R \times \mathcal{G}$ (on $\mathcal{T}_0 \times \mathcal{T}_\tau \times \mathcal{G}$) solution $\chi(t;t_0,x_0)$ of system (2.1.7) which satisfies the condition $\|x_0\| < \delta$, $v(t_0,x_0) > 0$ implies $\|\chi(t;t_0,x_0)\| < \varepsilon$ $\forall\, t \in R$ $(\forall\, t_0 \in \mathcal{T}_\tau)$ the inequality*

$$|v(t,\, \chi(t;t_0,x_0),\, y)| \geq v(t_0,x_0,y) \exp\left(\int_{t_0}^{t} \lambda(s)\, ds\right)$$

does not hold for all $t \geq t_0$, $t_0 \in R$ $(t_0 \in \mathcal{T}_\tau)$, $t \in \mathcal{T}_0$.

Then and only then the state $x = 0$ of system (2.1.7) is unstable (on $\mathcal{T}_\tau$).

PROOF. *Necessity.* Let the state $x = 0$ of system (2.1.7) be unstable (on $\mathcal{T}_\tau$). We construct two functions v and $\widetilde{\theta}$ satisfying the conditions of Theorem 2.3.7. The instability (on $\mathcal{T}_\tau$) of state $x = 0$ of system (2.1.7) yields the existence of an $\varepsilon^* > 0$ such that for any $\delta > 0$ a x_0 and a t_0, $t_0 \in R$ $(t_0 \in \mathcal{T}_\tau)$ can be taken so that the inequality

$$\|\chi(t;t_0,x_0)\| < \varepsilon^* \tag{2.3.9}$$

does not hold for all $t \geq 0$ in spite of the fact that $\|x_0\| < \delta$, $t_0 \geq 0$.

Let $t = t(t_0,x_0)$ be the next time after t_0 when inequality (2.3.9) is violated. The set of points $\Pi = \{(t_0,x_0)\colon \|x_0\| < \delta,\ t_0 \geq 0\}$ is divided conventionally into sets Π_1 and Π_2 such that

(A) for $(t_0,x_0) \in \Pi_1$ the solutions $\chi(t;t_0,x_0)$ of system (2.1.7) satisfy condition (2.3.9) provided all $t \geq t_0$.
(B) for $(t_0,x_0) \in \Pi_2$ the solutions $\chi(t;t_0,x_0)$ of system (2.1.7) intersect the surface $\|x\| = \varepsilon^*$ when the time increases.

We set in case (A) $y = (1, 1, \dots, 1) \in R^m$ and

$$y^{\mathrm{T}}U(t,x)y = v(t,x) \equiv 0 \qquad \forall\, t \in \mathcal{T}_0. \tag{2.3.10}$$

For the case (B) we set $y = (1, 1, \dots, 1) \in R^m$ and

$$y^{\mathrm{T}}U(t_0,x_0)y = v(t_0,x_0) = \exp\,(t_0 - t(t_0,x_0))\,. \tag{2.3.11}$$

It is clear that $v(t,x_0) \equiv 0$ for function (2.3.10) when $t \geq t_0$, and

$$v(t,x_0) = \exp\,(t_0 - t(t_0,x_0))$$

for function (2.3.11).

Hence, we get $dv/dt = v$. Comparing this result with condition (2) of Theorem 2.3.7 we obtain $\lambda = 1$ and $\widetilde{\theta} = 0$.

Function v is strictly positive semi-definite (on $\mathcal{T}_\tau$) and bounded, $\int\limits_{t_0}^{t} \lambda(s)\,ds$ diverges as $t - t_0 \to \infty$, since $\lambda \equiv 1$. Therefore, condition (3) of the Theorem 2.3.7 is also satisfied.

Sufficiency. Let all hypotheses of Theorem 2.3.7 be satisfied.

We are going to show that the state $x = 0$ of system (2.1.7) is unstable (on $\mathcal{T}_\tau$). If not, then using $\varepsilon > 0$ a $\delta > 0$ can be taken so that

$$\|\chi(t;t_0,x_0)\| < \varepsilon \qquad \forall\, t \in \mathcal{T}_0, \quad \forall\, t \in R \quad (\forall\, t_0 \in \mathcal{T}_\tau)\,, \tag{2.3.12}$$

when $\|x_0\| < \delta$.

According to condition (2) of Theorem 2.3.7 we take t_0 and x_0 so that $v(t_0,x_0,y) > 0$ and consider along the solution $\chi(t;t_0,x_0)$ of system (2.1.7) the correlation

$$\frac{dQ}{dt} = \lambda Q(t) + P(t) \qquad \forall\, t \in R \quad (\forall\, t \in \mathcal{T}_\tau)\,, \tag{2.3.13}$$

where $Q(t) = v\,(t,\, \chi(t;t_0,x_0))$ and $P(t) = \widetilde{\theta}\,(\chi(t;t_0,x_0))$.

In view of $P(t) \geq 0$ for all $t \in R$ we find from correlation (2.3.13)

$$Q(t) \geq Q(t_0)\,\exp\left(\int\limits_{t_0}^{t} \lambda(s)\,ds\right) \qquad \forall\, t \in R \quad (\forall\, t \in \mathcal{T}_\tau)\,.$$

This inequality contradicts condition (3) of Theorem 2.3.7 and, therefore, inequality (2.3.9) can be satisfied for all $t \in \mathcal{T}_0$, i.e. the state $x = 0$ of system (2.1.7) is unstable (on $\mathcal{T}_\tau$).

COROLLARY 2.3.1. If conditions (1) and (2) of Theorem 2.3.7 are satisfied and

(1) the function $v(t, x, y) = y^{\mathrm{T}} U(t, x) y$ is bounded (on $\mathcal{T}_\tau$);

(2) $\int\limits_{t_0}^{t} \lambda(s)\, ds \to +\infty$ as $t \to +\infty$.

Then the state $x = 0$ of system (2.1.7) is unstable (on $\mathcal{T}_\tau$).

COROLLARY 2.3.2. If conditions (1) and (2) of Theorem 2.3.7 are satisfied and

(1) the function $v(t, x, y) = y^{\mathrm{T}} U(t, x) y$ is bounded (on $\mathcal{T}_\tau$);

(2) the function λ is a positive constant.

Then the state $x = 0$ of system (2.1.7) is unstable (on $\mathcal{T}_\tau$).

REMARK 2.3.4. Corollary 2.3.2 is a new version of Liapunov's theorem on instability (cf. Liapunov [101], Theorem III, pp. 68).

COROLLARY 2.3.3. If conditions (1) and (2) of Theorem 2.3.7 are satisfied and

(1) $\dfrac{dv}{dt} = \widetilde{\theta}(x) \quad \forall\, t \in \mathcal{T}_0 \ (\forall\, t \in \mathcal{T}_\tau) \ \ \forall\, x \in \mathcal{G}$;

(2) using number $\varepsilon > 0$ and $\delta > 0$ can be taken so that $\widetilde{\theta}(x) > 0$ for $v(t, x, y) > \varepsilon$.

Then the state $x = 0$ of system (2.1.7) is unstable (on $\mathcal{T}_\tau$).

REMARK 2.3.5. Corollary 2.3.3 is a new version of Chetaev's theorem on instability (cf. Chetaev [19], pp. 33).

2.4 On Comparison Method

The concept of the matrix Liapunov function together with the theory of differential inequalities provides a very general comparison principle under much less restrictive assumptions. In this set up, the matrix Liapunov function may be viewed as a transformation that reduces the study of a given complicated differential system to the study of relatively simpler scalar differential equations.

2.4.1 Differential inequalities

Let us consider the following scalar differential equation

$$\frac{du}{dt} = g(t,u), \qquad u(t_0) = u_0 \geq 0, \quad t_0 \in R \ (t_0 \in \mathcal{T}_\tau), \tag{2.4.1}$$

where $g \in C(R \times R, R)$ (or $g \in C(\mathcal{T}_\tau \times R, R)$) and $g(t,0) = 0 \ \forall t \in \mathcal{T}_0$).

DEFINITION 2.4.1. Let $\gamma(t)$ be a solution of (2.3.11) existing on some interval $J = [t_0, t_0 + \alpha)$, $0 < \alpha \leq +\infty$, $t_0 \in R$ $(t_0 \in \mathcal{T}_\tau)$. Then $\gamma(t)$ is said to be *the maximal solution of* (2.4.1) if for every solution $u(t) = u(t; t_0, x_0)$ of (2.4.1) existing on J, the following inequalities hold

$$u(t) \leq \gamma(t), \qquad t \in \mathcal{G}, \quad t_0 \in R \ (t_0 \in \mathcal{T}_\tau). \tag{2.4.2}$$

A minimal solution is defined similarly by reversing the inequality (2.4.2).

We need the following known results for our discussion the proof of which may be found in (see e.g. Olech and Opial [150], Yoshizawa [174], and Lakshmikantham, Leela and Martynyuk [94]).

PROPOSITION 2.4.1. *Let $g \in C(R \times R, R)$ (or $g \in C(\mathcal{T}_\tau \times R, R)$) and $\gamma(t) = \gamma(t; t_0, x_0)$ be the maximal solution of (2.4.1) existing on J. Suppose that $m \in C(R, R_+)$ $(m \in C(\mathcal{T}_\tau, R_+))$ and*

$$D^* m(t) \leq g\left(t, m(t)\right), \qquad t \in J, \tag{2.4.3}$$

where D^* is any fixed Dini derivative.

Then $m(t_0) \leq u_0$ implies

$$m(t) \leq \gamma(t), \qquad \forall t \in J. \tag{2.4.4}$$

PROPOSITION 2.4.2. *Let $g \in C(R \times R, R)$ (or $g \in C(\mathcal{T}_\tau \times R, R)$) and $\rho(t) = \rho(t; t_0, x_0)$ be the minimal solution of (2.4.1) existing on J. Suppose that $m \in C(R, R_+)$ $(m \in C(\mathcal{T}_\tau, R_+))$ and*

$$D^* m(t) \geq g\left(t, m(t)\right), \qquad t \in J. \tag{2.4.5}$$

Then $m(t_0) \geq u_0$ implies

$$m(t) \geq \rho(t), \qquad \forall t \in J. \tag{2.4.6}$$

PROPOSITION 2.4.3. *Let for system (1.2.7) there exist*

(1) *a matrix-valued function* $U \in C(R \times R^n, R^{m\times m})$ $(U \in C(\mathcal{T}_\tau \times R^n, R^{m\times m}))$ *and a vector* $y \in R^m$ *such that the function* $v(t,x,y) = y^{\mathrm{T}}U(t,x)y$ *is locally Lipschitzian in* x *for every* $t \in R$ $(t \in \mathcal{T}_\tau)$:
(2) *a majorizing function* $g \in C(R \times R_+, R)$ $g \in C(\mathcal{T}_\tau \times R_+, R)$, $g(t,0) = 0$ $\forall\, t \in \mathcal{T}_0$ $(\forall\, t \in \mathcal{T}_\tau)$ *such that*

$$D^+v(t,x,y) \le g\,(t,\, v(t,x,y))$$
$$\forall\,(t,x,y) \in R \times R^n \times R^m \quad (\mathcal{T}_\tau \times R^n \times R^m)\,;$$

(3) *a maximal solution* $\gamma(t) = \gamma(t;t_0,u_0)$ *of comparison equation (2.4.1) on* J.

Then along any solution $\chi(t;t_0,x_0)$ *of system (1.2.7) existing on* $J_1 \subseteq J$ *the estimate*

$$v(t_0,x_0,u_0) \le u_0, \qquad t_0 \in R\ (t_0 \in \mathcal{T}_\tau) \tag{2.4.7}$$

implies the inequality

$$v\,(t,\, \chi(t;t_0,x_0), y) \le \gamma(t) \qquad \forall\, t \in J_1 \cap J. \tag{2.4.8}$$

PROOF. Let $m(t) = v\,(t,\, \chi(t;t_0,x_0), y)$ and $\chi(t;t_0,x_0)$ being a solution of (1.2.7) such that (2.4.7). Since $v(t,x,y)$ is locally Lipschitzian in x, we get, by (1.2.7) and (2.4.1), the differential inequality

$$D^+m(t) \le g\,(t,\, m(t))\,, \qquad m(t_0) \le u_0, \quad t_0 \in R\ (t_0 \in \mathcal{T}_\tau), \quad t \in J,$$

and Proposition 2.4.1 gives the desired result (2.4.8).

A comparison result analogous to Proposition 2.4.3 which yields lower bounds is the following.

PROPOSITION 2.4.4. *If in Proposition 2.4.3, assumption (2) is reversed to*

$$D^+v(t,x,y) \ge g\,(t,\, v(t,x,y))$$
$$\forall\,(t,x,y) \in R \times R^n \times R^m \quad (\mathcal{T}_\tau \times R^n \times R^m)$$

and $\rho(t) = \rho(t;t_0,u_0)$ *is the minimal solution of (2.4.1) existing for* $t \ge t_0$, *then*

$$v\,(t,\, \chi(t;t_0,x_0), y) \ge \rho(t) \qquad \forall\, t \in J, \quad t_0 \in R\ (t_0 \in \mathcal{T}_\tau)$$

whenever $v(t_0,x_0,y) \ge u_0$.

PROOF is similar to proof of Proposition 2.4.3.

In some situations, estimating $D^+v(t,x,y)$ as a function of t, x and $v(t,x,y)$ is more natural (see e.g. Matrosov [134], Hatvani [71], and Grujić, Martynyuk and Ribbens-Pavella [57]).

PROPOSITION 2.4.5. *Let for system (1.2.7) there exist*

(1) *a matrix-valued function* $U \in C(R\times R^n,\ R^{m\times m})$ $(U \in C(\mathcal{T}_\tau \times R^n, R^{m\times m}))$ *and a vector* $y \in R^m$ *such that the function* $v(t,x,y) = y^{\mathrm{T}}U(t,x)y$ *is locally Lipschitzian in* x *for every* $t \in R$ $(t \in \mathcal{T}_\tau)$*:*

(2) *a majorizing function* $g \in C(R \times R^n \times R_+, R)$ $g \in C(\mathcal{T}_\tau \times R^n \times R_+, R)$, $g(t,x,u)$ *nondecreasing in* u, $g(t,0,0) = 0$ $\forall\, t \in \mathcal{T}_0$ *such that*

$$D^+v(t,x,y) \le g\,(t,x,\ v(t,x,y)) \qquad \forall\,(t,x,y) \in R \times R^n \times R^m$$
$$(\forall\,(t,x,y) \in \mathcal{T}_\tau \times R^n \times R^m)\,;$$

(3) *a maximal solution* $r(t) = r(t;t_0,u_0,x_0)$ *of comparison equation*

$$\frac{du}{dt} = g(t,x(t),u), \qquad u(t_0) = u_0 \ge 0$$

exist for all $t \ge t_0$, $t_0 \in R$ $(t_0 \in \mathcal{T}_\tau)$.

Then $v(t_0,x_0,y) \le u_0$ *implies*

$$v\,(t,\chi(t;t_0,x_0),y) \le r(t;t_0,u_0,x_0) \qquad t \in J.$$

PROOF is similar to proof of Proposition 2.4.3.

COROLLARY 2.4.1. If in conditions of Proposition 2.4.3

(i) $g(t,u) \equiv 0$ $\ \forall\, t \in R$ $(\forall\, t \in \mathcal{T}_\tau)$
then

$$v\,(t,\chi(t;t_0,x_0),y) \le v(t_0,x_0,y) \qquad \forall\, t \in R\ (\forall\, t \in \mathcal{T}_\tau);$$

(ii) $g(t,u) = \lambda(t)u$
then

$$v\,(t,\chi(t;t_0,x_0),y) \le u_0 \exp\left(\int_{t_0}^{t} \lambda(s)\,ds\right),$$
$$t \ge t_0, \quad t_0 \in R\ (t_0 \in \mathcal{T}_\tau);$$

(iii) $g(t,u) = a \exp\,[-ku] + \varphi(t) - a$, where $\varphi\colon R \to R$; a, $k =$ const, then

$$v\left(t,\, \chi(t; t_0, x_0), y\right) \le \psi(t) + k^{-1} \ln \Big\{ \exp\,(ku_0)$$
$$+ak \int\limits_0^t \exp\,(-k\psi(s))\; ds \Big\}, \qquad t \ge t_0;$$

(iv) $g(t,u) = -c(u),\ c \in K$
then

$$v\left(t,\, \chi(t; t_0, x_0), y\right) \le G^{-1}\left[G(u_0) - (t - t_0)\right], \qquad t \ge t_0$$

where $G(u) = \int\limits_0^u \dfrac{ds}{c(s)}$ if $\int\limits_0^u \dfrac{ds}{c(s)} < +\infty$ and
$G(u) = \int\limits_\delta^u \dfrac{ds}{c(s)}$ if $\int\limits_0^u \dfrac{ds}{c(s)} = \infty$; $\delta = \text{const} > 0$,
G^{-1} is a function converse to the function G.

2.4.2 Theorems on stability via matrix Liapunov functions and scalar comparison equations

The estimates of function $v(t,x,y)$ found in Propositions 2.4.3 – 2.4.4 allow the reduction of the problem on stability of state $x = 0$ of system (1.2.7) to the stability investigation of solution $u = 0$ of equation (2.4.1). Let us formulate first stability definitions for solution $u = 0$ of equation (2.4.1).

DEFINITION 2.4.2. The state $u = 0$ of the equation (2.4.1) is:

(i) *stable with respect to* $\mathcal{T}_i$ iff for any $t_0 \in \mathcal{T}_i$ and any $\varepsilon \in (0, \infty)$ there exists $\delta(t_0, \varepsilon)$ such that for any u_0, $0 \le u_0 < \delta$ an estimation $\gamma(t; t_0 u_0) < \varepsilon$ is fulfilled for all $t \in \mathcal{T}_0$;

(ii) *uniformly stable with respect to* $\mathcal{T}_i$ iff conditions of the definition 2.4.2 (i) are fulfilled and for every $\varepsilon \in (0, \infty)$ the corresponding maximal δ denoted by $\delta_M(t, \varepsilon)$ obeys:

$$\inf\,(\delta_M(t,\varepsilon) \colon t \in \mathcal{T}_i) > 0;$$

(iii) *stable in the whole with respect to* $\mathcal{T}_i$ iff conditions of the definition 2.4.2 (i) are fulfilled and $\delta_M(t,\varepsilon) \to +\infty$, $\forall\, \varepsilon \to +\infty$, $\forall\, t \in \mathcal{T}_i$;

(iv) *uniformly stable in the whole with respect to* $\mathcal{T}_i$ iff conditions of the definition 2.4.2 (ii), (iii) are fulfilled.

DEFINITION 2.4.3. The state $u = 0$ for the equation (2.4.1) is:

(i) *attractive with respect to* $\mathcal{T}_i$ iff for any $t_0 \in \mathcal{T}_i$ there exists $\Delta(t_0) > 0$ and for any $\zeta > 0$ and $u_0, : 0 \le u_0 \le \Delta(t_0)$ there is $\tau(t_0, x_0, \zeta) \in [0, \infty)$ such that an estimation $\gamma(t; t_0, u_0) < \zeta$ is fulfilled for all $t \in (t_0 + \tau(t_0, x_0, \zeta) + \infty)$;

(ii) u_0 – *attractive with respect to* $\mathcal{T}_i$ iff conditions under (i) are fulfilled and for any $t_0 \in \mathcal{T}_i$ and any $\eta \in (0, +\infty)$ there exists $\Delta(t_0) > 0$ and $\tau_u(t_0, \Delta(t_0), \eta) \in [0, +\infty)$ such that

$$\sup\,(\tau_m(t, u_0, \eta) \colon 0 \le u_0 \le \Delta(t_0)) = \tau_u(t_0, \Delta(t_0), \eta);$$

(iii) t_0 – *uniformly attractive with respect to* $\mathcal{T}_i$ iff conditions of (i) are fulfilled and for any $\eta \in (0, +\infty)$ there exists $\Delta > 0$ and $\tau_u(u_0, \eta) \in [0, +\infty)$ such that

$$\sup\,(\tau_m(t_0, u_0, \eta) \colon (t_0, u_0 \in \mathcal{T}_i) = \tau_u(u_0, \eta);$$

(iv) *uniformly attractive with respect to* $\mathcal{T}_i$ iff conditions of the definitions 2.4.3 (i) – (iii) are fulfilled and for any $\eta \in (0, +\infty)$ there exists $\Delta > 0$ and $\tau_u(\Delta, \eta) \in [0, +\infty)$ such that

$$\sup\,(\tau_m(t_0, u_0, \eta) \colon (t_0, u_0) \in \mathcal{T}_i \times [0 \le u_0 \le \Delta]) = \tau_u(\Delta, \eta).$$

DEFINITION 2.4.4. The state $u = 0$ of the equation (2.4.1) is:

(i) *asymptotically stable with respect to* $\mathcal{T}_i$ iff it is stable with respect to $\mathcal{T}_i$ and attractive with respect to $\mathcal{T}_i$;

(ii) *equi-asymptotically stable with respect to* $\mathcal{T}_i$ iff it is stable with respect to $\mathcal{T}_i$ and u_0 - uniformly attractive with respect to $\mathcal{T}_i$;

(iii) *quasi-asymptotically stable with respect to* $\mathcal{T}_i$ iff it is uniformly stable with respect to $\mathcal{T}_i$ and t_0 - uniformly attractive with respect to $\mathcal{T}_i$;

(iv) *exponentially stable with respect to* $\mathcal{T}_i$ iff there exists $\Delta > 0$ and real values $\alpha \ge 1$ such that for $0 \le u_0 \le \Delta$ the inequality

$$u(t; t_0, u_0) \le \alpha u_0 \exp\,(-\beta(t - t_0)), \qquad \forall\, t \in \mathcal{T}_0, \quad \forall\, t_0 \in \mathcal{T}_i$$

is valid.

Definitions 2.4.4 (i) – (iv) become the corresponding definitions of asymptotic stability in the whole provided both the corresponding type of stability in the whole and attraction in the whole.

In Definitions 2.4.2 – 2.4.4, the expression "with respect to $\mathcal{T}_i$" can be omitted iff $\mathcal{T}_i = \mathcal{R}$.

THEOREM 2.4.1. *Let vector-function f in system (1.2.7) be continuous on $R \times \mathcal{N}$ (on $\mathcal{T}_\tau \times \mathcal{N}$). If there exist*

(1) *an open connected time-invariant neighborhood $\mathcal{G} \subset \mathcal{N}$ of point $x = 0$;*
(2) *a matrix-valued function $U \in C(R \times \mathcal{G}, R^{m\times m})$ $(U \in C(\mathcal{T}_\tau \times \mathcal{G}, R^{m\times m}))$ and vector $y \in R^m$ such that the function $v(t,x,y) = y^{\mathrm{T}}U(t,x)y$ is locally Lipschitzian in x for every $t \in R$ $(t \in \mathcal{T}_\tau)$;*
(3) *a majorizing function $g \in C(R \times R_+, R)$ $(g \in C(\mathcal{T}_\tau \times R_+, R))$ $g(t,0) = 0$ $\forall\, t \in \mathcal{T}_0$ $(\forall\, t \in \mathcal{T}_\tau)$ such that*

$$D^+v(t,x,y) \le g(t, v(t,x,y)) \qquad \forall\,(t,x,y) \in R \times \mathcal{G} \times R^m$$
$$(\forall\,(t,x,y) \in \mathcal{T}_\tau \times \mathcal{G} \times R^m).$$

Then properties of the function

$$y^{\mathrm{T}}U(t,x)y = v(t,x,y) \tag{2.4.9}$$

and properties of the zero solution of the equation (2.4.1)

$$\frac{du}{dt} = g(t,u), \qquad u(t_0) = u_0 \ge 0$$

provide the corresponding properties of the state $x = 0$ of system (1.2.7).

REMARK 2.4.1. In condition (2) of Theorem 2.4.1 alongside the function defined by (2.4.9) another suitable function, such as

$$v(t,x) = \max\{u_{ij}(t,x)\colon (i,j) \in [1,m]\},$$
$$v(t,x,\eta) = \eta^{\mathrm{T}}U(t,x)\eta, \qquad \eta \in R^m_+, \quad \eta > 0,$$
$$\text{or} \quad v(t,x) = Q\,(U(t,x))$$

can be utilized, where $Q \in C\,(R^{m\times m}, R_+)$, $Q(u)$ is nondecreasing in u and $Q(0) = 0$.

We shall state some properties of the zero solution of equation (2.4.1) and function (2.4.9) and prove Theorem 2.4.1.

PROOF. *Case A.* Let $g(t,u) = 0$, solution $u = 0$ of equation (2.4.1) be stable with respect to $\mathcal{T}_i$ and function (2.4.9) be positive definite on $\mathcal{G}$ $(\mathcal{T}_\tau \times \mathcal{G})$. Then, by Theorem 2.3.1 the state $x = 0$ of system (1.2.7) is stable (on $\mathcal{T}_\tau$).

Case B. Let solution $u = 0$ of equation (2.4.1 be stable with respect to $\mathcal{T}_i$, u_0 – uniformly attractive on $\mathcal{T}_i$ and function (2.4.9) is decreasing on $\mathcal{G}$ $(\mathcal{T}_\tau \times \mathcal{G})$. We are going to show that in this case the state $x = 0$ of system (1.2.7) is equi-asymptotically stable (on $\mathcal{T}_\tau$). In fact, for the function $v(t, x, y)$ mentioned above there exist functions a, $b \in K$ such that

$$a(\|x\|) \le v(t,x,y) \le b(\|x\|) \qquad \forall\,(t,x,y) \in R \times \mathcal{G} \times R^m \tag{2.4.10}$$
$$\forall\,(t,x,y) \in \mathcal{T}_\tau \times \mathcal{G} \times R^m.$$

Let $\varepsilon \in (0, H)$ and $t_0 \in R$ $(t_0 \in \mathcal{T}_\tau)$ be prespecified. The fact that $u = 0$ is stable with respect to $\mathcal{T}_i$ implies that for $a(\varepsilon) > 0$ and $t_0 \in R$ $(t_0 \in \mathcal{T}_\tau)$ there exists a $\delta_1 = \delta_1(t_0, \varepsilon) > 0$ such that it follows from $u_0 < \delta_1$ that $u(t; t_0, u_0) < a(\varepsilon)$ for all $t \ge t_0$. We take $u_0 = v(t_0, x_0, y)$ and $\delta = \delta(t_0, \varepsilon) > 0$ so that

$$b(\delta) < \delta_1. \tag{2.4.11}$$

Let the solution $\chi(t; t_0, x_0)$ of system (1.2.7) start in domain: $t_0 \in R$ $(t_0 \in \mathcal{T}_\tau)$ and $\|x_0\| < \delta$. We claim that $\|\chi(t; t_0, x_0)\| < \varepsilon$ for all $t_0 \in \mathcal{T}_0$. If not, there exists other solution $\chi(t)$ with the initial conditions in the same domain and value $t_1 > t_0$ such that

$$\|\chi(t_1; t_0, x_0)\| = \varepsilon\alpha\|\chi(t; t_0, x_0)\| \le \varepsilon \qquad \forall\, t \in [t_0, t_1]. \tag{2.4.12}$$

By Proposition 2.4.3 we have the estimate

$$v(t, \chi(t; t_0, x_0), y) \le \gamma(t) \qquad \forall\, t \in [t_0, t_1], \tag{2.4.13}$$

where $\gamma(t)$ is the maximal solution of equation (2.4.1).

Seeing that

$$v(t_0, x_0, y) \le b(\|x_0\|) \le b(\delta) < \delta_1 \tag{2.4.14}$$

and in view of inequalities (2.4.11) – (2.4.13) we get

$$a(\varepsilon) \le v(t_1, \chi(t_1; t_0, x_0), y) \le \gamma(t_1) < a(\varepsilon).$$

This proves stability with respect to $\mathcal{T}_i$ of the state $x = 0$ of system (1.2.7).

Further it follows from the property of u_0 – uniform attraction of the solution $u = 0$ of equation (2.4.1) that, given $a(\eta) > 0$ and $t_0 \in R$

($t_0 \in \mathcal{T}_\tau$), $0 < \eta < H$ there exists a $\delta_1^* = \delta_1^*(t_0) > 0$ and $\tau = \tau(t_0, \eta) > 0$ such that $u_0 < \delta_1^*$ implies $u(t; t_0, u_0) < a(\eta)$ for all $t \geq t_0 + \tau$. We take $u_0 = v(t_0, x_0, y)$ the same as before and find $\delta_0^* = \delta_0^*(t_0) > 0$ such that

$$b(\delta_0^*) < \delta_1^*. \tag{2.4.15}$$

Let $\delta_0 = \min(\delta_1^*, \delta_0^*)$ and $\|x_0\| < \delta_0$. Proceeding as in the proof of stability (on $\mathcal{T}_\tau$) of the state $x = 0$ of system (1.2.7) we conclude that $\|\chi(t; t_0, x_0)\| < H$ for $t \geq t_0$. Assume that there exists a sequence $\{t_k\}$, $t_k \geq t_0 + \tau$, $t_k \to +\infty$ as $k \to +\infty$ and $\eta \leq |\chi(t; t_0, x_0)|$, where $\chi(t; t_0, x_0)$ is a solution of system (1.2.7) with the initial conditions $\|x_0\| < \delta_0$ and $t_0 \in R$ ($t_0 \in \mathcal{T}_\tau$). Taking into account estimates (2.4.8) and (2.4.15) we get

$$a(\eta) \leq v(t_k, \chi(t_k; t_0, x_0), y) \leq \gamma(t_k) < a(\eta).$$

This proves that the state $x = 0$ of system (1.2.7) is attractive (on $\mathcal{T}_\tau$). By Definition 2.4.4 (ii) the state $x = 0$ of system (1.2.7) is equi-asymptotically stable (on $\mathcal{T}_\tau$).

THEOREM 2.4.2. *Let vector-function f in system (1.2.7) be continuous on $R \times \mathcal{N}$ (on $\mathcal{T}_\tau \times \mathcal{N}$). If*

(1) *conditions (1)–(2) of Theorem 2.4.1 are satisfied and*
(2) *there exists a majorizing function G such that $G \in C(R \times R_+, R)$ ($G \in C(\mathcal{T}_\tau \times R_+, R)$) $G(t, 0) = 0$ $\forall\, t \in \mathcal{T}_0$ ($\forall\, t \in \mathcal{T}_\tau$) such that*

$$D^+ v(t, x, y) \geq G(t, v(t, x, y)) \qquad \forall\,(t, x, y) \in R \times \mathcal{G} \times R^m$$
$$(\forall\,(t, x, y) \in \mathcal{T}_\tau \times \mathcal{G} \times R^m).$$

Then properties of the function

$$y^{\mathrm{T}} U(t, x) y = v(t, x, y) \tag{2.4.16}$$

and instability properties of the zero solution of the equation

$$\frac{du}{dt} = G(t, u), \qquad u(t_0) = u_0 \geq 0 \tag{2.4.17}$$

imply instability (on $\mathcal{T}_\tau$) of the state $x = 0$ of the system (1.2.7).

PROOF. In order to prove Theorem 2.4.2 we shall state some properties of the function (2.4.16). Namely, assume positive definite on $\mathcal{G}$ (on $\mathcal{T}_\tau \times \mathcal{G}$)

function, and for any $\delta > 0$ and $t_0 \in R$ $(t_0 \in \mathcal{T}_\tau)$ an x_0 is found, $\|x_0\| < \delta$, such that $v(t_0, x_0, y) > 0$ $\forall y \in R^m$. Instability of the zero solution of equation (2.4.17) ensures that given $\varepsilon^* > 0$ and $t_0 \in R$ $(t_0 \in \mathcal{T}_\tau)$, for any $\delta^* > 0$ a $u_0 \colon 0 \le u_0 < \delta^*$ can be found so that $\rho(t; t_0, u_0) \ge \varepsilon^*$. Since function (2.4.16) is positive definite on $\mathcal{G}$ (on $\mathcal{T}_\tau \times \mathcal{G}$), a function $a \in K$ can be taken so that

$$a(\|x\|) \le v(t, x, y) \qquad \forall\, (t, x, y) \in R \times \mathcal{G} \times R^m$$
$$\forall\, (t, x, y) \in \mathcal{T}_\tau \times \mathcal{G} \times R^m.$$

We take $\varepsilon > 0$ so that

$$a(\varepsilon) < \varepsilon^*. \tag{2.4.18}$$

This is possible due to assumptions on function $v(t, x, y)$.

Now we determine $u_0 \le \delta^*$ and $t \ge t_0$ so that $\rho(t; t_0, u_0) \ge \varepsilon^*$. If x_0 is taken in accordance with (2.4.18) and $t \notin \mathcal{T}_0$ $(t \notin \mathcal{T}_\tau)$ then the theorem is proved, since the solution $\chi(t; t_0, x_0)$ cannot cease its existence without leaving the domain $\|x\| < \varepsilon$. Let $t_0 \in \mathcal{T}_0$ $(t_0 \in \mathcal{T}_\tau)$. Then we get according to Proposition 2.4.4

$$a\left(\|\chi(t; t_0, x_0)\|\right) \ge v(t, \chi(t), y) \ge \rho(t) \ge \varepsilon^* > a(\varepsilon).$$

Consequently, $\|\chi(t; t_0, x_0)\| > \varepsilon$ and the state $x = 0$ of system (1.2.7) is unstable (on $\mathcal{T}_\tau$).

2.5 Method of Matrix Liapunov Functions

As already mentioned in the introduction the application of matrix Liapunov functions make it possible to establish easily verified stability conditions for the state $x = 0$ of system (1.2.7) in terms of the property having a fixed sign of special matrices. The results presented in this section demonstrate the opportunities of the matrix Liapunov functions technique.

2.5.1 Nonautonomous systems

General Theorems 2.3.1 – 2.3.7 allows sufficient stability conditions for the state $x = 0$ of system (1.2.7) to be constructed as follows.

THEOREM 2.5.1. *Let the vector-function f in system (1.2.7) be continuous on $R \times \mathcal{N}$ (on $\mathcal{T}_\tau \times \mathcal{N}$). If there exist*

(1) *an open connected time-invariant neighborhood $\mathcal{G} \subset \mathcal{N}$ of the point $x = 0$;*
(2) *a matrix-valued function $U \in C(R \times \mathcal{N}, R^{m\times m})$ and a vector $y \in R^m$ such that the function $v(t,x,y) = y^{\mathrm{T}} U(t,x) y$ is locally Lipschitzian in x for all $t \in R$ $(t \in \mathcal{T}_\tau)$;*
(3) *functions $\psi_{i1}, \psi_{i2}, \psi_{i3} \in K$, $\widetilde{\psi}_{i2} \in CK$, $i = 1, 2, \dots, m$;*
(4) *$m \times m$ matrices $A_j(y)$, $j = 1, 2, 3$, $\widetilde{A}_2(y)$ such that*

(a)
$$\psi_1^{\mathrm{T}}(\|x\|) A_1(y) \psi_1(\|x\|) \le v(t,x,y) \le \widetilde{\psi}_2^{\mathrm{T}}(t,\|x\|) \widetilde{A}_2(y) \widetilde{\psi}_2(t,\|x\|)$$
$$\forall (t,x,y) \in R \times \mathcal{G} \times R^m \quad (\forall (t,x,y) \in \mathcal{T}_\tau \times \mathcal{G} \times R^m);$$

(b)
$$\psi_1^{\mathrm{T}}(\|x\|) A_1(y) \psi_1(\|x\|) \le v(t,x,y) \le \psi_2^{\mathrm{T}}(\|x\|) A_2(y) \psi_2(\|x\|)$$
$$\forall (t,x,y) \in R \times \mathcal{G} \times R^m \quad (\forall (t,x,y) \in \mathcal{T}_\tau \times \mathcal{G} \times R^m);$$

(c)
$$D^+ v(t,x,y) \le \psi_3^{\mathrm{T}}(\|x\|) A_3(y) \psi_3(\|x\|)$$
$$\forall (t,x,y) \in R \times \mathcal{G} \times R^m \quad (\forall (t,x,y) \in \mathcal{T}_\tau \times \mathcal{G} \times R^m).$$

Then, if the matrices $A_1(y)$, $A_2(y)$, $\widetilde{A}_2(y)$, $(y \neq 0) \in R^m$ are positive definite and $A_3(y)$ is negative semi-definite, then

(a) *the state $x = 0$ of system (1.2.7) is stable (on $\mathcal{T}_\tau$), provided condition (4)(a) is satisfied;*
(b) *the state $x = 0$ of system (1.2.7) is uniformly stable (on $\mathcal{T}_\tau$), provided condition (4)(b) is satisfied.*

PROOF. We shall prove assertion (a). Since matrices $A_1(y)$ and $\widetilde{A}_2(y)$ $\forall (y \neq 0) \in R^m$ are positive definite, then $\lambda_m(A_1) > 0$ and $\lambda_M(\widetilde{A}_2) > 0$, where $\lambda_m(\cdot)$ and $\lambda_M(\cdot)$ are minimal and maximal eigenvalues of matrices $A_1(y)$ and $\widetilde{A}_2(y)$ respectively.

Condition (3) of Theorem 2.5.1 provides the existence of functions $\pi \in K$ and $\rho \in CK$ such that

$$\pi(\|x\|) \le \psi_1^{\mathrm{T}}(\|x\|) \psi_1(\|x\|)$$

and

$$\rho(t,\|x\|) \ge \widetilde{\psi}_2^{\mathrm{T}}(t,\|x\|) \widetilde{\psi}_2(t,\|x\|).$$

Consequently,

$$\lambda_m(A_1)\pi(\|x\|) \le v(t,x,y) \qquad \forall\,(t,x,y) \in R \times \mathcal{G} \times R^m \\ (\forall\,(t,x,y) \in \mathcal{T}_\tau \times \mathcal{G} \times R^m) \tag{2.5.1}$$

and

$$v(t,x,y) \le \lambda_M(\widetilde{A}_2)\rho(t,\|x\|) \qquad \forall\,(t,x,y) \in R \times \mathcal{G} \times R^m \\ (\forall\,(t,x,y) \in \mathcal{T}_\tau \times \mathcal{G} \times R^m)\,. \tag{2.5.2}$$

Since matrix $A_3(y)$ is negative semi-definite, then

$$D^+v(t,x,y) \le 0 \qquad \forall\,(t,x,y \ne 0) \in R \times \mathcal{G} \times R^m \\ (\forall\,(t,x,y \ne 0) \in \mathcal{T}_\tau \times \mathcal{G} \times R^m)\,. \tag{2.5.3}$$

Taking into account (2.5.1) – (2.5.3) one can easily see that all conditions of Theorem 2.3.1 are satisfied and the state $x = 0$ of system (1.2.7) is stable (on $\mathcal{T}_\tau$).

The proof of assertion (b) of the Theorem 2.5.1 is the same, seeing that $\psi_{i2} \in K$.

THEOREM 2.5.2. *Let the vector-function f in system (1.2.7) be continuous on $R \times R^n$ (on $\mathcal{T}_\tau \times R^n$). If there exist*

(1) *a matrix-valued function $U \in C\,(R \times R^n, R^{m\times m})$ $(U \in C(\mathcal{T}_\tau \times R^n, R^{m\times m}))$ and a vector $y \in R^m$ such that the function $v(t,x,y) = y^{\mathrm{T}}U(t,x)y$ is locally Lipschitzian in x for all $t \in R$ $(t \in \mathcal{T}_\tau)$;*

(2) *functions $\varphi_{1i}, \varphi_{2i}, \varphi_{3i} \in KR$, $\widetilde{\varphi}_{2i} \in CKR$, $i = 1, 2, \dots, m$;*

(3) *$m \times m$ matrices $B_j(y)$, $j = 1,2,3$, $\widetilde{B}_2(y)$ such that*

(a) $$\varphi_1^{\mathrm{T}}(\|x\|)B_1(y)\varphi_1(\|x\|) \le v(t,x,y) \le \widetilde{\varphi}_2^{\mathrm{T}}(t,\|x\|)\widetilde{B}_2(y)\widetilde{\varphi}_2(t,\|x\|) \\ \forall\,(t,x,y) \in R \times R^n \times R^m \quad (\forall\,(t,x,y) \in \mathcal{T}_\tau \times R^n \times R^m)\,;$$

(b) $$\varphi_1^{\mathrm{T}}(\|x\|)B_1(y)\varphi_1(\|x\|) \le v(t,x,y) \le \varphi_2^{\mathrm{T}}(\|x\|)B_2(y)\varphi_2(\|x\|) \\ \forall\,(t,x,y) \in R \times R^n \times R^m \quad (\forall\,(t,x,y) \in \mathcal{T}_\tau \times R^n \times R^m)\,;$$

(c) $$D^+v(t,x,y) \le \varphi_3^{\mathrm{T}}(\|x\|)B_3(y)\varphi_3(\|x\|) \\ \forall\,(t,x,y) \in R \times R^n \times R^m \quad (\forall\,(t,x,y) \in \mathcal{T}_\tau \times R^n \times R^m)\,.$$

Then, provided that matrices $B_1(y)$, $B_2(y)$ *and* $\widetilde{B}_2(y)$, $\forall\,(y \neq 0) \in R^m$ *are positive definite and matrix* $B_3(y)$ *is negative definite,*

(a) *under condition (3)(a) the state* $x = 0$ *of system (1.2.7) is stable in the whole (on* $\mathcal{T}_\tau$*);*

(b) *under condition (3)(b) the state* $x = 0$ *of system (1.2.7) is uniformly stable in the whole (on* $\mathcal{T}_\tau$*).*

PROOF. Under conditions (1) – (3)(a) of Theorem 2.5.2 the function $v(t,x,y)$ is radially unbounded positive definite in the whole (on $\mathcal{T}_\tau$) and weakly decreasing in the whole (on $\mathcal{T}_\tau$). Since the matix $B_3(y)$, $\forall\,(y \neq 0) \in R^m$ is negaitive semi-definite, then we have in consequence of condition (3)(c) of Theorem 2.5.2

$$D^+v(t,x,y) \le 0 \qquad \forall\,(t,x,y \neq 0) \in R \times R^n \times R^m$$
$$(\forall\,(t,x,y \neq 0) \in \mathcal{T}_\tau \times R^n \times R^m)\,.$$

According to Theorem 2.3.2 the state $x = 0$ of system (1.2.7) is stable in the same manner taking into account conditions (1) – (3)(b) and (3)(c).

THEOREM 2.5.3. *Let the vector-function* f *in system (1.2.7) be continuous on* $R \times \mathcal{N}$ *(on* $\mathcal{T}_\tau \times \mathcal{N}$*). If there exist*

(1) *an open connected time-invariant neighborhood* $\mathcal{G} \subset \mathcal{N}$ *of the point* $x = 0$*;*

(2) *a matrix-valued function* $U \in C(R \times \mathcal{N},\, R^{m\times m})$ $(U \in C(\mathcal{T}_\tau \times \mathcal{N},\, R^{m\times m}))$ *and a vector* $y \in R^m$ *such that the function* $v(t,x,y) = y^{\mathrm{T}}U(t,x)y$ *is locally Lipschitzian in* x *for all* $t \in R$ $(t \in \mathcal{T}_\tau)$*;*

(3) *functions* $\eta_{1i}, \eta_{2i}, \eta_{3i} \in K$, $\widetilde{\eta}_{2i} \in CK$, $i = 1, 2, \dots, m$*;*

(4) $m \times m$ *matrices* $C_j(y)$, $j = 1,2,3$, $\widetilde{C}_2(y)$ *such that*

(a)
$$\eta_1^{\mathrm{T}}(\|x\|)C_1(y)\eta_1(\|x\|) \le v(t,x,y) \le \widetilde{\eta}_2^{\mathrm{T}}(t,\|x\|)\widetilde{C}_2(y)\widetilde{\eta}_2(t,\|x\|)$$
$$\forall\,(t,x,y) \in R \times \mathcal{G} \times R^m \quad (\forall\,(t,x,y) \in \mathcal{T}_\tau \times \mathcal{G} \times R^m)\,;$$

(b)
$$\eta_1^{\mathrm{T}}(\|x\|)C_1(y)\eta_1(\|x\|) \le v(t,x,y) \le \eta_2^{\mathrm{T}}(\|x\|)C_2(y)\eta_2(\|x\|)$$
$$\forall\,(t,x,y) \in R \times \mathcal{G} \times R^m \quad (\forall\,(t,x,y) \in \mathcal{T}_\tau \times \mathcal{G} \times R^m)\,;$$

(c)
$$D^*v(t,x,y) \le \eta_3^{\mathrm{T}}(\|x\|)C_3(y)\eta_3(\|x\|) + m\,(t,\, \eta_3(\|x\|))$$
$$\forall\,(t,x,y) \in R \times \mathcal{G} \times R^m \quad (\forall\,(t,x,y) \in \mathcal{T}_\tau \times \mathcal{G} \times R^m)\,,$$

where function $m(t,\cdot)$ *satisfies the condition*

$$\lim \frac{|m\,(t,\eta_3(\|x\|))\,|}{\|\eta_3\|} = 0 \quad as \quad \|\eta_3\| \to 0$$

uniformly in $t \in R$ $(t \in \mathcal{T}_\tau)$.

Then, provided the matrices $C_1(y)$, $C_2(y)$, $\widetilde{C}_2(y)$ *are positive definite and matrix* $C_3(y)$ $(y \neq 0) \in R^m$ *is negative definite, then*

(a) *under condition (4)(a) the state* $x = 0$ *of the system (1.2.7) is asymptotically stable (on* $\mathcal{T}_\tau$*);*

(b) *under condition (4)(b) the state* $x = 0$ *of the system (1.2.7) is uniformly asymptotically stable (on* $\mathcal{T}_\tau$*).*

PROOF. Following the arguments from the proof of Theorem 2.5.1 under conditions (1) – (4)(a) the function $v(t,x,y)$ is positive definite on $\mathcal{G}$ (on $\mathcal{T}_\tau \times \mathcal{G}$) and weakly decreasing on $\mathcal{G}$ (on $\mathcal{T}_\tau \times \mathcal{G}$). Consider condition (4)(c). Since $\eta_{3i} \in K$, $i = 1, 2, \dots, m$ there exists a function $\omega \in K$ such that

$$\omega(\|x\|) \geq \eta_3^{\mathrm{T}}(\|x\|)\eta_3(\|x\|). \tag{2.5.4}$$

Due to matrix $C_3(y)$ $(y \neq 0) \in R^m$ being negative definite all its eigenvalues are negative so that $\lambda_M(C_3) < 0$. Therefore, we get in view of (2.5.4)

$$\begin{gathered} D^*v(t,x,y) \leq \lambda_M(C_3)\omega(\|x\|) + m\,(t,\eta_3(\|x\|)) \\ \forall\,(t,x,y\neq 0) \in R \times \mathcal{G} \times R^m \quad (\forall\,(t,x,y \neq 0) \in \mathcal{T}_\tau \times \mathcal{G} \times R^m)\,. \end{gathered} \tag{2.5.5}$$

Under condition (2.5.2) for the given neighborhood $\mathcal{G} \subset \mathcal{N}$ of point $x = 0$ a $0 < \mu < 1$ can be taken so that

$$\begin{gathered} |m(t,\,\eta(\|x\|)| < -\mu\lambda_M(C_3)\eta_3^{\mathrm{T}}(\|x\|)\eta_3(\|x\|) \\ \forall\,(t,x,y\neq 0) \in R \times \mathcal{G} \times R^m \quad (\forall\,(t,x,y \neq 0) \in \mathcal{T}_\tau \times \mathcal{G} \times R^m)\,. \end{gathered} \tag{2.5.6}$$

Together with inequalities (2.5.5) condition (2.5.6) yields the estimate

$$D^*v(t,x,y) \leq (1-\mu)\lambda_M(C_3)\omega(\|x\|), \qquad \lambda_M(C_3) < 0.$$

Thus, function $D^*v(t,x,y)$ is negative definite on $\mathcal{G}$ (on $\mathcal{T}_\tau \times \mathcal{G}$). Therefore, all conditions of Theorem 2.3.3 are satisfied and the state $x = 0$ of the system (1.2.7) is asymptotically stable (on $\mathcal{T}_\tau$).

Assertion (b) of Theorem 2.5.3 is proved in the same manner taking into account that condition (4)(b) ensures function $v(t,x,y)$ decreasing on $\mathcal{G}$ (on $\mathcal{T}_\tau \times \mathcal{G}$).

THEOREM 2.5.4. *Let the vector-function f in system (1.2.7) be continuous on $R \times R^n$ (on $\mathcal{T}_\tau \times R^n$) and conditions (1) – (3) of Theorem 2.5.2 are satisfied.*

Then, provided that matrices $B_1(y)$, $B_2(y)$ and $\widetilde{B}_2(y)$ are positive definite and matrix $B_3(y)$ $\forall\,(y \neq 0) \in R^m$ is negative definite,

(a) *under condition (3)(a) of Theorem 2.5.2 the state $x = 0$ of system (1.2.7) is asymptotically stable in the whole (on $\mathcal{T}_\tau$);*
(b) *under condition (3)(b) of Theorem 2.5.2 the state $x = 0$ of system (1.2.7) is uniformly asymptotically stable in the whole (on $\mathcal{T}_\tau$).*

PROOF. Under conditions (1) – (3)(a) of Theorem 2.5.2 the function $v(t,x,y)$ is radially unbounded positive definite in the whole (on $\mathcal{T}_\tau$).

Because matrix $B_3(y)$ $\forall\,(y \neq 0) \in R^m$ is negative definite, proceeding as in the proof of Theorem 2.5.2 we arrive at the estimate

$$D^*v(t,x,y) \leq \lambda_M(B_3)\varphi_3^{\mathrm{T}}(\|x\|)\varphi_3(\|x\|)$$
$$\forall\,(t,x,y) \in R \times R^n \times R^m \quad (\forall\,(t,x,y) \in \mathcal{T}_\tau \times R^n \times R^m)\,.$$

Since $\varphi_{3i} \in CK$, $i = 1,2,\ldots,m$, there exist a function $\theta(\|x\|) \in KR$ such that

$$\theta(\|x\|) \geq \varphi_3^{\mathrm{T}}(\|x\|)\varphi_3(\|x\|).$$

Therefore,

$$D^*v(t,x,y) \leq \lambda_M(B_3)\theta(\|x\|), \qquad \lambda_M(B_3) < 0$$
$$\forall\,(t,x,y \neq 0) \in R \times R^n \times R^m \quad (\forall\,(t,x,y \neq 0) \in \mathcal{T}_\tau \times R^n \times R^m)\,.$$

Thus, function $D^*v(t,x,y)$ is negative definite in the whole (on $\mathcal{T}_\tau$).

According to Theorem 2.3.4 the state $x = 0$ of system (1.2.7) is asymptotically stable in the whole (on $\mathcal{T}_\tau$).

The proof of assertion (b) of Theorem 2.5.4 is similar to the above and takes into account the fact that by conditions (2) and (3) of Theorem 2.5.2 the function $v(t,x,y)$ is radially unbounded positive definite and decreasing in the whole (on $\mathcal{T}_\tau$).

THEOREM 2.5.5. *Let the vector-function f in system (1.2.7) be continuous on $R \times \mathcal{N}$ (on $\mathcal{T}_\tau \times \mathcal{N}$). If there exist*

(1) *an open connected time-invariant neighborhood $\mathcal{G} \subset \mathcal{N}$ of the point $x = 0$;*

(2) *a matrix-valued function* $U \in C(R \times \mathcal{N}, R^{m\times m})$ *and a vector* $y \in R^m$ *such that the function* $v(t,x,y) = y^{\mathrm{T}}U(t,x)y$ *is locally Lipschitzian in* x *for all* $t \in R$ $(t \in \mathcal{T}_\tau)$;

(3) *functions* σ_{2i}, $\sigma_{3i} \in K$, $i = 1, 2, \ldots, m$, *a positive real number* Δ_1 *and positive integer* p, $m \times m$ *matrices* $F_2(y)$, $F_3(y)$ *such that*

$$\text{(a)}\quad \begin{gathered}\Delta_1\|x\|^p \le v(t,x,y) \le \sigma_2^{\mathrm{T}}(\|x\|)F_2(y)\sigma_2(\|x\|)\\ \forall\,(t,x,y\ne 0) \in R\times\mathcal{G}\times R^m \quad (\forall\,(t,x,y\ne 0)\in \mathcal{T}_\tau\times\mathcal{G}\times R^m)\,;\end{gathered}$$

$$\text{(b)}\quad \begin{gathered}D^*v(t,x,y) \le \sigma_3^{\mathrm{T}}(\|x\|)F_3(y)\sigma_3(\|x\|)\\ \forall\,(t,x,y\ne 0) \in R\times\mathcal{G}\times R^m \quad (\forall\,(t,x,y\ne 0)\in \mathcal{T}_\tau\times\mathcal{G}\times R^m)\,.\end{gathered}$$

Then, provided that the matrices $F_2(y)$ $(y \ne 0) \in R^m$ *are positive definite, the matrix* $F_3(y)$ $(y \ne 0) \in R^m$ *is negative definite and functions* σ_{2i}, σ_{3i} *are the same magnitude, then the state* $x = 0$ *of system (1.2.7) is exponentially stable (on* $\mathcal{T}_\tau$*).*

PROOF. Under conditions (1) – (4)(a) function $v(t,x,y)$ is positive definite and decreasing (on $\mathcal{T}_\tau$). In fact, we have the estimate

$$\begin{gathered}v(t,x,y) \le \lambda_M(F_2)\sigma_2^{\mathrm{T}}(\|x\|)\sigma_2(\|x\|), \qquad \lambda(F_2) > 0\\ \forall\,(t,x,y) \in R\times\mathcal{G}\times R^m \quad (\forall\,(t,x,y\ne 0)\in \mathcal{T}_\tau\times\mathcal{G}\times R^m)\,.\end{gathered}$$

Since the functions $\sigma_{3i} \in K$, $i = 1, 2, \ldots, m$, there exists a function $\varkappa \in K$ such that

$$\varkappa(\|x\|) \ge \sigma_2^{\mathrm{T}}(\|x\|)\sigma_2(\|x\|).$$

Therefore

$$\begin{gathered}\Delta_1\|x\|^p \le v(t,x,y) \le \lambda_M(F_2)\varkappa(\|x\|), \qquad \lambda_M(F_2) > 0\\ \forall\,(t,x,y) \in R\times\mathcal{G}\times R^m \quad (\forall\,(t,x,y\ne 0)\in \mathcal{T}_\tau\times\mathcal{G}\times R^m)\,.\end{gathered} \tag{2.5.7}$$

We reduce condition (4)(b) of Theorem 2.5.5 to the form

$$\begin{gathered}D^*v(t,x,y) \le \lambda_M(F_3)\pi(\|x\|), \qquad \lambda_M(F_3) < 0\\ \forall\,(t,x,y) \in R\times\mathcal{G}\times R^m \quad (\forall\,(t,x,y)\in \mathcal{T}_\tau\times\mathcal{G}\times R^m)\,,\end{gathered} \tag{2.5.8}$$

where $\pi \in K$ is such that

$$\pi(\|x\|) \ge \sigma_3^{\mathrm{T}}(\|x\|)\sigma_3(\|x\|).$$

Since functions $\varkappa$ and π are of the same magnitude, there exist constants $k_1 > 0$ and $k_2 > 0$ such that

$$k_1\varkappa(\|x\|) \le \pi(\|x\|) \le k_2\varkappa(\|x\|).$$

We get from inequalities (2.5.7) and (2.5.8)

$$\tag{2.5.9} \begin{gathered} D^*v(t,x,y) \le \lambda v(t,x,y) \\ \forall\,(t,x,y\ne 0)\in R\times\mathcal{G}\times R^m \quad (\forall\,(t,x,y\ne 0)\in\mathcal{T}_\tau\times\mathcal{G}\times R^m)\,, \end{gathered}$$

where $\lambda = \lambda_M(F_3)\lambda_M^{-1}(F_2)$, $\lambda < 0$.

In view of the estimate from the left in (2.5.7) we obtain from (2.5.9)

$$v(t,x,y) \le v(t_0,x_0,y)\,\exp\left(\lambda(t-t_0)\right)$$

and

$$\tag{2.5.10} \|\chi(t;t_0x_0)\| \le \Delta_1^{-\frac{1}{p}}\lambda_M^{\frac{1}{p}}(F_2)\varkappa^{\frac{1}{p}}(\|x_0\|)\,\exp\left(\frac{\lambda}{p}(t-t_0)\right).$$

We designate according to Definition 1.2.3 (vi)

$$\alpha = \Delta_1^{-\frac{1}{p}}\lambda_M^{\frac{1}{p}}(F_2), \quad \beta = \frac{\lambda}{p}, \quad \beta < 0.$$

From (2.5.10) we obtain

$$\|\chi(t;t_0x_0)\| \le \alpha\varkappa^{\frac{1}{p}}(\|x_0\|)\,\exp\left(\beta(t-t_0)\right) \qquad \forall\,t\in\mathcal{T}_0, \quad \forall\,t_0\in\mathcal{T}_i.$$

This proves Theorem 2.5.5.

THEOREM 2.5.6. *Let the vector-function f in system (1.2.7) be continuous on $R\times R^n$ (on $\mathcal{T}_\tau\times R^n$). If there exist*

(1) *a matrix-valued function $U\in C(R\times R^n, R^{m\times m})$ $(U\in C(\mathcal{T}_\tau\times R^n, R^{m\times m}))$ and a vector $y\in R^m$ such that the function $v(t,x,y) = y^{\mathrm{T}}U(t,x)y$ is locally Lipschitzian in x for all $t\in R$ $(\forall\,t\in\mathcal{T}_\tau)$;*

(2) *functions ν_{2i}, $\nu_{3i}\in KR$, $i = 1,2,\dots,m$, a positive real number $\Delta_2 > 0$ and a positive integer q;*

(3) *$m\times m$ matrices H_2, H_3 such that*

$$\text{(a)} \quad \begin{gathered} \Delta_2\|x\|^q \le v(t,x,y) \le \nu_2^{\mathrm{T}}(\|x\|)H_2(y)\nu_2(\|x\|) \\ \forall\,(t,x,y\ne 0)\in R\times R^n\times R^m \quad (\forall\,(t,x,y)\in\mathcal{T}_\tau\times R^n\times R^m)\,; \end{gathered}$$

$$\text{(b)} \quad \begin{gathered} D^*v(t,x,y) \le \nu_3^{\mathrm{T}}(\|x\|)H_3(y)\nu_3(\|x\|) \\ \forall\,(t,x,y\ne 0)\in R\times R^n\times R^m \quad (\forall\,(t,x,y\ne 0)\in\mathcal{T}_\tau\times R^n\times R^m)\,. \end{gathered}$$

Then, if the matrix $H_2(y)$ $\forall\,(y \neq 0) \in R^m$ is positive definite, the matrix $H_3(y)$ $\forall\,(y \neq 0) \in R^m$ is negative definite and functions ν_{2i}, ν_{3i} are of the same magnitude, the state $x = 0$ of system (1.2.7) is exponentially stable in the whole (on $\mathcal{T}_\tau$).

PROOF of this Theorem is similar to that of Theorem 2.5.5 taking into account the fact that under conditions of Theorem 2.5.6 the function $v(t,x,y)$ is radially unbounded (on $\mathcal{T}_\tau$). Inequality (2.5.10) is replaced by

$$\|\chi(t;t_0x_0)\| \leq \Delta_2^{-\frac{1}{q}}\lambda_M^{\frac{1}{q}}(H_2)g^{\frac{1}{q}}(\|x_0\|)\exp\left(\frac{\lambda_1}{q}(t-t_0)\right)$$

where $g(\|x\|) \in KR$ and $g(\|x\|) \geq \nu_2^{\mathrm{T}}(\|x\|)\nu_2(\|x\|)$,

$$\lambda_1 = \lambda_M(H_3)k_1\lambda_M^{-1}(H_2), \qquad k_1 > 0, \quad \lambda_1 < 0.$$

We designate $\beta = \lambda_1 q^{-1}$ and define function $\Phi(\Delta) = \Delta_1^{-\frac{1}{q}}\lambda_M^{\frac{1}{q}}(H_2)g^{\frac{1}{q}}(\Delta)$ whenever $\|x_0\| < \Delta$, $\Delta = +\infty$. Then

$$\|\chi(t;t_0x_0)\| \leq \Phi(\Delta)\exp\left(\beta(t-t_0)\right), \qquad \beta < 0 \quad \forall\, t \in \mathcal{T}_0, \quad \forall\, t_0 \in \mathcal{T}_i.$$

This proves Theorem 2.5.6.

THEOREM 2.5.7. *Let the vector-function f in system (1.2.7) be continuous on $R \times \mathcal{N}$ (on $\mathcal{T}_\tau \times \mathcal{N}$). If there exist*

(1) *an open connected time-invariant neighborhood $\mathcal{G} \subset \mathcal{N}$ of the point $x = 0$;*
(2) *a matrix-valued function $U \in C^1(R \times \mathcal{N}, R^{m\times m})$ $(U \in C^1(\mathcal{T}_\tau \times \mathcal{N}, R^{m\times m}))$ and a vector $y \in R^m$;*
(3) *functions ψ_{1i}, ψ_{2i}, $\psi_{3i} \in K$, $i = 1,2,\dots,m$, $m\times m$ matrices $A_1(y)$, $A_2(y)$, $G(y)$ and a constant $\Delta > 0$ such that*

$$\text{(a)}\quad \begin{gathered}\psi_1^{\mathrm{T}}(\|x\|)A_1(y)\psi_1(\|x\|) \leq v(t,x,y) \leq \psi_2^{\mathrm{T}}(\|x\|)A_2(y)\psi_2(\|x\|)\\ \forall\,(t,x,y) \in R\times\mathcal{G}\times R^m \quad (\forall\,(t,x,y) \in \mathcal{T}_\tau\times\mathcal{G}\times R^m);\end{gathered}$$

$$\text{(b)}\quad \begin{gathered}Dv(t,x,y) \geq \psi_3^{\mathrm{T}}(\|x\|)G(y)\psi_3(\|x\|)\\ \forall\,(t,x,y) \in R\times\mathcal{G}\times R^m \quad (\forall\,(t,x,y) \in \mathcal{T}_\tau\times\mathcal{G}\times R^m);\end{gathered}$$

(4) *point $x = 0$ belong to $\partial\mathcal{G}$;*
(5) *$v(t,x,y) = 0$ on $\mathcal{T}_0 \times (\partial\mathcal{G} \cap B_\Delta)$, where $B_\Delta = \{x\colon \|x\| < \Delta\}$.*

Then, if matrices $A_1(y)$, $A_2(y)$ *and* $G(y)$ $\forall\,(y \neq 0) \in R^m$ *are positive definite, the state* $x = 0$ *of system (1.2.7) is unstable (on* $\mathcal{T}_\tau$*).*

PROOF. Under conditions (1) – (3)(a) of Theorem 2.5.7 it is easy to obtain for function $v(t,x,y)$ the estimate

$$\begin{gathered}\lambda_m(A_1)\gamma(\|x\|) \leq v(t,x,y) \leq \lambda_M(A_2)\zeta(\|x\|)\\ \forall\,(t,x,y) \in R \times \mathcal{G} \times R^m \quad (\forall\,(t,x,y) \in \mathcal{T}_\tau \times \mathcal{G} \times R^m)\,.\end{gathered} \tag{2.5.11}$$

Here $\gamma \in K$ and $\gamma(\|x\|) \leq \psi_1^{\mathrm{T}}(\|x\|)\psi_1(\|x\|)$, $\zeta \in K$ and $\zeta(\|x\|) \geq \psi_2^{\mathrm{T}}(\|x\|)\psi_2(\|x\|)$.

Since $\lambda_M(A_1) > 0$, $\lambda_M(A_2) > 0$, then by estimate (2.5.11) function $v(t,x,y)$ is positive and bounded (on $\mathcal{T}_\tau$). Hence, for every $\delta > 0$ an $x_0 \in \mathcal{G} \cap B_\Delta$ and a $a > 0$ can be found such that $a \geq v(t_0,x_0,y) > 0$ $\forall\,(y \neq 0) \in R^m$.

Condition (3)(b) of Theorem 2.5.7 is reduced to the form

$$\begin{gathered}Dv(t,x,y) \geq \lambda_m(G)\xi(\|x\|), \qquad \lambda_m(G) > 0,\\ \forall\,(t,x,y \neq 0) \in R \times \mathcal{G} \times R^m \quad (\forall\,(t,x,y) \in \mathcal{T}_\tau \times \mathcal{G} \times R^m)\,.\end{gathered} \tag{2.5.12}$$

Here $\xi \in K$ and $\xi \leq \psi_3^{\mathrm{T}}(\|x\|)\psi_3(\|x\|)$.

In view of (2.5.11) and (2.5.12) we have for $\chi(t;t_0,x_0 \in \mathcal{G}$

$$\begin{aligned}a &\geq v(t,\chi(t;t_0,x_)),y) = v(t_0,x_0,y) + \int_{t_0}^{t} Dv(\tau,\chi(\tau;t_0,x_0)y)\,d\tau\\ &\geq v(t_0,x_0,y) + \lambda_m(G)\xi(\|x_0\|)(t-t_0) \qquad \forall\,t \in \mathcal{T}_0 \quad (\forall\,t \in \mathcal{T}_\tau).\end{aligned}$$

Hence, it follows that the solution $\chi(t;t_0,x_0)$ must leave neighborhood $\mathcal{G}$ some time later. But because of condition (5) it cannot leave $\mathcal{G}$ through $\partial\mathcal{G} \in B_\Delta$. Consequently, $\chi(t;t_0,x_0)$ leaves the domain B_Δ and the state $x = 0$ of system (1.2.7) is unstable (on $\mathcal{T}_\tau$).

2.5.2 Autonomous systems

2.5.2.1 Definitions of stability domains and their estimates. For a while our attention will be focused on the difference between the notions "domain" and "region".

Referring to LaSalle and Lefschetz [98] a "region" is an open connected set. However, Santalo [163] defined "domain" as an open and connected set, and "region" as the union of a domain with some, none, or all its boundary points.

We want to emphasize that, for stability analysis of nonlinear systems, only a neighborhood (either open or closed or neither open nor closed) of the origin is of interest herein. Hahn [66] used "domain" in this sense. The reason for using a neighborhood that can be closed is that the domain of asymptotic stability of an equilibrium of a nonlinear system can be closed.

We accept:

DEFINITION 2.5.1. A set D_s, $D_s \subseteq R^m$, is *the domain of the equilibrium state* $x = 0$ defined by

$$D_s = \bigcup \left[D_s(\varepsilon) \colon \varepsilon \in \overset{\circ}{R}_+ \right],$$

where $D_s(\varepsilon)$ is such a neighborhood of $x = 0$ that $\|\chi(t; 0, x_0)\| < \varepsilon$ $\forall\, t \in R_+$, holds provided only that $x_0 \in D_s(\varepsilon)$ for every $\varepsilon \in \overset{\circ}{R}_+$.

The next definition has been commonly used (see Krasovskii [89], Hahn [66], LaSalle and Lefschetz [98]).

DEFINITION 2.5.2. A set D_a, $D_a \subseteq R^m$, is *the domain of attraction* of the equilibrium state $x = 0$ of the system (1.2.10) if and only if it is such a neighborhood of $x = 0$ that

$$\lim \left[\, \|\chi(t; 0, x_0)\| \colon t \to +\infty \right] = 0$$

holds provided only that $x_0 \in D_a$.

It is now natural to accept the definition of the domain of asymptotic stability of $x = 0$ in the form.

DEFINITION 2.5.3. A set D, $D \subseteq R^m$, is *the domain of asymptotic stability* of $x = 0$ of the system (1.2.10) if and only if it is both a neighborhood of $x = 0$ and the intersection of its domain of stability and domain of attraction, that is, that $D = D_s \cap D_a$ is a neighborhood of $x = 0$.

The exact determination of the domain of asymptotic stability has great engineering and theoretical importance. Unfortinately, we can realize it only in special cases. For these reasons we investigate its estimate E defined as follows.

DEFINITION 2.5.4. A set E, $E \subseteq R^m$, is *an estimate set* (in brief, *estimate*) of the asymptotic stability domain D of $x = 0$ of the system (1.2.10) if and only if

(i) E is a neighborhood of $x = 0$,
(ii) $E \subseteq D$

and

(iii) E is positively invariant set of the system (1.2.10), that is, that $x_0 \in E$ implies $\chi(t; 0, x_0) \in E$ for every $t \in R_+$.

2.5.2.2 System description and decomposition. Suppose autonomous system (1.2.10) to be decomposed into m interconnected subsystems

$$\frac{dx_i}{dt} = g_i(x_i) + h_i(x) \tag{2.5.13}$$

with individual subsystems

$$\frac{dx_i}{dt} = g_i(x_i), \qquad x_i(0) = x_{i0}, \quad i = 1, 2, \dots, m, \tag{2.5.14}$$

where $x_i \in R^{n_i}$, $g = (g_1^{\mathrm{T}}, g_2^{\mathrm{T}}, \dots, g_m^{\mathrm{T}})^{\mathrm{T}}$, $x = (x_1^{\mathrm{T}}, \dots, x_m^{\mathrm{T}})^{\mathrm{T}}$. Besides $g_i \in C(R^{n_i}, R^{n_i})$, $h_i \in C(R^n, R^{n_i})$ and $g_i(0) = 0$, $h_i(0) = 0$ $\forall i = 1, 2, \dots, m$.

ASSUMPTION 2.5.1. There are connected neighborhoods N_i of $x_i = 0$ $\forall i = 1, 2, \dots, m$ such that both

(i) motions $x_i(t, x_{i0})$ of (2.5.14) are continuous in $(t, x_{i0}) \in R_+ \times N_i$, where $x_i(0, x_{i0}) \equiv 0$ $\forall i = 1, 2, \dots, m$;

and

(ii) motions $x(t, x_0)$ of (1.2.10) (or (2.5.13)) are continuous in $(t, x_0) \in R_+ \times N$, where $N = N_1 \times N_2 \times \dots \times N_m$ and $x(0, x_0) \equiv 0$.

Let $U \colon R^n \to R^{m \times m}$ be the matrix-valued function with elements $u_{ij} \in C(R^n, R)$ for $i \neq j$ and $u_{ij} \in C(R^n, R_+)$ for $i = j$.

Let us construct the function

$$v(x, y) = y^{\mathrm{T}} U(x) y$$

by means vector $y \in R^m$ which was used above. We shall use expressions of one of Dini derivatives of function U

$$D^+U(x) = \limsup \{[U(x(t+\theta), x) - U(x)]\,\theta^{-1} \colon \theta \to 0^+\},$$
$$D_+U(x) = \liminf \{[U(x(t+\theta), x) - U(x)]\,\theta^{-1} \colon \theta \to 0^+\}$$

with the function $U(x)$. We shall denote by symbol $D^*U(x)$ the possibility of utilizing any of functions $D^+U(x)$ or $D_+U(x)$.

ASSUMPTION 2.5.2. Matrix-valued function U is radially increasing on N, that is, the following inequality holds elementwise

$$U(\lambda_1 x) < U(\lambda_2 x) \qquad \forall\,(x \neq 0) \in N, \quad \lambda_i \in (0, +\infty), \quad i = 1, 2, \quad \lambda_1 < \lambda_2.$$

Let K be the elementwise greatest $m \times m$ matrix, $K = (k_{ij})$ satisfying

$$U_K(x) \subseteq \operatorname{int} N \tag{2.5.15}$$

for the set $U_K(x)$ defined by

$$U_K(x) = \{x \colon U(x) < K\}, \tag{2.5.16}$$

where int N is the interior of N. In case N is unbounded then $k_{ij} = +\infty$ is possible for some $(i, j) \in [1, m]$.

Let

$$\begin{gathered} E = \bigcup \{E_{ij} \colon (i, j) \in [1, m]\}, \qquad E_{ij} = \{x \colon u_{ij}(x) < k_{ij}\}, \\ k_{ij} < +\infty, \quad \partial E_{ij} = \{x \colon u_{ij}(x) = k_{ij}\}. \end{gathered} \tag{2.5.17}$$

2.5.2.3 A metric aggregation form. Metric on R^n will be introduced by the Euclidean norm $\|\cdot\|$. A metric aggregation form is determined by

ASSUMPTION 2.5.3. There are $U \in C(R^n, R^{m\times m})$, $w \in C(R^n, R^m)$ and real number α_{ij} such that

(i) $w(x) = 0$ for $x \in N$ iff $x = 0$;

(ii) $U(x) \in C(N, R^{m\times m})$;

(iii) $$D^*U(x) \leq \begin{pmatrix} \alpha_{11}\|w(x)\|^2 & \dots & \alpha_{1m}\|w(x)\|^2 \\ \multicolumn{3}{c}{\dotfill} \\ \alpha_{m1}\|w(x)\|^2 & \dots & \alpha_{mm}\|w(x)\|^2 \end{pmatrix} \qquad \forall x \in N.$$

THEOREM 2.5.8. *Let Assumptions 2.5.1 – 2.5.3 hold. In order for the set E (2.5.17) to be an estimate of D it is sufficient that U is positive definite on N, $\alpha_{ij} < 0$ and $u_{ij}(x)$ is radially unbounded in case N is unbounded, $\forall\, i, j = 1, 2, \dots, m$.*

PROOF. Positive definiteness of $U(x)$ on N implies positive definiteness of $u_{ij}(x)$ on N $\forall (i, j) \in [1, m]$. The conditions (iii) of Assumption 2.5.3 proves

$$D^*u_{ij}(x) \leq \alpha_{ij}\|w(x)\|^2 \qquad \forall x \in N \quad \forall\,(i, j) \in [1, m]. \tag{2.5.18}$$

Since $\alpha_{ij} < 0$ and $w(x) = 0$ iff $x = 0$ due to (i) of Assumption 2.5.3 then Assumption 2.5.1, (2.5.18) and Assumption 2.5.2 prove that E_{ij} (2.5.17) is an estimate of D $\forall (i,j) \in [1,m]$. Hence E (2.5.17) is also an estimate of D.

Let $v_i\colon R^n \to R^m$ be defined by

$$v_i = (u_{i1}, u_{i2}, \dots, u_{im})^{\mathrm{T}}, \qquad \forall\, i = 1, 2, \dots, m. \tag{2.5.19}$$

ASSUMPTION 2.5.4. Vector $b \in R^m$ is elementwise positive, $b^{\mathrm{T}}v_i$ is positive definite on N and radially unbounded in case N is unbounded, and $k_i \in (0, +\infty)$ is such that a set

$$V_i = \{x\colon b^{\mathrm{T}}v_i(x) < k_i\}, \qquad k_i < +\infty \to \partial V_i = \{x\colon b^{\mathrm{T}}v_i(x) = k_i\}, \quad \forall\, i \in [1,m] \tag{2.5.20}$$

is the largest connected neighborhood of $x = 0$ in N determined by $b^{\mathrm{T}}v_i(x)$.

THEOREM 2.5.9. *Let Assumptions 2.5.1, 2.5.3 and 2.5.4 hold. In order for the set E (2.5.21)*

$$E = \bigcup \{V_i\colon i \in [1,m]\} \tag{2.5.21}$$

to be estimate of D it is sufficient that the matrix $A = (\alpha_{ij})$ and the vector b obey elementwise $Ab < 0$.

PROOF. From (iii) Assumption 2.5.3 and $b > 0$ (Assumption 2.5.4) it results

$$D^*U(x)b \le Ab\|w(x)\|^2 \qquad \forall\, x \in N. \tag{2.5.22}$$

The condition $Ab < 0$, (i) of Assumption 2.5.3 and (2.5.22) prove $D^*U(x)b < 0$ elementwise on N, $x \neq 0$. This result, Assumption 2.5.4, and (2.5.19) prove that both $b^{\mathrm{T}}v_i$ is positive definite and $D^*U(x)b$ elementwise negative ($x \neq 0$) on the closure $\overline{E}_i$ (2.5.21), $\forall i \in [1,m]$. These facts and Assumption 2.5.1 prove that $E_i = V_i$ is an estimate of D. Since this holds for every $i \in [1,m]$, then E (2.5.20), (2.5.21) is an estimate of D.

Let k be the greatest number or the symbol $+\infty$ such that the set V_k

$$V_k = \{x\colon b^{\mathrm{T}}U(x)b < k\}, \qquad k < +\infty \to \partial V_k = \{x\colon b^{\mathrm{T}}U(x)b = k\} \tag{2.5.23}$$

is the largest connected neighborhood of $x = 0$ in N determined by b and V.

THEOREM 2.5.10. *Let Assumptions 2.5.1 and 2.5.3 hold. In order for the set* $E = V_k$ *(2.5.23) to be an estimate of* D *of* $x = 0$ *of (1.2.10) it is sufficient that* U *is positive definite on* N, $v(x) = b^{\mathrm{T}}U(x)b$ *is radially unbounded in case* N *is unbounded, the vector* b *is elementwise positive and the scalar* $b^{\mathrm{T}}Ab$ *is negative for* $A = (\alpha_{ij})$.

PROOF. Positive definiteness of U on N means that $v(x) = b^{\mathrm{T}}U(x)b$ is positive definite on N. Condition (iii) of Assumption 2.5.3 and $b > 0$ imply

$$D^*v(x) = b^{\mathrm{T}}D^*U(x)b \le (b^{\mathrm{T}}Ab)\|w(x)\| \qquad \forall\, x \in N.$$

These results, $b^{\mathrm{T}}Ab < 0$, the condition (i) of Assumption 2.5.3 and Assumption 2.5.1 prove that $E = V_k$ (2.5.21) is an estimate of D.

2.5.2.4 A quadratic aggregation form. A generalized quadratic aggregation form is this setting introduced by

ASSUMPTION 2.5.5. There are $U \in C(R^n, R^{m\times m})$, $w \in C(R^n, R^m)$ and matrices $A_{ij} \in R^{m\times m}$ such that

(i) $w(x) = 0$ for $x \in N$ iff $x = 0$;

(ii) $U(x) = 0$ for $x \in N$ iff $x = 0$;

(iii) $$D^*U(x) \le \begin{pmatrix} w^{\mathrm{T}}(x)A_{11}w(x) & \dots & w^{\mathrm{T}}(x)A_{1m}w(x) \\ \dots & \dots & \dots \\ w^{\mathrm{T}}(x)A_{m1}w(x) & \dots & w^{\mathrm{T}}(x)A_{mm}w(x) \end{pmatrix} \qquad \forall\, x \in N.$$

THEOREM 2.5.11. *Let Assumptions 2.5.1, 2.5.2 and 2.5.5 hold. In order for the set* E *(2.5.17) to be an estimate of* D *it is sufficient that* U *is positive definite on* N, $u_{ij}(x)$ *is radially unbounded in case* N *is unbounded* $\forall\,(i,j) \in [1,m]$, *and the matrix* $(A_{ij} + A_{ij}^{\mathrm{T}})$ *is negative definite* $\forall\,(i,j) \in [1,m]$.

PROOF. Let $\lambda_M(A_{ij} + A_{ij}^{\mathrm{T}})$ be the maximal eigenvalue of $(A_{ij} + A_{ij}^{\mathrm{T}})$ and

$$\alpha_{ij} = \frac{1}{2}\lambda_m(A_{ij} + A_{ij}^{\mathrm{T}}).2.5.24$$

Negative definiteness of $(A_{ij} + A_{ij}^{\mathrm{T}})$ implies $\alpha_{ij} < 0$ $\forall\,(i,j) \in [1,m]$. This resullt and the conditions of Theorem 2.5.11 satisfy all the requirements of Theorem 2.5.8, which proves the statement of Theorem 2.5.11.

THEOREM 2.5.12. *Let Assumptions 2.5.1, 2.5.4 and 2.5.5 hold. In order for the set E (2.5.21), (2.5.22) to be an estimate of D it is sufficient that the matrix* $\sum_{j=1}^{m}[b_j(A_{ij} + A_{ij}^{\mathrm{T}})]$ *is negative definite for all* $i \in [1, m]$.

PROOF. Using $b > 0$ elementwise (Assumption 2.5.4) we derive

$$D^{+}U(x)b \leq \begin{pmatrix} w^{\mathrm{T}}(x)\Big(\sum_{j=1}^{m} b_j A_{1j}\Big)w(x) \\ \cdots\cdots\cdots\cdots\cdots\cdots\cdots \\ w^{\mathrm{T}}(x)\Big(\sum_{j=1}^{m} b_j A_{mj}\Big)w(x) \end{pmatrix} \quad \forall x \in N \tag{2.5.25}$$

from (2.5.24). Negative definiteness of $\sum_{j=1}^{m}[b_j(A_{ij} + A_{ij}^{\mathrm{T}})]$, and the conditions (i) and (iii) of Assumption 2.5.5 prove negativeness of $b^{\mathrm{T}}D^{*}v_i(x)$ for every $(x \neq 0) \in N$ $\forall i \in [1, m]$, due to (2.5.25). This result, and Assumption 2.5.1, positive definiteness of $b^{\mathrm{T}}v_i$ on N $\forall i \in [1, m]$ prove that E (2.5.21), (2.5.22) is an estimate of D.

THEOREM 2.5.13. *Let Assumptions 2.5.1 and 2.5.5 hold. In order for the set* $E = V_k$ *(2.5.23) to be an estimate of D it is sufficient that U is positive definite on N,* $v(x) = b^{\mathrm{T}}U(x)b$ *is radially unbounded in case N is unbounded, the vector b is elementwise positive and the matrix* $\sum_{i,j=1}^{m}[b_i b_j(A_{ij} + A_{ij}^{\mathrm{T}})]$ *is negative definite.*

PROOF. Function $v(x)$, $v(x) = b^{\mathrm{T}}U(x)b$, is positive definite on N due to positive definiteness of $U(x)$ on N. Its derivative $D^{*}v(x)$ is negative for every $(x \neq 0) \in N$ in view of (i) and (iii) of Assumption 2.5.5,

$$D^{*}v(x) \leq \frac{1}{2}\,w^{\mathrm{T}}(x)\left[\sum_{i,j=1}^{m}[b_i b_j(A_{ij} + A_{ij}^{\mathrm{T}})]\right]w(x)$$

and negative definiteness of $\sum_{i,j=1}^{m}[b_i b_j(A_{ij} + A_{ij}^{\mathrm{T}})]$. These results and Assumption 2.5.1 prove that $E = V_k$ (2.5.23) is an estimate of D.

2.5.2.5 Generalized Michel's aggregation form. The aggregation form will be generalized by referring to Grujić, Martynyuk and Ribbens-Pavella [57] and Michel [141] as follows:

ASSUMPTION 2.5.6. There are $U \in C(R^n, R^{m\times m})$, $w \in C(R^n, R^m)$, $w(x) = [w_1(x), \dots, w_m(x)]^{\mathrm{T}}$, and vector $a_{ij} \in R^s$ such that

(i) $w_{ij}(x) = 0$ for $x \in N$ iff $x = 0$:

(ii) the matrix-valued function $U(x)$ is continuous on N, $U \in C(N, R^{m\times m})$;

(iii) the matrix-valued function $U(x)$, the vector function w and the vector a_{ij} obey (2.5.26)

$$D^*U(x) \le \begin{pmatrix} w_1(x)a_{11}^{\mathrm{T}}w(x) & \dots & w_1(x)a_{1s}^{\mathrm{T}}w(x) \\ \dots & \dots & \dots \\ w_s(x)a_{s1}^{\mathrm{T}}w(x) & \dots & w_s(x)a_{ss}^{\mathrm{T}}w(x) \end{pmatrix} \quad \forall x \in N. \tag{2.5.26}$$

Let

$$A^j = (a_{1j}, a_{2j}, \dots, a_{mj})^{\mathrm{T}}, \qquad V^j = (u_{1j}, u_{2j}, \dots, u_{sj})^{\mathrm{T}},$$
$$K^j = (k_{1j}, k_{2j}, \dots, k_{mj})^{\mathrm{T}}.$$

Let

$$E^j = (e_{ki}^j) \in R^{m\times m}, \qquad e_{ki}^j = [r_{ki}^1\delta_{ki} + r_{ki}^2(1-\delta_{ki})]a_{ki}^j,$$

where $\delta_{ki} = 1$ for $k = i$, $\delta_{ki} = 0$ for $k \ne i$ and

$$r_{ki}^1 = \inf\,[w_i(x) \colon x \in \partial E_{ki}], \qquad r_{ki}^2 = \sup\,[w_i(x) \colon x \in \overline{E}_{ki}].$$

THEOREM 2.5.14. *Let Assumptions 2.5.1, 2.5.2 and 2.5.6 hold. In order for the set E (2.5.17) and its closure $\overline{E}$ to be estimates of D it is sufficient that $U(x)$ is positive definite on N, $u_{ij}(x)$ is radially unbounded in case N is unbounded $\forall\, k = 1, 2, \dots, m$, of A^j is non-negative and the vector $E^j\mathbf{1}$ is negative elementwise $\forall\, j \in [1, m]$.*

PROOF. Since $U(x) = [v^1(x), v^2(x), \dots, v^m(x)]$ then (2.5.26) can be rewritten as

$$\begin{aligned} &D^*U(x) \le W(x)\,[A^1w(x),\ A^2w(x),\ \dots,\ A^mw(x)], \\ &W(x) = \operatorname{diag}\{w_1(x),\ w_2(x),\ \dots,\ w_m(x)\}. \end{aligned} \tag{2.5.27}$$

Let $j \in [1, m]$ be arbitrarily chosen. Positive definiteness of $U(x)$ on N implies positive definiteness of $u_{ij}(x)$ on N $\forall\,(i,j) \in [1, m]$. From $E^j\mathbf{1} < 0$, the definitions of E^j and V^j, (2.5.27) and Assumption 2.5.6 it follows that $D^*u_{ij}(x) < 0$ $\forall x \in \partial E_{ij}$ $\forall i \in [1, m]$. This result, Assumption 2.5.1 and

Assumption 2.5.3 prove positive invariance of $\overline{E}_{ij}$ with respect to motions of (1.2.10). The definitions of A^j and E^j imply $A^j \le E^j$ elementwise. Hence $E^j\mathbf{1} < 0$ implies $A^j\mathbf{1} < 0$.

Since $a^j_{ki} \ge 0$, $k \ne j$, then there is positive diagonal $D^j = \operatorname{diag}\{d_{1j}, d_{2j}, \dots, d_{mj}\}$ such that $[(A^j)^{\mathrm{T}}D^j + D^jA^j]$ is negative definite. Hence a function v^j, $v^j(x) = (d^j)^{\mathrm{T}}v^j(x)$ for $d^j = (d_{1j}, d_{2j}, \dots, d_{mj})^{\mathrm{T}}$, is positive definite and

$$D^*v^j(x) \le \frac{1}{2}\, w^{\mathrm{T}}(x)\,[(A^j)^{\mathrm{T}}D^j + D^jA^j]\,w(x) < 0$$

$\forall\,(x \ne 0) \in N$ due to negative definiteness of the matrix $[(A^j)^{\mathrm{T}}D^j + D^jA^j]$ and (i) of Assumption 2.5.6. These results, Assumption 2.5.3 together with positive invariance of all $\overline{E}_{ij}$ prove that E_{ij} and $\overline{E}_{ij}$ are estimates of D. Since this holds for every $(i, j) \in [1, m]$, then E and $\overline{E}$ are estimates of D of $x = 0$ of (1.2.10).

Let

$$A_j = (a_{j1}, a_{j2}, \dots, a_{jm})^{\mathrm{T}}, \qquad b = \operatorname{diag}(b_1, b_2, \dots, b_m),$$
$$A(b) = (A_1^{\mathrm{T}}b, A_2^{\mathrm{T}}b, \dots, A_m^{\mathrm{T}}b)^{\mathrm{T}}.$$

THEOREM 2.5.15. *Let Assumptions 2.5.1, 2.5.4 and 2.5.6 hold. In order for the set E (2.5.20), (2.5.21) to be an estimate of D it is sufficient that the vector $A_i^{\mathrm{T}}b$ is negative elementwise $\forall\, i \in [1, m]$.*

PROOF. Since $b > 0$ (Assumption 2.5.4) then (2.5.19) and (2.5.27) yield (2.5.28) due to (iii) of Assumption 2.5.6,

(2.5.28)
$$D^*U(x)b = \begin{pmatrix} b^{\mathrm{T}}D^*v_1(x) \\ \cdots\cdots\cdots \\ b^{\mathrm{T}}D^*v_m(x) \end{pmatrix} \le \begin{pmatrix} w_1(x)\, b^{\mathrm{T}}A_1 w(x) \\ \cdots\cdots\cdots\cdots \\ w_m(x)\, b^{\mathrm{T}}A_m w(x) \end{pmatrix} \qquad \forall\, x \in N.$$

Elementwise negativeness of $A_i^{\mathrm{T}}b_i$ $\forall\, i \in [1, m]$, (i) of Assumption 2.5.6 and (2.5.28) imply $b^{\mathrm{T}}D^*v_i(x) < 0$ $\forall\,(x \ne 0) \in N$. Hence, Assumption 2.5.1 and Assumption 2.5.4 prove that E (2.5.20), (2.5.21) is an estimate of D.

THEOREM 2.5.16. *Let Assumptions 2.5.1 and 2.5.6 hold. In order for the set $E = V_k$ (2.5.23) to be an estimate of D it is sufficient that $U(x)$ is positive definite on N, $v(x) = b^{\mathrm{T}}U(x)b$ is radially unbounded in case N is unbounded and the matrix $[A^{\mathrm{T}}(b)B + BA(b)]$ is negative definite for the elementwise positive vector b.*

PROOF. Theorem 2.5.16 is proved in the same way as Theorem 2.5.13. In order to achive this the matrix $\sum_{i,j=1}^{m} [b_i b_j (A_{ij} + A_{ij}^{\mathrm{T}})]$ should be replaced by the matrix $[A^{\mathrm{T}}(b)B + BA(b)]$ in the proof of Theorem 2.5.13.

2.5.2.6 Grujić–Šiljak's aggregation form. The aggregation form can be applied to matrix-valued function aggregation of (1.2.10) as follows:

ASSUMPTION 2.5.7. There are $U \in C(R^n, R^{m\times m})$, $w \in C(R^n, R^m)$, and vectors $a_{ij} \in R^m$ such that

(i) $w(x) = 0$ for $x \in N$ iff $x = 0$;

(ii) $U(x) \in C(R^n, R^{m\times m})$:

(iii) $D^*U(x) \le \begin{pmatrix} a_{11}^{\mathrm{T}} w(x) & \dots & a_{1m}^{\mathrm{T}} w(x) \\ \dots & \dots & \dots \\ a_{m1}^{\mathrm{T}} w(x) & \dots & a_{mm}^{\mathrm{T}} w(x) \end{pmatrix} \quad \forall x \in N.$

THEOREM 2.5.17. *Let Assumptions 2.5.1, 2.5.3 and 2.5.7 hold. In order for the set E (2.5.17) and its closure $\overline{E}$ to be estimates of D it is sufficient that $U(x)$ is positive definite on N, $u_{ij}(x)$ is radially unbounded in case N is unbounded $\forall (i,j) \in [1,m]$, off-diagonal element a_{ki}^j ($k \ne i$, $k,i = 1,2,\dots,m$) of A^j is nonnegative and the vector $E^j\mathbf{1}$ is negative elementwise $\forall j \in [1,m]$.*

PROOF. The condition (iii) of Assumption 2.5.7 can be set in the form

$$D^*U(x) \le [A^1 w(x), A^2 w(x), \dots, A^m w(x)] \qquad \forall x \in N. \tag{2.5.29}$$

We consider now $v^j(x) = \mathbf{1}^{\mathrm{T}} v^j(x)$. Positive definiteness of $U(x)$ on N implies positive definiteness of all $u_{ij}(x)$, hence of all $u^j(x)$, on N. Radial unboundedness of all $u_{ij}(x)$ implies radial unboundedness of all $v^j(x)$ in case N is unbounded. Assumption 2.5.2 implies radial increasing of all $v^j(x)$. From (2.5.29) and (i) of Assumption 2.5.7 it follows that $D^*v^j(x) < 0$ $\forall (x \ne 0) \in N$, $\forall j \in [1,m]$. The definition of E^j and $E^j\mathbf{1} < 0$ prove positive invariance of E_{ij} $\forall (i,j) \in [1,m]$. These results and Assumption 2.5.1 prove that both E (2.5.17) and $\overline{E}$ are estimates of D.

THEOREM 2.5.18. *Let Assumptions 2.5.1, 2.5.4 and 2.5.6 hold. In order for the set E (2.5.20), (2.5.21) to be an estimate of D it is sufficient that the vector $A_i^{\mathrm{T}} b$ is negative elementwise $\forall i \in [1,m]$.*

PROOF. Since $b > 0$ (Assumption 2.5.4) then (2.5.19) and (2.5.29) in view of (iii) of Assumption 2.5.7 yield

$$D^*U(x)\,b = \begin{pmatrix} b^{\mathrm{T}}D^*v_1(x) \\ \cdots\cdots\cdots \\ b^{\mathrm{T}}D^*v_m(x) \end{pmatrix} \le \begin{pmatrix} b^{\mathrm{T}}A_1w(x) \\ \cdots\cdots\cdots \\ b^{\mathrm{T}}A_mw(x) \end{pmatrix} \qquad \forall\, x \in N. \tag{2.5.30}$$

Now, $A_i^{\mathrm{T}}b < 0$ elementwise, (i) of Assumption 2.5.7 and (2.5.30) imply $D^*v_i(x) < 0\ \forall\,(x \ne 0) \in N$ for $v_i(x) = b^{\mathrm{T}}V_i(x)\ \forall i \in [1,m]$. This result, Assumption 2.5.1 and Assumption 2.5.4 prove that E (2.5.20), (2.5.21) is an estimate of D.

THEOREM 2.5.19. *Let Assumptions 2.5.1 and 2.5.7 hold. In order for the set $E = V_k$ (2.5.23) to be an estimate of D it is sufficient that $U(x)$ is positive definite on N, $v(x) = b^{\mathrm{T}}U(x)b$ is radially unbounded in case N is unbounded and the vector $A^{\mathrm{T}}(b)b$ is negative elementwise for the elementwise positive vector b.*

PROOF. Since $U(x)$ is positive definite on N for $y = b \in R^m$, then v, $v = b^{\mathrm{T}}U(x)b$, is also positive definite on N. From $b > 0$ and (iii) of Assumption 2.5.7 we derive

$$D^*v(x) \le b^{\mathrm{T}}A(b)w(x) \qquad \forall\, x \in N$$

so that

$$D^*v(x) < 0 \qquad \forall\,(x \ne 0) \in N$$

due to (i) of Assumption 2.5.7 and $A^{\mathrm{T}}(b)b < 0$. These results and Assumption 2.5.1 prove that $E = V_k$ (2.5.23) is an estimate of D.

2.5.2.7 L-aggregation form. L-aggregation form is being introduced in this framework by

ASSUMPTION 2.5.8. There are $U \in C(R^n, R^{m\times m})$, $w \in C(R^n, R^m)$, $b \in R^s$, $b = (b_1, b_2, \dots, b_m)^{\mathrm{T}}$ and $A \in R^{m\times m}$ such that

(i) $\|w(x)\| = 0$ for $x \in N$ iff $x = 0$;

(ii) $U(x) \in C(R^n, R^{m\times m})$;

(iii) $v(x) = b^{\mathrm{T}}U(x)b$ obeys

$$D^*v(x) \le w^{\mathrm{T}}(x)(A^{\mathrm{T}}B + BA)w(x) \qquad \forall\, x \in N$$

for $B = \mathrm{diag}\,(b_1, b_2, \dots, b_m)$.

THEOREM 2.5.20. *Let Assumptions 2.5.1 and 2.5.8 hold. In order for the set* $E = V_k$ *(2.5.23) to be an estimate of* D *it is sufficient that* $U(x)$ *is positive definite on* N, $v(x)$ *is radially unbounded in case* N *is unbounded and the matrix* $(A^{\mathrm{T}}B + BA)$ *is negative definite.*

PROOF. The function $U(x)$ is positive definite on N due to positive definiteness of $V(x)$ on N. Negative definiteness of $(A^{\mathrm{T}}B + BA)$ and conditions (i) and (iii) of Assumption 2.5.8 imply $D^*v(x) < 0 \;\; \forall\, (x \neq 0) \in N$. These results and Assumption 2.5.1 prove that $E = V_k$ (2.5.23) is an estimate of D.

2.6 On Multistability of Motion

As is well known, stability analysis of nonlinear systems is made under the assumption of the "equality" of all solutions coordinates with respect to dynamical properties as it is accepted in classical papers by Liapunov [101] and his adherents. The exeption is made for stability with respect to a part of variables. In the problem, phase vector of variables is divided into two subvectors, the norm of one of which is said to be "nonincreasing" to infinity for the finite time.

2.6.1 General problem on multistability

A large-scale system of dimension n is governed by

$$\frac{dx_i}{dt} = f_i(t, x_1, \dots, x_s), \qquad x_i(t_0) = x_{i0}, \tag{2.6.1}$$

where $x_i \in R^{n_i}$, $t \in \mathcal{T}_\tau$, $\mathcal{T}_\tau = [\tau, +\infty)$, $\tau \in R$, $t_0 \in \mathcal{T}_i$, $\mathcal{T}_i \subset R$, $f_i \colon \mathcal{T}_\tau \times R^{n_1} \times \dots \times R^{n_s} \to R^{n_i}$ and it is assumed that $f_i(t, x_1, \dots, x_s) = 0$ for all $t \in \mathcal{T}_\tau$ iff $x_1 = x_2 = \dots = x_s = 0$. Together with (2.6.1) we shall show in vector notion the system (1.6.1)

$$\frac{dx}{dt} = f(t, x), \qquad x(t_0) = x_0, \tag{2.6.2}$$

where $x \in R^n$, $n = \sum\limits_{i=1}^{s} n_i$; $f \colon \mathcal{T}_\tau \times R^n \to R^n$, $x_0 = \left(x_{10}^{\mathrm{T}}, \dots, x_{s0}^{\mathrm{T}}\right)^{\mathrm{T}}$. It is clear that $f(t, x) = 0$ for all $t \in \mathcal{T}_\tau$ iff $x = 0$.

DEFINITION 2.6.1. System (2.6.1) is called *multistability (on $\mathcal{T}_\tau$)* iff its zero solution $(x_1^{\mathrm{T}}, \dots, x_s^{\mathrm{T}})^{\mathrm{T}} = 0$ is stable in some type (on $\mathcal{T}_\tau$) and attractive (on $\mathcal{T}_\tau$) with respect to groups of variables $\{x_i^{\mathrm{T}}\}$, $i = 1, 2, \dots, s$ (with respect to totality of groups of variables $\{x_1^{\mathrm{T}}, \dots, x_l^{\mathrm{T}}\}$, $l < s$).

REMARK 2.6.1. When multistability of solution $x = 0$ of (2.6.1) is discussed with respect to all groups of variables $\{x_1^{\mathrm{T}}, \dots, x_s^{\mathrm{T}}\}$ system (2.6.2) is defined in domain $B(\rho) = \Big\{x_i \colon \sum\limits_i \|x_i\| < \rho\Big\}$ or in R^n as usual.

REMARK 2.6.2. If multistability of solution $x = 0$ of (2.6.1) is discussed with respect to a group of variables $\{x_1^{\mathrm{T}}, \dots, x_l^{\mathrm{T}}\}$, $l < s$ then it is sufficient to define system (2.6.1) in the domain

$$B_{(\cdot)}(\rho) = \{x_i^{\mathrm{T}} \colon \| \left(x_1^{\mathrm{T}}, \dots, x_l^{\mathrm{T}}\right)^{\mathrm{T}} \| < \rho\}, \qquad \rho = \text{const},$$
$$D_{(\cdot)}(\rho) = \{x_i^{\mathrm{T}} \colon 0 < \| \left(x_{l+1}^{\mathrm{T}}, \dots, x_s^{\mathrm{T}}\right)^{\mathrm{T}} \| < +\infty\},$$

here solution $x(t, \cdot) = \left(x_1^{\mathrm{T}}(t, \cdot), \dots, x_l^{\mathrm{T}}(t, \cdot)\right)^{\mathrm{T}}$ of the system (2.6.1) is assumed to be continuable along $\left(x_{l+1}^{\mathrm{T}}, \dots, x_s^{\mathrm{T}}\right)^{\mathrm{T}}$, i.e. solution $\left(x_1^{\mathrm{T}}, \dots, x_s^{\mathrm{T}}\right)^{\mathrm{T}}$ is definite for all $t \in \mathcal{T}_\tau$ for which $\| \left(x_1^{\mathrm{T}}(t), \dots, x_l^{\mathrm{T}}(t)\right)^{\mathrm{T}} \| \le \rho$.

The construction of sufficient (and necessary) conditions ensuring multistability of zero solutions of (2.6.1) in terms of Definition 2.6.1 makes the general problem on multistability of motion.

2.6.2 On the relationship of the definition of multistability with the other notions of stability of motion

We shall recall the well known definition with reference to system (2.6.1).

DEFINITION 2.6.2. The zero solution $x_1 = x_2 = \dots = x_s = 0$ of system (2.6.1) is

(i) *stable relatively $\mathcal{T}_i$* if for any $\varepsilon > 0$ and $t_0 \in \mathcal{T}_i$ there exist $\delta(t_0, \varepsilon) > 0$ such that $\sum\limits_i^s \|x_i(t; t_0, x_0)\| < \varepsilon$ for all $\sum\limits_i^s \|x_{i0}\| < \delta$ and all $t \ge t_0$;

(ii) *asymptotically stable relatively $\mathcal{T}_i$* if the conditions of Definition 2.6.2 (i) are satisfied and $\sum\limits_{i=1}^s \|x_i(t; t_0, x_0)\| \to 0$ as $t \to +\infty$.

Having compared Definition 2.6.1 with Definition 2.6.2, we see that if all subvectors x_i in system (2.6.1) are homogeneous with respect to the dynamical properties and in Definition 2.6.1 of stability one Euclidean norm $\|x\| = \sum_{i=1}^{s} \|x_i\|$ is used, Definition 2.6.1 of multistability degenerates into Definition 2.6.2 of stability in the sense of Liapunov of the zero solution of the system (2.6.1) iff $\mathcal{T}_i = R$.

DEFINITION 2.6.3. The zero solution of the system (2.6.1) is

(i) *stable relatively the subvectors* $x_1, \dots, x_k$ $(k < s)$ *and respect to* $\mathcal{T}_i$, if for every $\varepsilon > 0$ and $t_0 \in \mathcal{T}_i$ there exist $\delta_1(t_0, \varepsilon) > 0$ and $\delta_2(t_0, \varepsilon) > 0$ such that $\sum_{i=1}^{k} \|x_i(t; t_0, x_{i0})\| < \varepsilon$ for $\sum_{i=1}^{k} \|x_{i0}\| < \delta_1$ and $\sum_{i=k+1}^{s} \|x_{i0}\| < \delta_2$ for all $t \ge t_0$;

(ii) *asymptotically stable with respect to the subvectors* $x_1, \dots, x_k$ $(k < s)$ *relatively* $\mathcal{T}_i$ if under conditions (i) of Definition 2.6.3 the relation $\sum_{i=1}^{k} \|x_i(t; t_0, x_{i0})\| \to 0$ holds for all $t \to +\infty$.

The comparison of Definition 2.6.1 and 2.6.3 shows that if the subvectors x_i, $i < k$ are homogeneous relatively the dynamical properties and the solution of the system (2.6.1) is continuable relatively $x_{k+1}, \dots, x_s$, the Definition of multistability with respect to a part of the variables implies Definition 2.6.1.

According as Movchan [147], Lakshmikantham and Salvadori [93], Lakshmikantham, Leela and Martynyuk [94] we consider the classes of functions

$$M = \{\rho \in C(R_+ \times R^n, R_+) \colon \inf_{(t,x)} \rho(t, x) = 0\},$$
$$M_0 = \{\rho \in M \colon \inf_{x} \rho(t, x) = 0 \text{ for all } t \in R_+\}.$$

DEFINITION 2.6.4. System (2.6.1) is

(i) (ρ_0, ρ)*-stable with respect to* $\mathcal{T}_i$, if for any $\varepsilon > 0$ and $t_0 \in \mathcal{T}_i$ there exists a positive function $\delta(t_0, \varepsilon)$, being continuous in $t_0 \in \mathcal{T}_i$ for every $\varepsilon > 0$ and such that $\rho_0(t_0, x_0) < \delta$ implies $\rho(t, x(t)) < \varepsilon$ for all $t \le t_0$;

(ii) *asymptotically* (ρ_0, ρ)*-stable with respect to* $\mathcal{T}_i$ if under the conditions of Definition 2.6.4 (i) $\rho(t, x(t)) \to 0$ as $t \to +\infty$.

The comparison of Definitions 2.6.1 and 2.6.4 yields that the Definition 2.6.4 provides the general characteristics of the dynamical properties of the subvectors x_i, $i = 1, 2, \dots, s$, without distinguishing between them.

Let us consider the system (2.6.1) and introduce the measures

$$\rho_0 = \rho_0(t, x_1, \dots, x_s) \in M_0;$$
$$\rho_1 = \rho_1(t, x_1, \dots, x_{s-1}) \in M;$$
$$\dots$$
$$\rho_{s-1} = \rho_{s-1}(t, x_1) \in M.$$

DEFINITION 2.6.5. System (2.6.1) is *multistable with respect to the measures* $(\rho_0, \dots, \rho_{s-1})$ *relatively* $\mathcal{T}_i$, iff it is $(\rho_0, \dots, \rho_k)$-stable in some type (on $\mathcal{T}_\tau$), $k < s-1$.

Thus, the examination of Definitions 2.6.2 – 2.6.5 indicates that only Definition 2.6.5 is a generalization of Definition 2.6.1, while the rest of the definitions follow from it.

2.6.3 Multistability investigation

In order to apply the method of matrix Liapunov function to the problem in question, we introduce classes of matrix-valued function with particular properties.

Together with (2.6.1) we consider a two-indexed system of functions

$$U(t,x) = [v_{ij}(t,x)], \qquad i, j \in [1,s] \tag{2.6.3}$$

with $v_{ii} \in C(\mathcal{T}_\tau \times R^n, R_+)$ and $v_{ij} \in C(\mathcal{T}_\tau \times R^n, R)$ for $i \neq j \in [1,s]$.

The notion of the definiteness of an auxiliary function (that is used in the direct Liapunov's method) is a main one, since this behaves as a scalar function having all norm properties.

DEFINITION 2.6.6. The matrix-valued function $U \colon \mathcal{T}_\tau \times B_{(1,l)} \times D_{(l+1,s)} \to R^{s\times s}$ is:

(i) *positive definite on* $\mathcal{T}_\tau$, $\tau \in R$, *with respect to variables* $(x_1^{\mathrm T}, \dots, x_l^{\mathrm T})$ iff there exist time-invariant connected neighborhoods $\mathcal{N}^*$, $\mathcal{N}^* \subset R^l$ of $x = 0$, a vector $\varphi \in R^s_+$, $\varphi > 0$ and a scalar positive definite in the sence of Liapunov function $w \colon \mathcal{N}^* \to R_+$ such that

(a) $U(t,x) \in C\left(\mathcal{T}_\tau \times \mathcal{N}^* \times D_{(l+1,s)}, R^{s\times s}\right)$;

(b) $U(t,x) = 0$ for all $t \in \mathcal{T}_\tau$ and $(x_1^{\mathrm T}, \dots, x_l^{\mathrm T}) = 0$;

(c) $\varphi^{\mathrm T} U(t,x)\varphi \geq w(x_{1,l}^{\mathrm T})$ for all $(t, x \neq 0, \varphi \neq 0) \in \mathcal{T}_\tau \times \mathcal{N}^* \times D_{(l+1,s)} \times R^s_+$, $x_{1,l}^{\mathrm T} = (x_1^{\mathrm T}, \dots, x_l^{\mathrm T})$;

(ii) *positive definite on* $\mathcal{T}_\tau \times \mathcal{G}^*$ *with respect to variables* $(x_1^{\mathrm{T}}, \dots, x_l^{\mathrm{T}})$ if conditions of Definition 2.6.6 (i) hold for $B_{(1,l)} = \mathcal{G}^*$;

(iii) *positive definite in the whole (on* $\mathcal{T}_\tau$*) with respect to variables* $(x_{1,l}^{\mathrm{T}})$ if condition of Definition 2.6.6 (i) hold for $B_{(1,l)} = R^l$;

(iv) *negative definite (in the whole) on* $\mathcal{T}_\tau \times B_{(1,l)}$ *(on* $\mathcal{T}_\tau$*) with respect to variables* $(x_{1,l}^{\mathrm{T}})$, iff $(-U)$ is positive definite (in the whole) on $\mathcal{T}_\tau \times B_{(1,l)}$ (on $\mathcal{T}_\tau$) with respect to variables $(x_{1,l}^{\mathrm{T}})$.

PROPOSITION 2.6.1. *The matrix-valued function* $U\colon \mathcal{T}_\tau \times R^n \to R^{s\times s}$ *is positive definite on* $\mathcal{T}_\tau$ *with respect to* $(x_1^{\mathrm{T}}, \dots, x_l^{\mathrm{T}})$, *iff it can be represented in the form*

$$\varphi^{\mathrm{T}} U(t,x)\varphi = \varphi^{\mathrm{T}} U_+(t,x)\varphi + w(x_1^{\mathrm{T}}, \dots, x_l^{\mathrm{T}}), \tag{2.6.4}$$

where $U_+(t,x)$ *is positive semi-definite with respect to all variables* $(x_1^{\mathrm{T}}, \dots, x_s^{\mathrm{T}})$ *and* w *is a function explicitly independent of* $t \in \mathcal{T}_\tau$ *and positive definite with respect to variables* $(x_1^{\mathrm{T}}, \dots, x_l^{\mathrm{T}})$, $l < s$.

PROOF. *Necessity.* Let the matrix-valued function $U(t,x)$ be $(x_1^{\mathrm{T}}, \dots, x_l^{\mathrm{T}})$ positive definite on $\mathcal{T}_\tau$. Then, by Definition 2.6.6 there exists a positive definite in the sense of Liapunov function $w(x_1^{\mathrm{T}}, \dots, x_l^{\mathrm{T}})$ such that on the domain $\mathcal{T}_\tau \times B_{(1,l)} \times D_{(l+1,s)} \times R_+^s$ condition (i) of Definition 2.6.6 is satisfied. We introduce the function

$$\varphi^{\mathrm{T}} U_+(t,x)\varphi = \varphi^{\mathrm{T}} U(t,x)\varphi - w(x_{1,l}^{\mathrm{T}})$$

which, is non-negative by condition 2.6.6 (c). Hence the function $\varphi^{\mathrm{T}} U(t,x)\varphi$ can be presented in the form (2.6.4).

Sufficiency. Let equality (2.6.4) be satisfied, where $\varphi^{\mathrm{T}} U_+(t,x)\varphi \geq 0$ and $w(x_{1,l}^{\mathrm{T}})$ is a positive definite function with respect to the variables $(x_1^{\mathrm{T}}, \dots, x_l^{\mathrm{T}})$. Then equality (2.6.4) implies

$$\varphi^{\mathrm{T}} U(t,x)\varphi - w(x_{1,l}^{\mathrm{T}}) = \varphi^{\mathrm{T}} U_+(t,x)\varphi \geq 0.$$

Hence condition 2.6.6 (c) for the function $\varphi^{\mathrm{T}} U(t,x)\varphi$ holds. This proves the Proposition 2.6.1.

PROPOSITION 2.6.2. *The matrix-valued function* $U\colon \mathcal{T}_\tau \times R^n \to R^{s\times s}$ *is positive definite on* $\mathcal{T}_\tau$ *with respect to variables* $(x_1^{\mathrm{T}}, \dots, x_l^{\mathrm{T}})$ *(in the whole) iff there exist function* $a \in K(KR)$ *such that*

$$\varphi^{\mathrm{T}} U(t,x)\varphi \geq a\left(\|(x_{1,l}^{\mathrm{T}})^{\mathrm{T}}\|\right) \tag{2.6.5}$$

in the domain $\mathcal{T}_\tau \times \mathcal{N}^* \times D_{(l+1,s)} \times R_+^s$.

DEFINITION 2.6.7. The matrix-valued function $U\colon \mathcal{T}_\tau \times R^n \to R^{s\times s}$ is called

(i) *decreasing on $\mathcal{T}_\tau$ with respect to variables* $(x_1^{\mathrm T}, \dots, x_l^{\mathrm T})$ iff there exists time-invariant connected neighborhood $\mathcal{N}^* \subseteq R^l$ of $x = 0$, a positive definite function $w_2\colon \mathcal{N}^* \to R_+$ and a vector $\varphi \in R^s_+$, $\varphi > 0$ such that

  (a) conditions (a), (b) of Definition 2.6.6 hold and

  (b) $\varphi^{\mathrm T} U(t,x)\varphi \ge w_2(x_{1,l}^{\mathrm T})$ for all $(t,\, x \ne 0,\, \varphi \ne 0) \in \mathcal{T}_\tau \times \mathcal{N}^* \times D_{(l+1,s)} \times R^s_+$.

PROPOSITION 2.6.3. *The matrix-valued function $U\colon \mathcal{T}_\tau \times R^n \to R^{s\times s}$ is decreasing on $\mathcal{T}_\tau$ with respect to variables $(x_1^{\mathrm T}, \dots, x_l^{\mathrm T})$, iff it can be presented in the form*

$$\varphi^{\mathrm T} U(t,x)\varphi = \varphi^{\mathrm T} U_-(t,x)\varphi + w_2(x_1^{\mathrm T}, \dots, x_l^{\mathrm T}), \tag{2.6.6}$$

where $U_-(t,x)$ is negative semi-definite with respect to all of variables $(x_1^{\mathrm T}, \dots, x_s^{\mathrm T})$, and w_2 is independent of $t \in \mathcal{T}_\tau$ positive definite function of variables $(x_1^{\mathrm T}, \dots, x_l^{\mathrm T})$, $l < s$.

PROOF. Repeating the same argument as in Proposition 2.6.1, one can show there is a matrix-valued function $U(t,x)$ for which the condition (2.6.6) holds.

PROPOSITION 2.6.4. *The matrix-valued function $U\colon \mathcal{T}_\tau \times R^n \to R^{s\times s}$ is decreasing on $\mathcal{T}_\tau \times \mathcal{N}^*$ with respect to variables $(x_1^{\mathrm T}, \dots, x_l^{\mathrm T})$ iff there exist a function $b \in K_{[0,\alpha]}$, where $\alpha = \sup\{x_{1,l}^{\mathrm T} \in \mathcal{N}^*\}$ and estimate*

$$\varphi^{\mathrm T} U(t,x)\varphi \le b\left(\|(x_{1,l}^{\mathrm T})^{\mathrm T}\|\right) \tag{2.6.7}$$

holds for all $(t,x) \in \mathcal{T}_\tau \times \mathcal{N}^* \times D_{(l+1,s)} \times R^s_+$.

Let $U \in C(\mathcal{T}_\tau \times R^n,\, R^{s\times s})$. The right-hand upper Dini derivative of functions $U(t,x)$ along solutions of the system (2.6.1) are defined by

$$D^+U(t,x) = \left[D^+v_{ij}(t,x)\right] \qquad \forall\, i,\, j \in [1,s], \tag{2.6.8}$$

where

$$D^+v_{ij}(t,x) = \limsup\left\{[v_{ij}(t+\theta,\, x+\theta f(t,x)) - v_{ij}(t,x)]\,\theta^{-1} \colon \theta \to 0^+\right\}.$$

2.6.4 Principle of comparison and multistability

The investigation of multistability of the solution of systems of differential equations (2.6.1) via the comparison technique assumes the presence of the corresponding comparison theorems.

2.6.4.1 The functions of SL-class. All scalar functions of the type

$$v(t, x, a) = a^{\mathrm{T}} U(t, x) a, \tag{2.6.9}$$

where $U \in C(\mathcal{T}_\tau \times R^n,\, R^{s\times s})$ are attributed to the class SL.

The vector a can be defined as

(i) $a = y \in R^s,\ \ y \neq 0;$

(ii) $a = \psi \in C(R^n, R^s_+),\ \ \psi(0) = 0;$

(iii) $a = \theta \in C(\mathcal{T}_\tau \times R^n,\, R^s)\ \ \theta(t, 0) = 0,\ \ \forall\, (t, x) \in \mathcal{T}_\tau \times \mathcal{N};$

(iv) $a = \varphi \in R^s_+,\ \ \varphi > 0.$

Applying function (2.6.9) and quasimonotone nondecreasing in u for each t function $g\colon g \in C(R^2_+, R),\ g(t, 0) = 0$ we shall formulate the following comparison result.

PROPOSITION 2.6.5. *Let the function $U\colon \mathcal{T}_\tau \times R^n \to R^{s\times s}$ be locally Lipschitzian in x. Suppose that the function*

$$\varphi^{\mathrm{T}} D^+ U(t, x)\varphi \triangleq D^+ v(t, x, \varphi) \tag{2.6.10}$$

and the function $g \in C(R_+ \times R^n \times R_+,\, R)$ such that

$$D^+ v(t, x, \varphi) \le g(t, x,\, v(t, x, \varphi))$$

holds for $(t, x, \varphi) \in R_+ \times R^n \times R^s_+$. Let $x(t) = x(t; t_0, x_0)$ be a solution of (2.6.1) existing on $[t_0, \infty)$ and $r(t; t_0, x_0, u_0)$ be the maximal solution of

$$\frac{du}{dt} = g(t, x(t), u), \qquad u(t_0) = u_0 \ge 0 \tag{2.6.11}$$

existing for $t \ge t_0$. Then $v(t_0, x_0, \varphi) \le u_0$ implies

$$v(t, x(t), \varphi) \le r(t; t_0, x_0, u_0), \qquad t \ge t_0. \tag{2.6.12}$$

PROOF is similar to the proof of Proposition 2.4.3.

COROLLARY 2.6.1. If all conditions of Proposition 2.6.5 are satisfied and function $g(t,x,v)$ satisfies either of the conditions

C_1. $g(t,x,v)=0$ for all $t\ge t_0$;

C_2. $g(t,x,v)=\psi^{\mathrm{T}}A\psi$, where $\psi\in C(R^n,R^s_+)$, $\psi(0)=0$, A is a constant matrix $s\times s$;

C_3. $g(t,x,v)=w^{\mathrm{T}}Bw+r(t,w,\varphi)$, where $w\in C(R^n,R^s_+)$, B is a constant matrix $s\times s$, $r\in C(R_+\times R^s_+\times R^s_+,R)$ is a polynomial in power higher than two;

C_4. $g(t,x,v)=W(t,w,\varphi)+r^*(t,w,\varphi)$, where $W\in C(R_+\times R^s_+\times R^s_+,R)$ is atleast a second-power polynomial, and r^* is the same polynomial as in case C_3;

C_5. $g(t,x,v)=w^{\mathrm{T}}(x)\left[A^{\mathrm{T}}B+BA\right]w(x)$, where $w\in C(R^n,R^s)$, $A\in R^{s\times s}$, $b\in R^s$, $B=\operatorname{diag}(b_1,\dots,b_s)$, then estimate (2.6.12) is satisfied, and the investigation of comparison equation (2.6.11) is simplified.

2.6.4.2 The functions of VL-class. All vector functions of the type

$$L(t,x,b)=AU(t,x)b, \tag{2.6.13}$$

where $U\in C(\mathcal{T}_\tau\times R^n,\,R^{s\times s})$, A is a constant matrix $s\times s$, and vector b is defined according to (i) – (iv) similarly to the definition of the vector a.

For any function $U(t,x)$, which is associated with system (2.6.1) we shall define the function

$$D^+L(t,x,\varphi)=AD^+U(t,x)\varphi \tag{2.6.14}$$

for all $(t,x,\varphi)\in\mathcal{T}_\tau\times R^n\times R^s_+$.

PROPOSITION 2.6.6. *Let there exist*

(1) *a matrix-valued function $U\in C(\mathcal{T}_\tau\times R^n,\,R^{s\times s})$ such that $U(t,x)$ is locally Lipschitzian in x;*

(2) *a constant $s\times s$ matrix A, a vector $\varphi\in R^s_+$ and vector $y\in R^n$ such that*

$$y^{\mathrm{T}}L(t,x,\varphi)\ge a(\|x\|),$$

where $a\in K$;

(3) *a vector function* $G \in C(\mathcal{T}_\tau \times R^n \times R_+^s,\ R^s)$ *such that* $G(t,x,u)$ *is quasimonotone nondecreasing in* u *for every* $t \in R_+$ *such that the estimate*

$$D^+L(t,x,\varphi) \le G\,(t,x,L(t,x,\varphi)) \tag{2.6.15}$$

holds;

(4) *let* $x(t;t_0,x_0)$ *be any solution of (2.6.1) existing on* $[t_0,\infty)$ *and* $w(t;t_0,w_0,x_0)$ *be the maximal solution of*

$$\frac{du}{dt} = G(t,x,u), \qquad u(t_0) = w_0 \ge 0 \tag{2.6.16}$$

existing for $t \ge t_0$. *Then* $L(t_0,x_0,\varphi) \le w_0$ *implies*

$$L(t,x(t),\varphi) \le w(t;t_0,w_0,x_0) \qquad t \ge t_0. \tag{2.6.17}$$

PROOF. It is proved in a standard way by the comparison method (see e.g. Lakshmikantham, Leela and Martynyuk [94]).

COROLLARY 2.6.2. Let conditions (1) and (2) of Proposition 2.6.6 be satisfied and in conditions (3) and (4) the function $G \in C(\mathcal{T}_\tau \times R^n, R^s)$. Then, estimate (2.6.17) is satisfied for the maximal solution $w^*(t;t_0,w_0)$ of the comparison system

$$\frac{du}{dt} = G(t,u), \qquad u(t_0) = u_0 \ge 0. \tag{2.6.18}$$

COROLLARY 2.6.3. Let conditions (1) and (2) Proposition 2.6.6 be satisfied and the function $G(t,x,L)$ have the form

$$G(t,x,L(t,x,\varphi)) = PL(t,x,\varphi) + m(t,L(t,x,\varphi)),$$

where $P = [p_{ij}]$ is a $s \times s$ matrix with elements $p_{ij} \ge 0$ $(i \ne j)$ and $m \in C(\mathcal{T}_\tau \times R^s, R^s)$ is quasimonotone in L and

$$\lim_{\|L\|\to 0} \frac{\|m(t,L)\|}{\|L\|} = 0$$

uniformly in $t \ge t_0$.

Then, estimate (2.6.18) is true for the maximal solution $w(t; t_0, w_0)$ of the comparison system

$$\frac{du}{dt} = Pu + m(t, u), \qquad u(t_0) = w_0 \geq 0.$$

2.6.4.3 The functions of ML-class. In order to formulate the theorem of comparison with matrix-valued Liapunov function relatively to arbitrary cone K in space R^n we shall need some auxiliary information. Following Lakshmikantham, Leela and Martynyuk [94] a proper subset $K \subset R^n$ is called a cone if the following properties hold:

$$\begin{gathered} \lambda K \subset K, \quad \lambda \geq 0, \quad K + K \subset K, \quad K = \overline{K} \\ K \cap \{-K\} = \{0\} \quad \text{and} \quad \text{int}\, K \neq \emptyset \end{gathered} \tag{2.6.19}$$

where $\overline{K}$ denotes the closure of K, int K is the interior of K. We shall denote by ∂K the boundary of K. The cone K induces the order relations on R^n defined by

$$\begin{gathered} x \overset{K}{\leq} y \quad \text{iff} \quad y - x \in K \quad \text{and} \\ x \overset{K}{<} y \quad \text{iff} \quad y - x \in \text{int}\, K. \end{gathered} \tag{2.6.20}$$

The set K^* defined by $K^* = \{\varphi \in R^n \colon \varphi(x) \geq 0 \text{ for all } x \in K\}$, where $\varphi(x)$ denotes the scalar product $\langle \varphi, x \rangle$, is called the adjoint cone and satisfies the properties (2.6.19).

We note that $K = (K^*)^*$, $x \in \text{int}\, K$ iff $\varphi(x) > 0$ for all $\varphi \in K_0^*$ and $x \in \partial K$ iff $\varphi(x) = 0$ for some $\varphi \in K_0^*$, where $K_0 = K - \{0\}$.

We can now define as quasimonotone property a function relative to the cone K.

A function $f \in C(R^n, R^n)$ is said to be quasimonotone nondecreasing relative to K if $x \overset{K}{\leq} y$ and $\varphi(x - y) = 0$ for some $\varphi \in K_0^*$ implies $\varphi(f(x) - f(y)) \leq 0$.

If f is linear, that is, $f(x) = Ax$ where A is an n by n matrix, the quasimonotone property of f means the following: $x \geq 0$ and $\varphi(x) = 0$ for some $\varphi \in K_0^*$ imply $\varphi(Ax) \geq 0$.

If $K = R^n_+$, the function f is said to be quasimonotone nondecreasing if $x < y$ and $x_i = y_i$ for some i, $1 \leq i \leq n$, implies $f_i(x) \leq f_i(y)$.

We consider the system

$$\frac{dz}{dt} = g(t, z), \qquad g(t, 0) = 0, \tag{2.6.21}$$

where $g \in C(R_+ \times R^n,\ R^n)$, $g(t, z)$ is a locally Lipschitzian in z. Let $z_1(t; t_0, z_{10})$, $z_2(t; t_0, z_{20})$ be solutions of the system (2.6.21) with the initial conditions (t_0, z_{10}) and (t_0, z_{20}) respectively.

DEFINITION 2.6.8. We shall say that system (2.6.21) has monotone (strictly) solutions, if

$$z_{20} - z_{10} \in K, \qquad z_{20} \neq z_{10}$$

imply the inclusions

$$z_2(t) - z_1(t) \in K \qquad (z_2(t) - z_1(t) \in \operatorname{int} K)$$

for all $t \geq t_0$ respectively.

DEFINITION 2.6.9. System (2.6.21) is said to belong the class $W_0(K)$ $(W_s(K))$ if $(z - y) \in \partial K$, $z \neq y$ implies the inequalities

$$g(t, z) - g(t, y) \geq 0 \qquad (g(t, z) - g(t, y) > 0)$$

respectively.

DEFINITION 2.6.10. The operator $p(t, z)$ is *positive on* $J_1 \times D$ if $z \in D$ implies $p(t, z) \geq 0$ for all $t \in J_1$, with respect to the cone K.

We shall formulate now a basic Proposition of the principle of comparison in the space, ordered by an arbitrary cone.

PROPOSITION 2.6.7. *Let*

(1) *there exists a function* $g(t, z) \in W_0(K)$ *continuous in open* (t, z) *set* $J_1 \times D$ *and satisfying the uniqueness conditions of solutions* $z(t; t_0, z_0)$ *of system*

$$\frac{dz}{dt} = g(t, z), \qquad z(t_0) = z_0; \tag{2.6.22}$$

(2) *there exists a function* $h(t, y)$ *continuous on open* (t, y) *set* $J_2 \times D \subset J_1 \times D$, $J_2 \subseteq J_1$ *such that* $g(t, y) - h(t, y) = p(t, y)$, *where* $p(t, y)$ *is a positive operator on set* $J_0 \times D$, *where* $J_0 = J_1 \cap J_2$.

Then satisfy the relation

$$z(t) - y(t) \in K$$

whenever $z_0 - y_0 \in K$, *where* $y(t)$ *is an arbitrary solution of the system*

$$\frac{dy}{dt} = h(t, y), \qquad y(t_0) = y_0. \tag{2.6.23}$$

PROOF. Together with system (2.6.21) consider a weakly perturbed system

$$\frac{dz}{dt} = g(t, z) + \varepsilon(z, u^*)u, \tag{2.6.24}$$

where $u \in K$, $u^* \in K^*$ and $\varepsilon \in (0, \varepsilon^*)$, the solution $z(t, \varepsilon) = z(t; t_0, z_0, \varepsilon)$ of which exists on $[t_0, \tau]$, where $\tau \in J_0$ and $\lim\limits_{\varepsilon \to 0} z(t, \varepsilon) = z(t)$ uniform on $[t_0, \tau)$, where $z(t)$ is a solution of system (2.6.22). Let $z(t, \varepsilon) - y(t) \notin K$ for all $t \in [t_0, \tau)$. Then there exists a $t^* \in [t_0, \tau]$ such that

$$z(t, \varepsilon) - y(t) \in K \quad \text{for all} \quad t \in [t_0, t^*)$$

and $z(t, \varepsilon) - y(t) \notin K$ for the values $t > t^*$ arbitrarily close to t^*. For $t = t^*$ the inclusion

$$z(t^*, \varepsilon) - y(t^*) \in \partial K \tag{2.6.25}$$

and the condition

$$z(t^*, \varepsilon) \neq y(t^*) \tag{2.6.26}$$

are satisfied. For the function $m(t, \varepsilon) = z(t, \varepsilon) - y(t)$ we make the differential equation, in view of the system of equations (2.6.22) and (2.6.23). Namely

$$\begin{aligned} \frac{dm}{dt} &= g(t, z) + \varepsilon(z, u^*)u - h(t, y) \\ &= g(t, z) - g(t, y) + g(t, y) - h(t, y) + \varepsilon(z, u^*)u. \end{aligned}$$

By condition (2) of Proposition 2.6.7

$$\frac{dm}{dt} = g(t, z) - g(t, y) + p(t, y) + \varepsilon(z, u^*)u, \tag{2.6.27}$$

where $p(t, y)$ is a positive operator. By conditions (1) and (2) of Proposition 2.6.7 $p(t^*, y) \geq 0$ and $g(t^*, z) - g(t^*, y) \geq 0$ whenever $(z - y) \in \partial K$. The last condition is satisfied due to (2.6.25) and (2.6.26). The item $\varepsilon(z, u^*)u$ is also non-negative, since $u \in K$ and $u^* \in K^*$.

We confront with the set of point m from the boundary of cone K the indicatory function $\delta(\cdot \mid K)$, setting

$$\delta(m \mid K) = \begin{cases} 0, & \text{if } m \in K; \\ +\infty, & \text{if } m \notin K. \end{cases}$$

For the indicatory function $\delta(m \mid K)$ we compute the subgradient $\gamma(m)$ and scalar multiply the right and left side of the equation by $\gamma(m)$. We get

$$\left(\gamma, \frac{dm}{dt}\right) < -\alpha, \qquad \alpha = \text{const} > 0$$

at point $t = t^*$. Therefore, $m(t, \varepsilon)$ will not leave the cone K for all $t > t^*$ as $\varepsilon \to 0$. The proof is complete.

2.6.5 The system (2.6.1) analysis for s = 2

For the system (2.6.1) we construct a matrix-valued function

$$U(t, x) = [v_{ij}(t, x)], \qquad i, j = 1, 2, \tag{2.6.28}$$

where $x \in R^{N_0}$, $N_0 = n_1 + n_2$ and v_{ij} is locally Lipschitzian in x. With the aid of vector $y \in R^2$, $y \neq 0$ we construct a scalar function

$$v(t, x, y) = y^{\mathrm{T}} U(t, x) y. \tag{2.6.29}$$

Function (2.6.29) allows us to investigate multistability of the system under definite conditions.

2.6.5.1 Direct application of matrix-valued function. Suppose that system (2.6.1) is defined in domain

$$\mathcal{T}_\tau \times B_1(\rho) \times B_2(\rho), \qquad \rho = \text{const} > 0 \tag{2.6.30}$$

and the following stability definition holds true for it.

DEFINITION 2.6.11. System (2.6.1) is called *multistable (on $\mathcal{T}_\tau$)* if its zero solution $(x_1^{\mathrm{T}}, x_2^{\mathrm{T}})^{\mathrm{T}} = 0$ is

(i) uniformly $(x_1^{\mathrm{T}}, x_2^{\mathrm{T}})$-stable with respect to $\mathcal{T}_i$;
(ii) uniformly asymptotically x_2^{T}-stable with respect to $\mathcal{T}_i$.

THEOREM 2.6.1. *Let vector-function $f = (f_1^{\mathrm{T}}, f_2^{\mathrm{T}})^{\mathrm{T}}$ in (2.6.1) continuous on $R \times B_1(\rho) \times B_2(\rho)$ on $(\mathcal{T}_\tau \times B_1(\rho) \times B_2(\rho))$. If there exists*

(1) *open connected time-invariant neighborhood $\mathcal{G}$ of $x = 0$;*
(2) *matrix-valued function $U(t, x)$ is*
  (a) *positive definite on $\mathcal{G}$ (on $\mathcal{T}_\tau \times \mathcal{G}$);*
  (b) *decreasing on $\mathcal{G}$ (on $\mathcal{T}_\tau \times \mathcal{G}$);*
(3) *matrix-valued function $D^+U(t, x)$ is*
  (a) *negative semi-definite on $R \times \mathcal{G}$ (on $\mathcal{T}_\tau \times \mathcal{G}$);*
  (b) *x_2^{T}-negative definite on $R \times \mathcal{G}$ (on $\mathcal{T}_\tau \times \mathcal{G}$).*

Then system (2.6.1) is multistability (on $\mathcal{T}_\tau$) in the sense of Definition 2.6.11.

PROOF. If conditions (1), (2), (3)(a) of the Theorem 2.6.1 hold for system (2.6.1) with function (2.6.29), then all hypotheses of Theorem 2.5.1 are fulfilled and state $(x = 0) \in R^{N_0}$ is uniformly stable (on $\mathcal{T}_\tau$).

If conditions (1), (2), (3)(b) of Theorem 2.6.1 hold for system (2.6.1) with function (2.6.29), then all hypotheses of Theorem 2.5.3 are fulfilled and state $(x = 0) \in R^{N_0}$ is uniformly asymptotically x_2^{T}-stable (on $\mathcal{T}_\tau$).

The Theorem 2.6.1 is proved.

Further we suppose that multistability of (2.6.1) for $s = 2$ is investigated in the domain

$$\mathcal{T}_\tau \times B_1(\rho) \times D_2, \qquad D_2 = \{x_2 \colon 0 < \|x_2\| < +\infty\}. \tag{2.6.31}$$

The next result can be easily verified (see e.g. Martynyuk [122]).

THEOREM 2.6.2. *Let vector function $f = (f_1^{\mathrm{T}}, f_2^{\mathrm{T}})^{\mathrm{T}}$ in (2.6.1) be continuous on $R \times B_1(\rho) \times D_2$ (on $\mathcal{T}_\tau \times B_1(\rho) \times D_2$). If there exists*

(1) *an open connected time-invariant neighborhood $\mathcal{G}$ of $(x = 0) \in R^{n_1}$;*
(ii) *matrix-valued function $U(t, x)$ is*
  (a) *x_1^{T}-positive definite on $\mathcal{G}$ (on $\mathcal{T}_\tau \times \mathcal{G}$);*
  (b) *decreasing on $\mathcal{G}$ (on $\mathcal{T}_\tau \times \mathcal{G}$);*

(c) x_1^{T}-*decreasing on* $\mathcal{G}$ *(on* $\mathcal{T}_\tau \times \mathcal{G}$*);*

(3) *matrix-valued function* $D^+U(t,x)$ *is*

(a) *negative semi-definite on* $R \times \mathcal{G}$ *(on* $\mathcal{T}_\tau \times \mathcal{G}$*);*

(b) x_1^{T}-*negative definite on* $R \times \mathcal{G}$ *(on* $\mathcal{T}_\tau \times \mathcal{G}$*);*

(c) *negative definite on* $R \times \mathcal{G}$ *(on* $\mathcal{T}_\tau \times \mathcal{G}$*).*

Then, respectively

(a) *the conditions (1), (2)(a) and (3)(a) are sufficient for stability of state* $(x = 0) \in R^{N_0}$ *of (2.6.1) (on* $\mathcal{T}_\tau$*);*

(b) *the conditions (1), (2)(a), (2)(b) and (3)(a) are sufficient for uniform* x_1^{T}-*stability of state* $(x = 0) \in R^{N_0}$ *of (2.6.1) (on* $\mathcal{T}_\tau$*);*

(c) *the conditions (1), (2)(a) and (3)(c) are sufficient for asymptotic* x_1^{T}-*stability of state* $(x = 0) \in R^{N_0}$ *of (2.6.1) (on* $\mathcal{T}_\tau$*);*

2.6.5.2 The application of matrix-valued Liapunov function via transition to vector function. Basing on matrix-valued function $U(t,x)$ and vector $y \in R^s$, $y \neq 0$, $s = 2$ we construct a vector function

$$L(t,x,y) = AU(t,x)y \tag{2.6.32}$$

where A is a constant 2 by 2 matrix. Consider a system of comparison

$$\frac{du}{dt} = G(t,u), \qquad u(t_0) = u_0 \geq 0 \tag{2.6.33}$$

where $u \in R_+^2$, $G(t,u) = (g_1(t,u_1),\, g_2(t,u_1,u_2))^{\mathrm{T}}$, $g_1 \in C(\mathcal{T}_\tau \times R_+,\, R)$, $g_2 \in C(\mathcal{T}_\tau \times R_+ \times R_+,\, R)$

$$g_1(t,0) = g_2(t,0,0) = 0 \quad \text{for all} \quad t \in \mathcal{T}_\tau.$$

DEFINITION 2.6.12. A comparison system (2.6.33) is called *multistable (on* $\mathcal{T}_\tau$*)*, if its zero solution is

(i) u_1-stable with respect to $\mathcal{T}_\tau$;

(ii) uniformly u_2-stable with respect to $\mathcal{T}_\tau$.

THEOREM 2.6.3. *Let vector function* $f = (f_1^{\mathrm{T}}, f_2^{\mathrm{T}})^{\mathrm{T}}$ *in (2.6.1) be continuous on* $R \times B_1(\rho) \times D_2$ *(on* $\mathcal{T}_\tau \times B_1(\rho) \times D_2$*). If there exists*

(1) *a matrix-valued function* $U\colon \mathcal{T}_\tau \times B_1(\rho) \times D_2 \to R^{2\times 2}$, *vector* $y \in R^2$, $y \neq 0$ *and a constant matrix* A 2 *by* 2 *such that components* $L_i(t, x, y)$, $i = 1, 2$ *of vector function (2.6.32) are locally Lipschitzian in* x *and satisfy the conditions*
   (a) $L_1(t, 0, y) = 0 \ \forall\, t \in R \ (t \in \mathcal{T}_\tau)$;
   (b) $a(\|x_1\|) \leq L_2(t, x, y) \leq b(\|x_1\|) + b_1(L_1(t, x, y))$
   for all $(t, x) \in \mathcal{T}_\tau \times B_1(\rho) \times D_2 \cap B_1^c(\eta)$ *when each* $0 < \eta < \rho$;

(2) *a vector function* $g \in C(\mathcal{T}_\tau \times R_+^2, R^2)$, $G(t, u)$ *is quasimonotone nondecreasing with respect to* u *for the components of which*
   (a) $D^+ L_1(t, x, y) \leq g_1(t, L_1(t, x, y), 0)$
   hold for all $(t, x) \in \mathcal{T}_\tau \times B_1(\rho) \times D_2$, *and*
   (b) $D^+ L_2(t, x, y) \leq g_2(t, L_1(t, x, y), L_2(t, x, y))$
   hold for all $(t, x) \in \mathcal{T}_\tau \times B_1(\rho) \times D_2 \cap B_1^c(\eta)$ *for* $o < \eta < \rho$;

(3) *zero solution of system (2.6.33) is multistable (on* $\mathcal{T}_\tau$*)in the sense of Definition 2.6.12.*

Then the system (2.6.1) is x_1^{T}*-stable (on* $\mathcal{T}_\tau$*).*

PROOF. Let $(t_0, \varepsilon)\colon t_0 \in \mathcal{T}_i$ and $0 < \varepsilon$, ρ be given. It follows from condition (3) of the theorem that for given $\varepsilon_1, \varepsilon_2 > 0$ and $t_0 \in \mathcal{T}_i$ there exist $\delta_{10} = \delta_{10}(t_0, \varepsilon_1) > 0$ and $\delta_{20} = \delta_{20}(\varepsilon_2) > 0$ such that

(a) $\eta^{\mathrm{T}} u_0 < \delta_{10}$ implies that $u_1(t; t_0, u_0) < \varepsilon_1 \ \forall\, t \geq t_0$ and
(b) $\eta^{\mathrm{T}} u_0 < \delta_{20}$ implies that $u_2(t; t_0, u_0) < \varepsilon_2 \ \forall\, t \geq t_0$.

Let $\varepsilon_2 = a(\varepsilon)$ and $\varepsilon_1 = b_1^{-1}\left(\dfrac{1}{2}\delta_{20}\right)$. It follows from the continuity of the function $L_1(t; x, y)$ and condition (1)(a) that there exists $\delta_1 = \delta_1(t_0, \varepsilon) > 0$ such that

$$L_1(t_0, x_0, y) < \delta_{10} \quad \text{and} \quad \|x_0\| < \delta_1.$$

Let $\delta = \min(\delta_1, \delta_2)$. It is clear that δ depends on $t_0 \in \mathcal{T}_i$ and on $0 < \varepsilon < \rho$. For δ defined in this way, we can assert that the zero solution of (2.6.1) is x_1^{T}-stable (on $\mathcal{T}_\tau$)with respect to $\mathcal{T}_i$.

Assume the countary, i.e., that the zero solution of (2.6.1) is not x_1^{T}-stable (on $\mathcal{T}_\tau$)when all the conditions of Theorem 2.6.3 are fulfilled. Then for the solution $x(t;t_0,x_0)$ of (2.6.1) with initial conditions $t_0 \in \mathcal{T}_i$ and $\|x_0\| < \delta$ there exists a time $t_2 > t_1 > t_0$ such that

$$\begin{aligned} &\|x_1(t_2)\| = \varepsilon < \rho, \qquad \|x_1(t_1)\| = \delta_2(\varepsilon) \\ &x_2(t) \in \mathcal{N}_{(1,2)}(\rho) \cap \mathcal{N}^c_{(1,2)}(\eta), \qquad \eta = \delta_2(\varepsilon) > 0 \end{aligned} \tag{2.6.34}$$

at the same time that $\|x_2(t)\| < +\infty$.

Let $m(t) = L(t,x(t),y)$; in view of condition (2) of the theorem we obtain

$$D^+m_1(t) \le g_1(t,m_1(t),0), \qquad t_0 \le t \le t_2; \tag{2.6.35}$$

$$D^+m_2(t) \le g_2(t,m(t)), \qquad t_1 \le t \le t_2. \tag{2.6.36}$$

Let $u^*(t) = u(t;t_1,m(t_1)) \ge 0$ be the extension of $u(t)$ to the left from t_1 to t_0, and let $u^*(t_0) = u_0^*$. We assume that $L_1(t_0,x_0,y) = u_1(t_0)$ and that $u^*(t_0) = u_0$.

From the differential inequality

$$D^+m_1(t) \le g_1(t,m_1(t),u_2^*(t)), \qquad m_1(t_0) = u_1(t_0)$$

and the comparison theorem we have

$$m_1(t) \le u_1(t;t_0,u_0), \quad t_0 \le t \le t_1, \quad u_0 = (u_1(t_0),u_2(t_0))^{\mathrm{T}}. \tag{2.6.37}$$

From this it is clear that $u(t) = (u_1(t;t_0,u_0),\, u_2^*(t;t_1,m(t_1)))^{\mathrm{T}}$ is a solution of (2.6.33) on $[t_0,t_1]$. From condition (1) of Theorem 2.6.3 and inequalities (2.6.34), (2.6.35) and (2.6.37) we obtain

$$a(\varepsilon) = a(\|x_1(t_2)\|) < L_2(t_2,x(t_2),y) \le u_2(t_2;t_1,m(t_1)). \tag{2.6.38}$$

From the fact that

$$L_1(t_1,x(t_1),y) \le u_1(t_1;t_0,u_0) < b_1^{-1}\left(\frac{1}{2}\delta_{20}\right)$$

as soon as $\eta^{\mathrm{T}}u_0 < \delta_{10}$ and also from conditions (2.6.34), we have, by condition (1)(a)

$$\begin{aligned} &L_2(t_1,x(t_1),y) \le b(\|x_1(t)\|) + b_1(L_1(t_1,x(t_1),y)) \\ &\le b(\delta_2(\varepsilon)) + b_1\left(b_1^{-1}\left(\frac{1}{2}\delta_{20}\right)\right) < \frac{1}{2}\delta_{20} + \frac{1}{2}\delta_{20} = \delta_{20}. \end{aligned} \tag{2.6.39}$$

It follows from the uniform u_2-stability of the zero solution of (2.6.33) with respect to $\mathcal{T}_i$ that

$$u_2\left(t_2; t_1, m(t_1)\right) < a(\varepsilon). \tag{2.6.40}$$

Inequality (2.6.40) contradicts condition (2.6.37). This completes the proof of the theorem.

2.6.6 The system (2.6.1) analysis for s = 3

Suppose that for $s = 3$ the right-hand side of (2.6.1) are defined in the region

$$\mathcal{T}_\tau \times B_1(\rho) \times B_2(\rho) \times D_3, \qquad D_3 = \{x_3 \colon 0 < \|x_3\| < +\infty\}. \tag{2.6.41}$$

Consider a matrix-valued function

$$U(t,x) = [v_{ij}(t,x)], \qquad i,j = 1,2,3, \tag{2.6.42}$$

where $v_{ij} \in C(\mathcal{T}_\tau \times R^{N_1}, R)$, $v_{ij}(t,x)$ are locally Lipschitzian in x, $N_1 = n_1 + n_2 + n_3$. With the aid of vector $\varphi \in R^3_+$, $\varphi > 0$ and matrix-valued function (2.6.42) we construct the function

$$v(t,x,\varphi) = \varphi^{\mathrm{T}} U(t,x)\varphi, \tag{2.6.43}$$

where $v \in C(\mathcal{T}_\tau \times R^{N_1} \times R^3_+, R)$.

Function (2.6.42) is applied in two approaches as in Section 2.6.5.

2.6.6.1 Direct application of matrix-valued function.

DEFINITION 2.6.13. System (2.6.1) is *multistable (on $\mathcal{T}_\tau$)* if its zero solution is

(i) t_0-uniformly $(x_1^{\mathrm{T}}, x_2^{\mathrm{T}})$-stable in the whole (on $\mathcal{T}_\tau$);
(ii) asymptotically x_2^{T}-stable in the whole (on $\mathcal{T}_\tau$).

THEOREM 2.6.4. *Let vector function $f = \left(f_1^{\mathrm{T}}, f_2^{\mathrm{T}}, f_3^{\mathrm{T}}\right)^{\mathrm{T}}$ in (2.6.1) be continuous on $R \times B_1(\rho) \times B_2(\rho) \times D_3$ (on $\mathcal{T}_\tau \times B_1(\rho) \times B_2(\rho) \times D_3$). If there exists*

(1) *an open connected time-invariant neighborhood $\mathcal{G}^*$ of point $(x = 0) \in R^{N_0}$, $N_0 = n_1 + n_2$;*

(2) *a matrix-valued function* $U(t,x)$

(a) $(x_1^{\mathrm{T}}, x_2^{\mathrm{T}})$*-positive definite on* $\mathcal{G}^*$ *(on* $\mathcal{T}_\tau \times \mathcal{G}^* \times D_3$*);*

(b) *decreasing on* $\mathcal{G}^* \times D_3$ *(on* $\mathcal{T}_\tau \times \mathcal{G}^* \times D_3$*);*

(3) *a matrix-valued function* $D^+U(t,x)$ *is*

(a) *negative semi-definite on* $R \times \mathcal{G}^* \times D_3$ *(on* $\mathcal{T}_\tau \times \mathcal{G}^* \times D_3$*);*

(a) x_2^{T}*-negative semi-definite on* $R \times \mathcal{G}$ *(on* $\mathcal{T}_\tau \times \mathcal{G}^* \times D_3$*);*

(4) *a constant* $m > 0$ *for which*

$$\|f_2(t, x_1, x_2, x_3)\| \leq m \qquad \forall\, (t,x) \in \mathcal{T}_\tau \times B_1(\rho) \times B_2(\rho) \times D_3$$

Then, respectively

(a) *hypotheses (1), (2)(a) and (3)(a) are sufficient for* $(x_1^{\mathrm{T}}, x_2^{\mathrm{T}})$*-stability of* $(x = 0) \in R^{N_1}$, $N_1 = n_1 + n_2 + n_3$ *of the system (2.6.1) (on* $\mathcal{T}_\tau$*);*

(b) *hypotheses (1), (2)(a), (2)(b) and (3)(a) are sufficient for uniform* $(x_1^{\mathrm{T}}, x_2^{\mathrm{T}})$*-stability of* $(x = 0) \in R^{N_1}$ *of the system (2.6.1) (on* $\mathcal{T}_\tau$*);*

(c) *hypotheses (1), (2) and (3)(b) are sufficient for asymptotical* x_2^{T}*-stability in the whole of state* $(x = 0) \in R^{N_1}$ *of (2.6.1) (on* $\mathcal{T}_\tau$*).*

PROOF. We show that if all hypotheses of Theorem 2.6.4 are satisfied, then for $\|x_0\| < \Delta$, $\Delta < +\infty$ the correlation

$$\lim_{t \to +\infty} \|x_2(t; t_0, x_0)\| = 0 \tag{2.6.44}$$

is valid. Suppose on the contrary. That there exists a number $\delta^* > 0$, a point $x_0^*\colon \|x_0^*\| < \Delta$ and a sequence $t_k \to \infty$ such that inequality

$$\|x_2(t_k; t_0, x_0^*)\| \geq \delta^*, \qquad k = 1, 2, \dots \tag{2.6.45}$$

holds true.

Let $t_k - t_{k-1} \geq \alpha > 0$, $k = 1, 2, \dots$. We present x_2-component of solution $x(t) = \left(x_1^{\mathrm{T}}(t), x_2^{\mathrm{T}}(t), x_3^{\mathrm{T}}(t)\right)^{\mathrm{T}}$ of the system (2.6.1) in the neighborhood of $t = t_k$ in the form

$$x_2(t; t_0, x_0^*) = x_2(t_k; t_0, x_0^*) + \int_{t_k}^{t} f_2\left(s,\ x(s; t_0, x_0^*)\right) ds. \tag{2.6.46}$$

In view of (2.6.45) and (2.6.46) we have

$$\|x_2(t; t_0, x_0^*)\| \geq \delta^* - m(t - t_k), \qquad k = 1, 2, \dots.$$

Hence, there exists a β, $0 < \beta < \frac{1}{2}\alpha$ such that

$$\frac{1}{2}\delta^* \le \|x_2(t; t_0, x_0^*)\| \le \rho \qquad \forall\, t \in [t_k - \beta,\, t_k + \beta] \tag{2.6.47}$$

for $k = 1, 2, \ldots$. By force of Proposition 2.6.3 and hypotheses (3)(b) of the theorem we have

$$D^+ v(t, x, \varphi) \le -c(\|x_2\|) \tag{2.6.48}$$

where $c \in KR$.

From the Liapunov correlation for function $v(t, x, y)$ we have

$$0 \le v\,(t_k + \beta,\, x(t_k + \beta; t_0, x_0^*),\, \varphi) \le v(t_0, x_0^*, \varphi)$$
$$-\sum_{i=1}^{k} \int_{t_i-\beta}^{t_i+\beta} c\,(\|x_2(s; t_0, x_0^*)\|)\, ds \le v(t_0, x_0^*, \varphi) - 2k\beta\, c\left(\frac{\delta^*}{2}\right).$$

This shows that the condition $v\,(t_k + \beta,\, x(t_k + \beta; t_0, x_0^*),\, \varphi) \ge 0$ is violated for k being large enogh. Therefore, (2.6.44) is proved.

2.6.6.2 The application of matrix-valued function via transition to vector-function. Suppose $U\colon \mathcal{T}_\tau \times R^n \to R^{3\times 3}$, Q is a 3 by 3 constant matrix and $y \in R^3$. Construct a vector function

$$L(t, x, \varphi) = QU(t, x)y \tag{2.6.49}$$

where $L \in C(\mathcal{T}_\tau \times R^n \times R^3,\, R^3)$.

Let a, b be functions from classes K_1 and K_2, where $K_1 = \{a \in C\,(0, \rho), R_+)$ increases with u and $a(u) \to 0$ as $u \to 0\}$, and $K_2 = \{b \in C\,(0, 3\rho), R_+)$ increases with u and $b(u) \to 0$ as $u \to 0\}$.

Suppose that the components $L_1(t, x, y), \ldots, L_3(t, x, y)$ of the function (2.6.49) satisfy the following conditions:

(A) $L_1(t, 0, y) = 0$ for all $t \in R$ or $t \in \mathcal{T}_\tau$, and

$$L_1(t, x, y) \in C\left(\mathcal{T}_\tau \times B_1(\rho) \times B_2(\rho) \times D_3 \times R^3,\, R_+\right).$$

(B) There is a constant $0 < \eta_1 < \rho$ such that

$$L_2(t, x, y) \in C\left(\mathcal{T}_\tau \times B_1(\rho) \times B_2(\rho) \times D_3 \times R^3 \cap B_2^c(\eta_1),\, R_+\right)$$

and

$$a_2(\|x_2\|) \le L_2(t,x,y) \le b_2(\|x_1\| + \|x_2\| + \|x_3\|)$$
$$\forall\ (t,x,y) \in \mathcal{T}_\tau \times B_1(\rho) \times B_2(\rho) \times D_3 \times R^3 \cap B_2^c(\eta_1),$$

where $a_2 \in K_1$ and $b_2 \in K_2$.

(C) For any $0 < \eta_1 < \rho$ there exist an $\eta_2 < \rho$ such that

$$L_2(t,x,y) \in C\left(\mathcal{T}_\tau \times B_1(\rho) \times B_2(\rho) \times D_3 \times R^3 \cap B_1^c(\eta_2),\ R_+\right);$$
$$a_3(\|x_1\|) \le L_3(t,x,y) \le b_3(\|x_1\| + \|x_2\| + \|x_3\|)$$
$$\forall\ (t,x,y) \in \mathcal{T}_\tau \times B_1(\rho) \times B_2(\rho) \times D_3 \times R^3 \cap B_1^c(\eta_2),$$

where $a_3 \in K_1$ and $b_3 \in K_2$.

DEFINITION 2.6.14. The comparison system

$$\frac{dz}{dt} = G(t,z), \qquad z(t_0) = z_0 \ge 0, \tag{2.6.50}$$

where $z = (u,v,w)^{\mathrm{T}}$, $G = (g_1(t,u),\ g_2(t,v),\ g_3(t,w))^{\mathrm{T}}$ is *multistable (on $\mathcal{T}_\tau$)* if its zero solution is

(i) u-equistable (on $\mathcal{T}_\tau$), and
(ii) (v,w)-uniformly stable (on $\mathcal{T}_\tau$).

Following Lakshmikantham, Leela and Martynyuk [94], Martynyuk [118] and Koksal [88] the next result is obtained.

THEOREM 2.6.5. *Let vector function $f = \left(f_1^{\mathrm{T}}, f_2^{\mathrm{T}}, f_3^{\mathrm{T}}\right)^{\mathrm{T}}$ in (2.6.1) be continuous on $R \times B_1(\rho) \times B_2(\rho) \times D_3$ (on $\mathcal{T}_\tau \times B_1(\rho) \times B_2(\rho) \times D_3$). If there exists*

(1) *matrix-valued function $U(t,x)$, a vector $y \in R^3$ and a constant 3 by 3 matrix Q for which components L_1, L_2, L_3 of (2.6.49) the conditions (A) – (C) are satisfied;*
(2) *functions $g_k \in C(\mathcal{T}_\tau \times R_+, R)$, $g_k(t,0) = 0\ \forall\, t \in \mathcal{T}_\tau$ such that*
   (a) *The inequality*

$$D^+L_1(t,x,y) \le g_1\left(t,\ L_1(t,x,y)\right)$$

   holds in the domain $\mathcal{T}_\tau \times B_1(\rho) \times B_2(\rho) \times D_3 \times R^3$.

(b) *The inequality*

$$D^+L_1(t,x,y) + D^+L_2(t,x,y) \le g_2\,(t,\ L_1(t,x,y),\ L_2(t,x,y))$$

holds in the domain $\mathcal{T}_\tau \times B_1(\rho) \times B_2(\rho) \times D_3 \times R^3 \cap B_2^c(\eta_1)$.

(c) *The inequality*

$$D^+L_1(t,x,y) + D^+L_3(t,x,y) \le g_3\,(t,\ L_1(t,x,y),\ L_3(t,x,y))$$

holds in the domain $\mathcal{T}_\tau \times B_1(\rho) \times B_2(\rho) \times D_3 \times R^3 \cap B_1^c(\eta_2)$.

(3) *The zero solution of system (2.6.50) is multistable (on* $\mathcal{T}_\tau$*) in the sence of Definition 2.6.14.*

Then the zero solution of system (2.6.1) is $(x_1^{\mathrm{T}}, x_2^{\mathrm{T}})$*-stable (on* $\mathcal{T}_\tau$*).*

2.6.7 The system (2.6.1) analysis for s = 4

For $s = 4$ the system (2.6.1) is considered in region

$$\mathcal{T}_\tau \times B_1(\rho) \times B_2(\rho) \times B_3(\rho) \times D_4, \qquad D_4 = \{x_4 \colon 0 < \|x_4\| < +\infty\}.$$

Let $N_1 = n_1 + \cdots + n_4$.

DEFINITION 2.6.15. System (2.6.1) is *multistable (on* $\mathcal{T}_\tau$*)* if its zero solution $((x_1^{\mathrm{T}}, \dots, x_4^{\mathrm{T}}) = 0) \in R^{N_2}$ is

(i) t_0-uniformly $(x_1^{\mathrm{T}}, x_2^{\mathrm{T}}, x_3^{\mathrm{T}})$-stable (on $\mathcal{T}_\tau$);
(ii) asymptotically $(x_2^{\mathrm{T}}, x_3^{\mathrm{T}})$-stable (on $\mathcal{T}_\tau$);
(iii) practically x_3^{T}-stable (on $\mathcal{T}_\tau$), i.e. if given (λ, A) with $0 < \lambda < A$, the inequality $\|x_0\| < \lambda$ implies $\|x_3(t)\| < A$ for all $t \in \mathcal{T}_\tau$.

THEOREM 2.6.6. *Let vector function* $f = (f_1^{\mathrm{T}}, \dots, f_4^{\mathrm{T}})^{\mathrm{T}}$ *in (2.6.1) be continuous on* $R \times B_1(\rho) \times B_2(\rho) \times B_3(\rho) \times D_4$ *(on* $\mathcal{T}_\tau \times B_1(\rho) \times B_2(\rho) \times B_3(\rho) \times D_4$*). If there exists*

(1) *a matrix-valued function* $U(t,x)$ *which is*

(a) $(x_1^{\mathrm{T}}, x_2^{\mathrm{T}}, x_3^{\mathrm{T}})$*-positive definite (on* $\mathcal{T}_\tau$*);*
(b) $(x_1^{\mathrm{T}}, x_2^{\mathrm{T}}, x_3^{\mathrm{T}})$*-decreasing (on* $\mathcal{T}_\tau$*);*

(2) *the matrix-valued function* $D^+U(t,x)$ *which is*

(c) $(x_2^{\mathrm{T}}, x_3^{\mathrm{T}})$*-negative definite (on* $\mathcal{T}_\tau$*);*

(3) *a constant* $m_1 \in R_+$, $m_1 > 0$ *for which*

$$\| \left(f_2^{\mathrm{T}}(t,x),\ f_3^{\mathrm{T}}(t,x)\right)^{\mathrm{T}} \| \leq m_1$$

in the domain $\mathcal{T}_\tau \times B_1(\rho) \times B_2(\rho) \times B_3(\rho) \times D_4$*;*

(4) *a vector* $\varphi \in R_+^4$, $\varphi > 0$ *such that for all* $t \in \mathcal{T}_\tau$*, given* (λ, A) *inequality*

$$\sup\left(\varphi^{\mathrm{T}} U(t,x)\varphi \quad \text{for} \quad \|x\| < \lambda\right) < \inf\left(\varphi^{\mathrm{T}} U(t,x)\varphi \quad \text{for} \quad \|x_3\| = A\right)$$

holds true.

Then system (2.6.1) is multistable in the sense of Definition 2.6.15.

PROOF. Properties (i) and (ii) of the zero solution of the system (2.6.1) are implied by hypotheses (1) and (2) of the Theorem 2.6.6, when function $v(t,x,\varphi) = \varphi^{\mathrm{T}} U(t,x)\varphi$ and its derivative $D^+ v(t,x,\varphi)$ are considered along with solution of the system (2.6.1). To prove practical stability (on $\mathcal{T}_\tau$) of state $(x = 0) \in R^{N_2}$ with respect to variables of vectors x_3^{T} it is sufficient to make sure that when hypotheses of Theorem 2.6.6 hold, the value of norm $\|x_3(t; t_0, x_0)\|$ does not reach the value of A for all $t \in \mathcal{T}_\tau$ provided $\|x_0\| < \lambda$ for any $t_0 \in \mathcal{T}_i \subseteq R$. By hypotheses (2) of the Theorem 2.6.6 we have

$$D^+ v(t,x,\varphi) \leq 0$$

in the domain $\mathcal{T}_\tau \times B_1(\rho) \times B_2(\rho) \times B_3(\rho) \times D_4$. Hence

$$\begin{aligned} v(t,x,\varphi) &\leq v(t_0,x_0,\varphi) \\ &\leq \sup\left(\varphi^{\mathrm{T}} U(t,x)\varphi\right) \quad \text{for} \quad \|x\| < \lambda\right). \end{aligned} \tag{2.6.51}$$

Let hypotheses (3) of Theorem 2.6.6 be satisfied and inequality $\|x_3(t)\| < A$ be false for some $t \in \mathcal{T}_\tau$. If the violation of the inequality takes place at $t^* \in \mathcal{T}_\tau$, then

$$v(t^*,x,\varphi) \geq \inf\left(\varphi^{\mathrm{T}} U(t,x)\varphi\right) \quad \text{for} \quad \|x_3\| = A\right). \tag{2.6.51}$$

From (2.6.51) and (2.6.52) we get

$$\begin{aligned} &\sup\left(\varphi^{\mathrm{T}} U(t^*,x)\varphi \quad \text{for} \quad \|x\| < \lambda\right) \geq v(t^*,x,\varphi) \\ &\qquad \geq \inf\left(\varphi^{\mathrm{T}} U(t,x)\varphi \quad \text{for} \quad \|x_3\| = A\right). \end{aligned} \tag{2.6.53}$$

The inequality (2.6.53) contradicts hypothesis (4) of the Theorem 2.6.6 and proves that $(x = 0) \in R^{N_2}$ is practically stable (on $\mathcal{T}_\tau$) with respect to variable x_3^{T}.

2.7 Applications

In this section we present some applications of general theorems of matrix-valued Liapunov functions method to system of equations that model real engineering problems.

2.7.1 The problem of Lefschetz

We consider a problem on stability in a product space for a system of differential equations of the perturbed motion

$$\begin{aligned} \frac{dy}{dt} &= g(y) + G(y,z), \\ \frac{dz}{dt} &= h(z) + H(y,z). \end{aligned} \tag{2.7.1}$$

Here $y \in R^p$, $z \in R^q$, $g\colon R^p \to R^p$, $G\colon R^p \times R^q \to R^p$, $h\colon R^q \to R^q$, $H\colon R^p \times R^q \to R^q$. In addition, function g, G; h, H are continuous on R^p, R^q, $R^p \times R^q$ and they vanish for $y = z = 0$.

The problem itself is to point out the connection between the stability properties of equilibrium state $y = z = 0$ with respect to system (2.7.1) on $R^p \times R^q$ and its nonlinear approximation

$$\begin{aligned} \frac{dy}{dt} &= g(y), \\ \frac{dz}{dt} &= h(z). \end{aligned} \tag{2.7.2}$$

ASSUMPTION 2.7.1. Let there exist the time-invariant neighborhood $\mathcal{N}_y \subseteq R^p$ and $\mathcal{N}_z \subseteq R^q$ of the equilibrium state $y = 0$ and $z = 0$, respectively and let there exist a matrix-valued function

$$U(y,z) = \begin{pmatrix} v_{11}(y) & v_{12}(y,z) \\ v_{21}(y,z) & v_{22}(z) \end{pmatrix} \tag{2.7.3}$$

the element v_{ij} of which satisfy the estimations characteristic to the quadratic forms

$$\begin{aligned} v_{11}(y) &\geq c_{11}\|y\|^2 & &\forall\, (y \neq 0) \in \mathcal{N}_y; \\ v_{22}(z) &\geq c_{22}\|z\|^2 & &\forall\, (z \neq 0) \in \mathcal{N}_z; \\ v_{12}(y,z) = v_{21}(y,z) &\geq c_{12}\|y\|\|z\| & &\forall\, (y \neq 0,\ z \neq 0) \in \mathcal{N}_y \times \mathcal{N}_z. \end{aligned} \tag{2.7.4}$$

ASSUMPTION 2.7.2. Let there exist constants α_{ij}, $i = 1, 2$; $j = 1, 2, \ldots, 8$ such that

$$\begin{aligned}
&\left(\frac{\partial v_{11}}{\partial y}, g\right) \le \alpha_{11}||y||^2;\\
&\left(\frac{\partial v_{11}}{\partial y}, G\right) \le \alpha_{12}||y||^2 + \alpha_{13}||y||||z||;\\
&\left(\frac{\partial v_{22}}{\partial z}, h\right) \le \alpha_{21}||z||^2;\\
&\left(\frac{\partial v_{22}}{\partial z}, H\right) \le \alpha_{22}||z||^2 + \alpha_{23}||y||||z||;\\
&\left(\frac{\partial v_{12}}{\partial y}, g\right) \le \alpha_{14}||y||^2 + \alpha_{15}||y||||z||;\\
&\left(\frac{\partial v_{12}}{\partial y}, G\right) \le \alpha_{16}||y||^2 + \alpha_{17}||y||||z|| + \alpha_{18}||z||^2;\\
&\left(\frac{\partial v_{12}}{\partial z}, h\right) \le \alpha_{24}||z||^2 + \alpha_{25}||y||||z||;\\
&\left(\frac{\partial v_{12}}{\partial z}, H\right) \le \alpha_{26}||y||^2 + \alpha_{27}||y||||z|| + \alpha_{28}||z||^2.
\end{aligned} \tag{2.7.5}$$

THEOREM 2.7.1. *Suppose that*

(1) *all conditions of Assumptions 2.7.1, 2.7.2 are fulfilled;*
(2) *the matrix*

$$C = \begin{pmatrix} c_{11} & c_{12} \\ c_{21} & c_{22} \end{pmatrix}, \qquad c_{12} = c_{21}$$

be positive definite;
and

(3) *the matrix*

$$S = \begin{pmatrix} \sigma_{11} & \sigma_{12} \\ \sigma_{21} & \sigma_{22} \end{pmatrix}, \qquad \sigma_{12} = \sigma_{21}$$

be negative definite, where

$$\sigma_{11} = \eta_1^2(\alpha_{11} + \alpha_{12}) + 2\eta_1\eta_2(\alpha_{14} + \alpha 16 + \alpha_{26});$$
$$\sigma_{22} = \eta_2^2(\alpha_{21} + \alpha_{22}) + 2\eta_1\eta_2(\alpha_{18} + \alpha 24 + \alpha_{28});$$
$$\sigma_{12} = \frac{1}{2}\left(\eta_1^2\alpha_{13} + \alpha_{23}\eta_2^2\right) + \eta_1\eta_2(\alpha_{15} + \alpha 25 + \alpha_{17} + \alpha_{27}),$$

η_1, η_2 *being positive numbers.*

Then the state of equilibrium $y = z = 0$ *of the system (2.7.1) is uniformly asymptotically stable.*

If conditions of Assumptions 2.7.1, 2.7.2 are fulfilled for $\mathcal{N}_y = R^{\text{p}}$, $\mathcal{N}_y = R^{\text{q}}$ *and conditions (2), (3) of the theorem hold, then the equlibrium state* $y = z = 0$ *of the system (2.7.1) is uniformly asymptotically stable in the whole.*

PROOF. On the basis of estimations (2.7.4), it is not difficult to show that the function $v = \eta^{\text{T}} U(y, z) \eta$ satisfies the estimate

$$v \geq u^{\text{T}} \Phi^{\text{T}} C \Phi u, \tag{2.7.6}$$

where $u^{\text{T}} = (\|y\|, \|z\|)$, $\Phi = \operatorname{diag}[\eta_1, \eta_2]$.

Also, in view of Assumption 2.7.1 and the estimates (2.7.5), the derivative $Dv(y,z)$ defined by $Dv(y,z) = \eta^{\text{T}} DU(y,z)\eta$ satisfies

$$Dv(y,z) \leq u^{\text{T}} S u. \tag{2.7.7}$$

By virtue of (2) and (3) and the inequalities (2.7.6), (2.7.7), we see that all conditions of Theorem 2.5.3 are verified for the function $v(y,z)$ and its derivative. Hence the proof is complete.

If in estimate (2.7.5) we change the sign of inequality for the opposite one, then by means of the method similar to the given one we can obtain an estimate

$$Dv(y,z) \geq u^{\text{T}} \widetilde{S} u$$

which allows us to formulate instability conditions for the equilibrium state $y = z = 0$ of system (2.7.1) on the basis of Theorem 2.5.7.

The statement of Theorem 2.7.1 shows that asymptotic stability of the equilibrium state $y = z = 0$ of system (2.7.1) can hold even if the equilibrium state $y = z = 0$ of system (2.7.2) has no properties of asymptotic quasi-stability (cf. Lefschetz [100]).

2.7.2 Autonomous large scale systems

We consider a large scale systems be decomposed into three subsystems

$$\begin{aligned} \frac{dx}{dt} &= Ax + f(x,y,z), \\ \frac{dy}{dt} &= By + g(x,y,z), \\ \frac{dz}{dt} &= Cz + h(x,y,z), \end{aligned} \tag{2.7.8}$$

where $x \in R^{n_1}$, $y \in R^{n_2}$, $z \in R^{n_3}$, $n_1 + n_2 + n_3 = n$; A, B and C are constant matrices of the corresponding dimensions

$$f \in C\,(R^{n_1} \times R^{n_2} \times R^{n_3},\, R^{n_1});$$
$$g \in C\,(R^{n_1} \times R^{n_2} \times R^{n_3},\, R^{n_2});$$
$$h \in C\,(R^{n_1} \times R^{n_2} \times R^{n_3},\, R^{n_3}).$$

Moreover, the vector-functions f, g and h vanish for $x = y = z = 0$ and contain variables x, y and z in first power, i.e. the subsystems

$$\frac{dx}{dt} = Ax; \tag{2.7.9}$$

$$\frac{dy}{dt} = By; \tag{2.7.10}$$

$$\frac{dz}{dt} = Cz; \tag{2.7.11}$$

are not complete linear approximation of the system (2.7.8). Physically speaking this corresponds to the situation when the connections between subsystems (2.7.9) – (2.7.11) are carried out by time-invariant linear blocks. For different dynamical properties of subsystems (2.7.9) – (2.7.11) sufficient total stability conditions will be established for the state $x = y = z = 0$ of the system (2.7.8).

The solution algorithm for this problem is based on actual construction of the matrix-valued function

$$U(x, y, z) = [v_{ij}(\cdot)], \qquad v_{ij} = v_{ji} \quad \forall\,(i \neq j) \tag{2.7.12}$$

with the elements

$$\begin{aligned} v_{11}(x) &= x^{\mathrm{T}} P_{11} x, \\ v_{22}(y) &= y^{\mathrm{T}} P_{22} y, \\ v_{33}(z) &= z^{\mathrm{T}} P_{33} z; \\ v_{12}(x, y) &= x^{\mathrm{T}} P_{12} y, \\ v_{13}(x, z) &= x^{\mathrm{T}} P_{13} z, \\ v_{23}(y, z) &= y^{\mathrm{T}} P_{23} z, \end{aligned} \tag{2.7.13}$$

where P_{ii}, $i = 1, 2, 3$, are symmetrical and positive definite matrices, P_{12}, P_{13} and P_{23} are constant matrices. It can be easily verified that for the functions (2.7.13) there exist estimates

(2.7.14)

$$\begin{aligned}
v_{11}(x) &\geq \lambda_m(P_{11})\|x\|^2 && \forall\,(x \neq 0) \in \mathcal{N}_x;\\
v_{22}(y) &\geq \lambda_m(P_{22})\|y\|^2 && \forall\,(y \neq 0) \in \mathcal{N}_y;\\
v_{33}(z) &\geq \lambda_m(P_{33})\|z\|^2 && \forall\,(z \neq 0) \in \mathcal{N}_z;\\
v_{12}(x,y) &\geq -\lambda_M^{1/2}\left(P_{12}P_{12}^{\mathrm{T}}\right)\|x\|\|y\| && \forall\,(x \neq 0,\ y \neq 0) \in \mathcal{N}_x \times \mathcal{N}_y;\\
v_{13}(x,z) &\geq -\lambda_M^{1/2}\left(P_{13}P_{13}^{\mathrm{T}}\right)\|x\|\|z\| && \forall\,(x \neq 0,\ z \neq 0) \in \mathcal{N}_x \times \mathcal{N}_z;\\
v_{23}(y,z) &\geq -\lambda_M^{1/2}\left(P_{23}P_{23}^{\mathrm{T}}\right)\|y\|\|z\| && \forall\,(y \neq 0,\ z \neq 0) \in \mathcal{N}_y \times \mathcal{N}_z,
\end{aligned}$$

where $\lambda_m(P_{ii})$ are minimal eigenvalues of matrices P_{ii}, $i = 1,2,3$, $\lambda_M^{1/2}\left(P_{12}P_{12}^{\mathrm{T}}\right)$, $\lambda_M^{1/2}\left(P_{13}P_{13}^{\mathrm{T}}\right)$, $\lambda_M^{1/2}\left(P_{23}P_{23}^{\mathrm{T}}\right)$ are norms of matrices P_{12}, P_{13} and P_{23} respectively.

By means of the function

$$U(x,y,z) = \begin{pmatrix} v_{11}(x) & v_{12}(x,y) & v_{13}(x,z)\\ v_{12}(x,y) & v_{22}(y) & v_{23}(y,z)\\ v_{13}(x,z) & v_{23}(y,z) & v_{33}(z) \end{pmatrix}$$

and the vector $\eta \in R_+^3$, $\eta_i > 0$, $i = 1,2,3$ we introduce the function

(2.7.15) $$v(x,y,z,\eta) = \eta^{\mathrm{T}} U(x,y,z)\eta.$$

PROPOSITION 2.7.1. *Let for system (2.7.8) there exists matrix-valued function (2.7.12) with elements (2.7.13) and estimates (2.7.14). Then for function (2.7.15) the estimate*

(2.7.16) $$v(x,y,z,\eta) \geq u^{\mathrm{T}} H^{\mathrm{T}} P H u \quad \forall\,(x \neq 0,\ y \neq 0,\ z \neq 0) \in \mathcal{N}_x \times \mathcal{N}_y \times \mathcal{N}_z$$

is satisfied, where $u^{\mathrm{T}} = (\|x\|, \|y\|, \|z\|)$; $H = \operatorname{diag}[\eta_1, \eta_2, \eta_3]$,

(2.7.17) $$P = \begin{pmatrix} \lambda_m(P_{11}) & -\lambda_M^{1/2}\left(P_{12}P_{12}^{\mathrm{T}}\right) & -\lambda_M^{1/2}\left(P_{13}P_{13}^{\mathrm{T}}\right)\\ -\lambda_M^{1/2}\left(P_{12}P_{12}^{\mathrm{T}}\right) & \lambda_m(P_{22}) & -\lambda_M^{1/2}\left(P_{23}P_{23}^{\mathrm{T}}\right)\\ -\lambda_M^{1/2}\left(P_{13}P_{13}^{\mathrm{T}}\right) & -\lambda_M^{1/2}\left(P_{23}P_{23}^{\mathrm{T}}\right) & \lambda_m(P_{33}) \end{pmatrix}.$$

Together with function (2.7.15) we shall consider its total derivative

(2.7.18) $$Dv(x,y,z,\eta) = \eta^T DU(x,y,z)\eta$$

by virtue of system (2.7.8).

PROPOSITION 2.7.2. *Let for system (2.7.8) there exist matrix-valued function (2.7.12) with elements (2.7.13). For total derivatives of functions (2.7.13) by virtue of subsystems (2.7.9)–(2.7.11) the following estimates are satisfied*

(1) $(\nabla_x v_{11})^{\mathrm{T}} Ax \le \rho_{11}\|x\|^2 \quad \forall\, x \in \mathcal{N}_x$;

(2) $(\nabla_x v_{12})^{\mathrm{T}} Ax \le \rho_{12}\|x\|\|y\| \quad \forall\,(x,y) \in \mathcal{N}_x \times \mathcal{N}_y$;

(3) $(\nabla_x v_{13})^{\mathrm{T}} Ax \le \rho_{13}\|x\|\|z\| \quad \forall\,(x,z) \in \mathcal{N}_x \times \mathcal{N}_z$;

(4) $(\nabla_y v_{22})^{\mathrm{T}} By \le \rho_{21}\|y\|^2 \quad \forall\, y \in \mathcal{N}_y$;

(5) $(\nabla_y v_{21})^{\mathrm{T}} By \le \rho_{22}\|x\|\|y\| \quad \forall\,(x,y) \in \mathcal{N}_x \times \mathcal{N}_y$;

(6) $(\nabla_y v_{23})^{\mathrm{T}} By \le \rho_{23}\|y\|\|z\| \quad \forall\,(y,z) \in \mathcal{N}_y \times \mathcal{N}_z$;

(7) $(\nabla_z v_{33})^{\mathrm{T}} Cz \le \rho_{31}\|z\|^2 \quad \forall\, z \in \mathcal{N}_z$;

(8) $(\nabla_z v_{31})^{\mathrm{T}} Cz \le \rho_{32}\|x\|\|z\| \quad \forall\,(x,z) \in \mathcal{N}_x \times \mathcal{N}_z$;

(9) $(\nabla_z v_{32})^{\mathrm{T}} Cz \le \rho_{33}\|y\|\|z\| \quad \forall\,(y,z) \in \mathcal{N}_y \times \mathcal{N}_z$,

where $\nabla_u = \partial/\partial u$ *and*

$$
\begin{aligned}
\rho_{11} &= \lambda_{\max}\left[P_{11}A + A^{\mathrm{T}}P_{11}\right],\\
\rho_{21} &= \lambda_{\max}\left[P_{22}B + B^{\mathrm{T}}P_{22}\right],\\
\rho_{31} &= \lambda_{\max}\left[P_{33}C + C^{\mathrm{T}}P_{33}\right],\\
\rho_{12} &= \|A^{\mathrm{T}}P_{12}\|,\\
\rho_{13} &= \|A^{\mathrm{T}}P_{13}\|,\\
\rho_{22} &= \|P_{12}B\|,\\
\rho_{23} &= \|B^{\mathrm{T}}P_{23}\|,\\
\rho_{32} &= \|P_{13}C\|,\\
\rho_{33} &= \|P_{23}C\|
\end{aligned}
$$

respectively, ρ_{12}, ρ_{13}, ρ_{22}, ρ_{23}, ρ_{32}, ρ_{33} *are norms of matrices* $A^{\mathrm{T}}P_{12}$, $A^{\mathrm{T}}P_{13}$, $P_{12}B$, $B^{\mathrm{T}}P_{23}$, $P_{13}C$, $P_{23}C$.

ASSUMPTION 2.7.3. There exist constants ρ_{ij}, $i = 1,2,3$, $j = 4,5,\dots,12$, such that in open connected neighborhoods $\mathcal{N}_x \subseteq R^{n_1}$, $\mathcal{N}_y \subseteq R^{n_2}$, $\mathcal{N}_z \subseteq R^{n_3}$ or in its product there exist the estimates

(1') $(\nabla_x v_{11})^{\mathrm{T}} f \le \rho_{14}\|x\|^2 + \rho_{15}\|x\|\|y\| + \rho_{16}\|x\|\|z\|$;

(2') $(\nabla_x v_{12})^{\mathrm{T}} f \le \rho_{17}\|y\|^2 + \rho_{18}\|x\|\|y\| + \rho_{19}\|y\|\|z\|$;

(3') $(\nabla_x v_{13})^{\mathrm{T}} f \le \rho_{1.10}\|z\|^2 + \rho_{1.11}\|x\|\|z\| + \rho_{1.12}\|y\|\|z\|$;

(4') $(\nabla_y v_{22})^{\mathrm{T}} g \le \rho_{24}\|y\|^2 + \rho_{25}\|x\|\|y\| + \rho_{26}\|y\|\|z\|$;
(5') $(\nabla_y v_{21})^{\mathrm{T}} g \le \rho_{27}\|x\|^2 + \rho_{28}\|x\|\|y\| + \rho_{29}\|x\|\|z\|$;
(6') $(\nabla_y v_{23})^{\mathrm{T}} g \le \rho_{2.10}\|z\|^2 + \rho_{2.11}\|x\|\|z\| + \rho_{2.12}\|y\|\|z\|$;
(7') $(\nabla_z v_{33})^{\mathrm{T}} h \le \rho_{34}\|z\|^2 + \rho_{35}\|x\|\|z\| + \rho_{36}\|y\|\|z\|$;
(8') $(\nabla_z v_{13})^{\mathrm{T}} h \le \rho_{37}\|x\|^2 + \rho_{38}\|x\|\|y\| + \rho_{39}\|x\|\|z\|$;
(9') $(\nabla_z v_{23})^{\mathrm{T}} h \le \rho_{3.10}\|y\|^2 + \rho_{3.11}\|x\|\|y\| + \rho_{3.12}\|y\|\|z\|$.

PROPOSITION 2.7.3. *If estimates (1) - (9) and (1') - (9') are satisfied, then for all total derivatives of function (2.7.15) by virtue of system (2.7.8) the inequality*

$$Dv(x, y, z, \eta) \le u^{\mathrm{T}} S u \qquad \forall\ (x, y, z) \in \mathcal{N}_x \times \mathcal{N}_y \times \mathcal{N}_z \tag{2.7.19}$$

takes place, where

$$\begin{aligned}
S &= [\sigma_{ij}],\ \ \sigma_{ij} = \sigma_{ji}\ \ \forall\ (i,j) \in [1,3];\\
\sigma_{11} &= \eta_1^2(\rho_{11} + \rho_{14}) + 2\eta_1(\eta_2\rho_{27} + \eta_3\rho_{37});\\
\sigma_{22} &= \eta_2^2(\rho_{21} + \rho_{24}) + 2\eta_2(\eta_1\rho_{17} + \eta_3\rho_{3.10});\\
\sigma_{33} &= \eta_3^2(\rho_{31} + \rho_{34}) + 2\eta_3(\eta_1\rho_{1.10} + \eta_2\rho_{2.10});\\
\sigma_{12} &= \frac{1}{2}\eta_1^2\rho_{15} + \frac{1}{2}\eta_2^2\rho_{25} + \eta_1\eta_2(\rho_{12} + \rho_{22} + \rho_{18} + \rho_{28})\\
&\quad + \eta_3(\eta_1\rho_{38} + \eta_2\rho_{3.11});\\
\sigma_{13} &= \frac{1}{2}\eta_1^2\rho_{16} + \frac{1}{2}\eta_3^2\rho_{35} + \eta_1\eta_3(\rho_{13} + \rho_{32} + \rho_{1.11} + \rho_{39})\\
&\quad + \eta_2(\eta_1\rho_{29} + \eta_3\rho_{2.11});\\
\sigma_{23} &= \frac{1}{2}\eta_2^2\rho_{26} + \frac{1}{2}\eta_3^2\rho_{36} + \eta_2\eta_3(\rho_{23} + \rho_{33} + \rho_{2.12} + \rho_{3.12})\\
&\quad + \eta_1(\eta_2\rho_{19} + \eta_3\rho_{1.12}).
\end{aligned} \tag{2.7.20}$$

REMARK 2.7.1. The dynamical properties of subsystems (2.7.9)-(2.7.11) influence only the sign of coefficients ρ_{11}, ρ_{21} and ρ_{31}. The constants ρ_{12}, ρ_{13}, ρ_{22}, ρ_{23}, ρ_{32}, ρ_{33} can always be taken positive and the rest of the constants are independent of matrices A, B and C.

In view of the above remark we introduce the following designations

$$\begin{aligned}
c_{11} &= \eta_1^2\rho_{14} + 2\eta_1(\eta_2\rho_{27} + \eta_3\rho_{37});\\
c_{22} &= \eta_2^2\rho_{24} + 2\eta_2(\eta_1\rho_{17} + \eta_3\rho_{3.10});\\
c_{33} &= \eta_3^2\rho_{34} + 2\eta_3(\eta_1\rho_{1.10} + \eta_2\rho_{2.10}).
\end{aligned}$$

Hence we have

$$\sigma_{11} = \eta_1^2 \rho_{11} + c_{11}; \quad \sigma_{22} = \eta_2^2 \rho_{21} + c_{22}; \quad \sigma_{33} = \eta_3^2 \rho_{31} + c_{33}.$$

PROPOSITION 2.7.4. *The matrix S is negative definite if and only if*

(1) $\eta_1^2 \rho_{11} + c_{11} < 0$;

(2) $\eta_1^2 \eta_2^2 \rho_{11} \rho_{21} + \eta_1^2 \rho_{11} c_{22} + \eta_2^2 \rho_{21} c_{11} + c_{11} c_{22} - \sigma_{12}^2 > 0$;

(3) $\eta_1^2 \rho_{11} \left(\eta_2^2 \eta_3^2 \rho_{21} \rho_{31} + \eta_2^2 \rho_{21} c_{33} + \eta_3^2 \rho_{31} c_{22} + c_{22} c_{33} - \sigma_{23}^2\right) + \eta_2^2 \rho_{21} \times \left(\eta_3^2 \rho_{31} c_{11} + c_{11} c_{33} - \sigma_{13}^2\right) + \eta_3^2 \rho_{31} \left(c_{11} c_{22} - \sigma_{12}^2\right) + c_{11} c_{22} c_{33} + 2\sigma_{12} \times \sigma_{13} \rho_{23} - c_{11} \sigma_{23}^2 - c_{22} \sigma_{13}^2 - c_{33} \sigma_{12}^2 < 0$.

REMARK 2.7.2. If subsystems (2.7.9)–(2.7.11) are nonasymptotically stable, i.e. $\rho_{11} = \rho_{21} = \rho_{31} = 0$, the conditions of Proposition 2.7.4 become

(1') $c_{11} < 0$;
(2') $c_{11} c_{22} - \sigma_{12}^2 > 0$;
(3') $c_{11} c_{22} c_{33} + 2\sigma_{12} \sigma_{13} \sigma_{23} - c_{11} \sigma_{23}^2 - c_{22} \sigma_{13}^2 - c_{33} \sigma_{12}^2 < 0$.

REMARK 2.7.3. If subsystem (2.7.9) is nonasymptotically stable, subsystem (2.7.10) is asymptotically stable and (1.7.11) is unstable, i.e. $\rho_{11} = 0$, $\rho_{21} < 0$, $\rho_{31} > 0$, the conditions of Proposition 2.7.4 become

(1") $c_{11} < 0$;

(2") $\eta_2^2 \rho_{21} c_{11} + c_{11} c_{22} - \sigma_{12}^2 > 0$;

(3") $\eta_2^2 \rho_{21} \left(\eta_3^2 \rho_{31} c_{11} + c_{11} c_{33} - \sigma_{13}^2\right) + \eta_3^2 \rho_{31} \left(c_{11} c_{22} - \sigma_{12}^2\right) + c_{11} c_{22} c_{33} + 2\sigma_{12} \sigma_{13} \sigma_{23} - c_{11} \sigma_{23}^2 - c_{22} \sigma_{13}^2 - c_{33} \sigma_{12}^2 < 0$.

PROPOSITION 2.7.5. *Matrix S is negative semi-definite iff the inequality signs $<$ and $>$ in Proposition 2.7.4 are replaced by $\geqslant$ and $\leqslant$ correspondingly.*

Function (2.7.15) and its total derivative (2.7.18) together with estimates (2.7.16) and (2.7.19) allows us to establish sufficient conditions of stability (in the whole) and asymptotic stability (in the whole) for system (2.7.8).

THEOREM 2.7.2. *Suppose that the system (2.7.8) be such that*

(1) *in product $\mathcal{N} = \mathcal{N}_x \times \mathcal{N}_y \times \mathcal{N}_z$ there is the matrix-valued function $U: \mathcal{N} \to R^{3\times 3}$;*
(2) *there exist the vector $\eta \in R_+^3$, $\eta_i > 0$, $i \in [1,3]$;*
(3) *the matrix P is positive definite;*
(4) *the matrix S is negative semi-definite or equals to zero.*

Then the state $x = y = z = 0$ of the system (2.7.8) is uniformly stable.

If all estimates mentioned in conditions of Theorem 2.7.2 are satisfied for $\mathcal{N}_x = R^{n_1}$, $\mathcal{N}_y = R^{n_2}$, $\mathcal{N}_z = R^{n_3}$ and function (2.7.15) is radially unbounded, the state $x = y = z = 0$ of the system (2.7.8) is uniformly stable in the whole.

PROOF. Under all conditions of Theorem 2.7.2 the conditions of well-known Barbashin-Krasovskii's theorem are satisfied, and hence, the corresponding type of stability of state $x = y = z = 0$ of the system (2.7.8) takes place (see Theorem 2.5.2).

Let there exists the domain $\Omega = \{(x,y,z) \in \mathcal{N},\ 0 \le v(x,y,z,\eta) < a,\ a \in \overset{\circ}{R}_+\} \subset R^n$ where $Dv(x,y,z,\eta) \le 0$.

We designate by $\mathcal{M}$ the largest invariant set in Ω where

$$Dv(x,y,z,\eta) = 0.$$

THEOREM 2.7.3. *Suppose that the system (2.7.8) be such that*

(1) *the conditions (1) - (3) of Theorem 2.7.2 be satisfied;*
(2) *on the set Ω $Dv(x,y,z,\eta) \le 0$ i.e. the matrix S is negative semi-definite.*

Then the set $\mathcal{M}$ is attractive relative to the domain Ω, i.e. all motions of system (2.7.8) starting on set Ω tend to the set $\mathcal{M}$ as $t \to +\infty$.

Proof of this Theorem is similar to that of Theorem 26.1 by Hahn [66].

THEOREM 2.7.4. *Suppose that the system (2.7.8) is such that*

(1) *the conditions (1) - (3) of Theorem 2.7.2 are satisfied;*
(2) *the matrix S is negative semi-definite.*

Then the equilibrium state $x = y = z = 0$ of the system (2.7.8) is uniformly asymptotically stable.

If all estimates mentioned in conditions of Theorem 2.7.4 are satisfied for $\mathcal{N}_x = R^{n_1}$, $\mathcal{N}_y = R^{n_2}$, $\mathcal{N}_z = R^{n_3}$ and function (2.7.15) is radially unbounded, the state $x = y = z = 0$ of the system (2.7.8) is uniformly asymptotically stable in the whole.

The proof is similar to that of Theorem 25.2 by Hahn [66].

2.7.3 Large scale Lur'e-Postnikov system

We consider the system of equations

$$
\begin{aligned}
\frac{dx}{dt} &= A_{11}x + A_{12}y + A_{13}z + q_1 f_1(\sigma_1) \triangleq f_1^*; \\
\frac{dy}{dt} &= A_{21}x + A_{22}y + A_{23}z + q_2 f_2(\sigma_2) \triangleq f_2^*; \\
\frac{dz}{dt} &= A_{31}x + A_{32}y + A_{33}z + q_3 f_3(\sigma_3) \triangleq f_3^*,
\end{aligned}
\tag{2.7.21}
$$

where

$$
\sigma_i = c_{i1}^{\mathrm{T}} x + c_{i2}^{\mathrm{T}} y + c_{i3}^{\mathrm{T}} z,
$$

$$
f_i(\sigma_i)/\sigma_i \in [0, k_i], \qquad i = 1, 2, 3, \quad \sigma_i \in (-\infty, +\infty).
$$

Assume that for system (2.7.8) matrix-valued function (2.7.12) is constructed with elements (2.7.13) for which estimates (2.7.14) are satisfied, and matrix (2.7.17) is positive definite. It is easy to verify that for the total derivative of function (2.7.15) by virtue of system (2.7.21) the following estimate

$$
Dv(x, y, z, \eta) \le u^{\mathrm{T}} \widetilde{S} u
\tag{2.7.22}
$$

is satisfied, where $\widetilde{S} = [\widetilde{\sigma}_{ij}]$, $\widetilde{\sigma}_{ij} = \widetilde{\sigma}_{ji} \ \ \forall\,(i,j) \in [1,3]$ and

$$
\begin{aligned}
\sigma_{11} = \lambda_{\max}\big[&\eta_1^2\left(A_{11}^{\mathrm{T}} P_{11} + P_{11} A_{11} + P_{11}(q_1 k_1^* c_{11}^{\mathrm{T}}) + (q_1 k_1^* c_{11}^{\mathrm{T}})^{\mathrm{T}} P_{11}\right) \\
&+ 2\eta_1\eta_2\left(P_{12} A_{21} + P_{12}(q_2 k_2^* c_{21}^{\mathrm{T}})\right) + 2\eta_1\eta_3\left(P_{13} A_{31} + P_{13}(q_3 k_3^* c_{31}^{\mathrm{T}})\right)\big],
\end{aligned}
$$

$$
\begin{aligned}
\sigma_{22} = \lambda_{\max}\big[&\eta_2^2\left(A_{22}^{\mathrm{T}} P_{22} + P_{22} A_{22} + P_{22}(q_2 k_2^* c_{22}^{\mathrm{T}}) + (q_2 k_2^* c_{22}^{\mathrm{T}})^{\mathrm{T}} P_{22}\right) \\
&+ 2\eta_1\eta_2\left(A_{12}^{\mathrm{T}} P_{12} + (q_1 k_1^* c_{12}^{\mathrm{T}})^{\mathrm{T}} P_{12}\right) + 2\eta_1\eta_3\left(P_{23} A_{32} + P_{23}(q_3 k_3^* c_{22}^{\mathrm{T}})\right)\big],
\end{aligned}
$$

$$
\begin{aligned}
\sigma_{33} = \lambda_{\max}\big[&\eta_3^2\left(A_{33}^{\mathrm{T}} P_{33} + P_{33} A_{33} + P_{33}(q_3 k_3^* c_{33}^{\mathrm{T}}) + (q_3 k_3^* c_{33}^{\mathrm{T}})^{\mathrm{T}} P_{33}\right) \\
&+ 2\eta_1\eta_3\left(A_{13}^{\mathrm{T}} P_{13} + (q_1 k_1^* c_{13}^{\mathrm{T}})^{\mathrm{T}} P_{13}\right) + 2\eta_2\eta_3\left(A_{23}^{\mathrm{T}} P_{23} + (q_2 k_2^* c_{23}^{\mathrm{T}})^{\mathrm{T}} P_{23}\right)\big],
\end{aligned}
$$

σ_{ij}, $i \ne j$, $i, j \in [1,3]$ are norms of matrices:

$$\begin{aligned}\sigma_{12} = \big\|&\eta_1^2\left(P_{11}A_{12} + P_{11}(q_1k_1^*c_{12}^{\mathrm{T}})\right) + \eta_2^2\left(P_{22}A_{21} + P_{22}(q_2k_2^*c_{21}^{\mathrm{T}})\right)\\ &+ \eta_1\eta_2\left(A_{11}^{\mathrm{T}}P_{12} + P_{12}A_{22} + (q_1k_1^*c_{11}^{\mathrm{T}})^{\mathrm{T}}P_{12} + P_{12}(q_2k_2^*c_{22}^{\mathrm{T}})\right)\\ &+ \eta_1\eta_3\left(P_{13}A_{32} + P_{13}(q_3k_3^*c_{32}^{\mathrm{T}})\right) + \eta_2\eta_3\left(P_{23}A_{31} + P_{23}(q_3k_3^*c_{31}^{\mathrm{T}})\right)\big\|,\end{aligned}$$

$$\begin{aligned}\sigma_{13} = \big\|&\eta_1^2\left(P_{11}A_{13} + P_{11}(q_1k_1^*c_{13}^{\mathrm{T}})\right) + \eta_3^2\left(P_{33}A_{31} + P_{33}(q_3k_3^*c_{31}^{\mathrm{T}})\right)\\ &+ \eta_1\eta_2\left(P_{12}A_{23} + P_{12}(q_2k_2^*c_{23}^{\mathrm{T}})\right) + \eta_1\eta_3\left(P_{13}A_{33} + A_{11}^{\mathrm{T}}P_{13}\right.\\ &+ \left.P_{13}(q_3k_3^*c_{33}^{\mathrm{T}}) + (q_1k_1^*c_{11}^{\mathrm{T}})^{\mathrm{T}}P_{13}\right) + \eta_2\eta_3\left(A_{21}^{\mathrm{T}}P_{23} + (q_2k_2^*c_{21}^{\mathrm{T}})^{\mathrm{T}}P_{23}\right)\big\|,\end{aligned}$$

$$\begin{aligned}\sigma_{23} = \big\|&\eta_2^2\left(P_{22}A_{23} + P_{22}(q_2k_2^*c_{23}^{\mathrm{T}})\right) + \eta_3^2\left(P_{33}A_{32} + P_{33}(q_3k_3^*c_{32}^{\mathrm{T}})\right)\\ &+ \eta_1\eta_2\left(A_{13}^{\mathrm{T}}P_{12} + (q_1k_1^*c_{13}^{\mathrm{T}})^{\mathrm{T}}P_{12}\right) + \eta_1\eta_3\left(A_{12}^{\mathrm{T}}P_{13} + (q_1k_1^*c_{12}^{\mathrm{T}})P_{13}\right)\\ &+ \eta_2\eta_3\left(A_{22}^{\mathrm{T}}P_{23} + P_{23}A_{22} + (q_2k_2^*c_{22}^{\mathrm{T}})^{\mathrm{T}}P_{23} + P_{23}(q_3k_3^*c_{33}^{\mathrm{T}})\right)\big\|,\end{aligned}$$

$$k_i^* = \begin{cases} k_i & \text{for} \quad \sigma_i q_i^{\mathrm{T}} P_{ij} x > 0 \ (\text{or } \sigma_i q_i^{\mathrm{T}} P_{ij} y > 0 \\ & \qquad \text{or } \sigma_i q_i^{\mathrm{T}} P_{ij} z > 0); \\ 0 & \text{in other cases.} \end{cases}$$

Estimate (2.7.22) of total derivative $Dv(x, y, z, \eta)$ along solutions of system (2.7.21) makes possible the application of Theorems 2.7.2 – 2.7.4.

EXAMPLE 2.7.1. Let in system (2.7.21) matrices and vectors be defined as

$$A_{11} = \begin{pmatrix} -3 & 0 \\ 0 & -3 \end{pmatrix}; \quad A_{12} = \begin{pmatrix} -5 & 0 \\ -1 & -5 \end{pmatrix}; \quad A_{13} = \begin{pmatrix} 1 & 0 \\ 0 & 1 \end{pmatrix};$$

$$A_{21} = \begin{pmatrix} 5 & 0 \\ 1 & 5 \end{pmatrix}; \quad A_{22} = \begin{pmatrix} 0.1 & 0 \\ 0 & 0.1 \end{pmatrix}; \quad A_{23} = \begin{pmatrix} -2 & 0 \\ 0 & -2 \end{pmatrix};$$

$$A_{31} = \begin{pmatrix} -1 & 0 \\ 0 & -1 \end{pmatrix}; \quad A_{32} = \begin{pmatrix} 2.3 & 0 \\ 0 & 2.3 \end{pmatrix}; \quad A_{33} = \begin{pmatrix} 0 & 1 \\ -1 & 0 \end{pmatrix};$$

$$q_1 = \begin{pmatrix} 0.1 \\ 0 \end{pmatrix}; \quad q_2 = \begin{pmatrix} -0.1 \\ 0 \end{pmatrix}; \quad q_3 = \begin{pmatrix} 0 \\ 0.1 \end{pmatrix};$$

$$c_{11} = \begin{pmatrix} -0.1 \\ 0 \end{pmatrix}; \quad c_{12} = \begin{pmatrix} -0.01 \\ 0 \end{pmatrix}; \quad c_{13} = \begin{pmatrix} -0.1 \\ 0.1 \end{pmatrix};$$

$$c_{21} = \begin{pmatrix} 0.1 \\ 0 \end{pmatrix}; \quad c_{22} = \begin{pmatrix} 0.01 \\ 0 \end{pmatrix}; \quad c_{23} = \begin{pmatrix} 0.1 \\ 0.1 \end{pmatrix};$$

$$c_{31} = \begin{pmatrix} -0.1 \\ -0.1 \end{pmatrix}; \qquad c_{32} = \begin{pmatrix} 0 \\ -0.01 \end{pmatrix}; \qquad c_{33} = \begin{pmatrix} 0 \\ -0.1 \end{pmatrix};$$

$$k_i = 1, \qquad i = 1, 2, 3.$$

We take matrix-valued function $U(x)$ elements in the form

$$\begin{aligned} v_{11}(x) &= x^{\mathrm{T}} \operatorname{diag}(1,1)x; & v_{22}(y) &= y^{\mathrm{T}} \operatorname{diag}(1,1)y; \\ v_{33}(z) &= z^{\mathrm{T}} \operatorname{diag}(1,1)z; & v_{12}(x,y) &= x^{\mathrm{T}} \operatorname{diag}(0.1,0.1)y; \\ v_{13}(x,z) &= x^{\mathrm{T}} \operatorname{diag}(0.1,0.1)z; & v_{23}(y,z) &= y^{\mathrm{T}} \operatorname{diag}(0.1,0.1)z. \end{aligned}$$

For elements $v_{ij}(\cdot)$ estimates

$$\begin{gathered} v_{11}(x) \geq \|x\|^2, \quad v_{22}(y) \geq \|y\|^2, \quad v_{33}(z) \geq \|z\|^2, \\ v_{12}(x,y) \geq -0.1\|x\|\|y\|; \qquad v_{13}(x,z) \geq -0.1\|x\|\|z\|; \\ v_{23}(y,z) \geq -0.1\|y\|\|z\| \end{gathered}$$

are satisfied, and matrix $\widetilde{P}$ corresponding to matrix P in estimate (2.7.16)

$$\widetilde{P} = \begin{pmatrix} 1 & -0.1 & -0.1 \\ -0.1 & 1 & -0.1 \\ -0.1 & -0.1 & 1 \end{pmatrix}$$

is positive definite.

If $\eta = (1,1,1)^{\mathrm{T}}$ then, given choice of elements $v_{ij}(\cdot)$, $i, j \in [1,3]$ matrix-valued function $U(x,y,z)$, the matrix $\widetilde{S}$ takes the values

$$\widetilde{S} = \begin{cases} \begin{pmatrix} -5.2 & 0.16 & 0.2 \\ 0.16 & -0.34 & 0.15 \\ 0.2 & 0.15 & -0.2 \end{pmatrix} & \text{for } k_i^* = 0; \\ \begin{pmatrix} -5.202 & 0.18 & 0.03 \\ 0.18 & -0.34 & 0.012 \\ 0.03 & 0.012 & -0.202 \end{pmatrix} & \text{for } k_i^* = k_i = 1. \end{cases}$$

It easy to verify that in both cases matrix $\widetilde{S}$ is negative definite.

By Theorem 2.7.4 we find that the state $x = y = z = 0$ of system (2.7.21) with vectors and matrices defined in Example 2.7.1 is asymptotically stable in the whole (i.e., system (2.7.21) is absolutely stable).

2.7.4 A generalized Lotka-Volterra system

We consider a generalized Lotka-Volterra system of the form

$$\begin{aligned}\frac{dx_1}{dt} &= x_1\left(b_1 + a_{11}(x)x_1 + a_{12}(x)x_2\right),\\ \frac{dx_2}{dt} &= x_2\left(b_2 + a_{21}(x)x_1 + a_{22}(x)x_2\right),\end{aligned} \tag{2.7.23}$$

where $x_1, x_2 \in R_+$, $a_{ij} \in C\left(R_+^2, R\right)$, b_1, b_2 are constants, $x \in R_+^2$. The generalized Lotka-Volterra system (2.7.23) can have several equilibrium states x_e determined as solutions of

$$x_e = 0 \quad \text{or} \quad A(x)x_e = -b \tag{2.7.24}$$

when $b \neq 0$ and $\det A(x) = 0\ \forall x \in S(\rho)$, $S(\rho) \subseteq R_+^2$ or $b = 0$, $\det A(x) \neq 0\ \forall x \in S(\rho)$, in which case $x_e = 0$ is the unique equilibrium state of (2.7.23) which is a singular case.

Otherwise, the system (2.7.23) can have finitely many ($\det A(x) \neq 0$ $\forall x \in S(\rho)$, $b \neq 0$) or infinitely many ($\det A(x) = 0\ \forall x \in S(\rho)$, $b = 0$) equilibrium states. If we are interested in properties of $x \neq 0$, then we use the Liapunov transformation of the state variables,

$$y_1 = x_1 - x_{e1}, \qquad y_2 = x_2 - x_{e2} \tag{2.7.25}$$

and transform (2.7.23) into

$$\begin{aligned}\frac{dy_1}{dt} &= \left(a_{11}(x)y_1 + a_{12}(x)y_2\right)x_{e1} + \left(a_{11}(x)y_1 + a_{12}(x)y_2\right)y_1,\\ \frac{dy_2}{dt} &= \left(a_{21}(x)y_1 + a_{22}(x)y_2\right)x_{e2} + \left(a_{21}(x)y_1 + a_{22}(x)y_2\right)y_2.\end{aligned} \tag{2.7.26}$$

Together with equations (2.7.26) for $i = 1, 2$, we consider the real functions $v_{ij}(y_1, y_2)$ and matrix-valued function

$$U(y_1, y_2) = \begin{pmatrix} v_{11}(y_1) & v_{12}(y_1, y_2) \\ v_{12}(y_1, y_2) & v_{22}(y_2) \end{pmatrix} \tag{2.7.27}$$

with elements

$$\begin{aligned}&v_{11}(y_1) = \alpha y_1^2, \qquad v_{22}(y_2) = \beta y_2^2,\\ &v_{12}(y_1, y_2) = v_{21}(y_1, y_2) = -\gamma y_1 y_2\end{aligned} \tag{2.7.28}$$

α, $\beta > 0$ and γ a constant.

By means of the vector $\eta^{\mathrm{T}} = (\eta_1, \eta_2) \in R_+^2$, $\eta_i > 0$ we shall construct a scalar function

$$v(y) = \eta^{\mathrm{T}} U(y) \eta \tag{2.7.29}$$

for the generalized Lotka-Volterra system (2.7.23).

For all $y \in S(\rho)$ the inequality

$$v(y) \geq u^{\mathrm{T}} H^{\mathrm{T}} P H u \tag{2.7.30}$$

holds, where $u^{\mathrm{T}} = (|y_1|, |y_2|)$, $H = \operatorname{diag}(\eta_1, \eta_2)$,

$$P = \begin{pmatrix} \alpha & -\gamma \\ -\gamma & \beta \end{pmatrix}. \tag{2.7.31}$$

The total derivatives of the matrix-valued function (2.7.27) along solutions of (2.7.23) are given by

$$\begin{aligned}
\frac{dv_{11}}{dt} &\leq 2\alpha|a_{11}|x_{e1}|y_1|^2 + 2\alpha|a_{12}|x_{e1}|y_1||y_2| + 2\alpha|a_{11}||y_1|^3 \\
&\quad + 2\alpha|a_{12}|x_{e1}|y_1|^2|y_2|; \\
\frac{dv_{22}}{dt} &\leq 2\beta|a_{22}|x_{e2}|y_2|^2 + 2\beta|a_{21}|x_{e2}|y_1||y_2| + 2\beta|a_{22}||y_2|^3 \\
&\quad + 2\beta|a_{21}|x_{e2}|y_1||y_2|^2; \\
\frac{dv_{12}}{dt} &\leq \gamma|a_{21}|x_{e2}y_1^2 + \gamma|a_{12}|x_{e1}y_2^2 + |\gamma(a_{22}x_{e2} + a_{11}x_{e1})||y_1||y_2| \\
&\quad + |\gamma(a_{21} + a_{11})||y_1|^2|y_2| + |\gamma(a_{22} + a_{12})||y_1||y_2|^2,
\end{aligned}$$

and

$$\frac{dv}{dt} \leq u^{\mathrm{T}}(C + G(y_1, y_2))u, \tag{2.7.32}$$

where $u^{\mathrm{T}} = (|y_1|, |y_2|)$,

$$C = \begin{pmatrix} c_{11} & c_{12} \\ c_{12} & c_{22} \end{pmatrix}, \qquad G(y_1, y_2) = \begin{pmatrix} \sigma_{11}(y_1) & \sigma_{12}(y_1, y_2) \\ \sigma_{21}(y_1, y_2) & \sigma_{22}(y_2) \end{pmatrix}.$$

Here we have

$$\begin{aligned}
c_{11} &= 2\eta_1(\alpha\eta_1 a_{11}x_{e1} + \eta_2\gamma a_{21}x_{e2}); \\
c_{22} &= 2\eta_2(\beta\eta_2 a_{22}x_{e2} + \eta_1\gamma a_{12}x_{e1}); \\
c_{12} &= \alpha\eta_1^2|a_{12}|x_{e1} + \beta\eta_2^2|a_{21}|x_{e2} + \eta_1\eta_2|\gamma(a_{11}x_{e1} + a_{21}x_{e2}|
\end{aligned}$$

and

$$
\begin{aligned}
\sigma_{11}(y_1) &= 2\alpha\eta_1^2|a_{11}||y_1|,\\
\sigma_{22}(y_2) &= 2\beta\eta_2^2|a_{22}||y_2|,\\
\sigma_{12}(y_1,y_2) &= \left(\alpha\eta_1^2|a_{12}| + \eta_1\eta_2|\gamma(a_{21}+a_{11})|\right) y_1\\
&\quad + \left(\beta\eta_2^2|a_{21}| + \eta_1\eta_2|\gamma(a_{22}+a_{12})|\right) y_2.
\end{aligned}
$$

Inequalities (2.7.30) and (2.7.32) imply the following theorem, which is the main result of this section.

THEOREM 2.7.5. *The equilibrium x_e of the generalized Lotka-Volterra system (2.7.23) is asymptotically stable if*

(1) *the matrix P is positive definite;*
(2) *there exists a constant matrix $\overline{G}$ such that*

$$G(y_1,y_2) \le \overline{G} \quad \forall\ (y_1,y_2) \in S(\rho);$$

(3) *there exists a constant matrix $\overline{C}$ such that*

$$C(x_1,x_2) \le \overline{C} \quad \forall\ (x_1,x_2) \in S(\rho);$$

(4) *the matrix $\overline{C} + \overline{G}$ is negative definite.*

We believe that this result is the first of its kind for such generalized Lotka-Volterra systems.

2.8 Notes

2.1. The following is a summary of the formulation of the matrix Liapunov function method:

* discovery of double-index system of functions, as a structure suitable for constructing Liapunov functions (see Martynyuk and Gutowski [123]);
* formulation of the basic concepts of the MLMF on the basis of double-index system function (see Djordjević [27, 29], Grujić [47], Martynyuk [109, 112, 116].);
* formulation of the principle of invariance and investigation of autonomous systems (see Djordjević [28]; Grujić [47]; Grujić, Martynyuk and Ribbens-Pavella [57]; Martynyuk [116], etc.);

* development of methods for constructing matrix Liapunov functions (see Djordjević [30], Martynyuk and Krapivny [124], Grujić and Shaaban [61], etc.);
* construction of sufficient condition of stability for

  (a) systems with lumped parameters (see Djordjević [27 – 30], Grujić [47], Martynyuk [126], Martynyuk and Miladzhanov [125], etc.);

  (b) systems with a small parameter multiplying a derivative (see Martynyuk [114], Martynyuk and Miladzhanov [128], etc.);

  (c) systems with random parameters (see Azimov [7], Azimov and Martynyuk [8], Martynyuk [115], etc.);

2.2. The results in this section are due to Grujić [47], Martynyuk [116, 121]. Propositions 2.2.1 – 2.2.3 are new.

2.3. Theorems 2.3.1 – 2.3.4 uses the results of Liapunov [101], Persidskii [152], Yoshizawa [174], Zubov [178] and Grujić, Martynyuk and Ribbens-Pavella [57].

2.4. Theorems 2.4.1 – 2.4.2 are new. They generalize well-known theorems of comparison method in motion stability theory (see e.g. Lakshmikantham, Leela and Martynyuk [94]).

2.5. Theorems 2.5.1 – 2.5.7 of this section are new. The results of the investigation of autonomous system (Theorems 2.5.8 – 2.5.20) are presented based on those by Grujić [47] and Grujić, Martynyuk and Ribbens-Pavella [57].

2.6. The notion of multistability of motion is formulated in terms of refusal from "homogeneous" behavior of components of solutions for nonlinear system. This notion can be viewed as well as generalization of stability with respect to a part of variables (see e.g. Rumiantzev [160] and Aminov and Sirazetdinov [2]). The results of sections 2.6.1 - 2.6.4 are new. Theorem 2.6.3 is taken from Martynyuk [118]. Theorems 2.6.4 and 2.6.5 were published by Martynyuk [117] and Theorem 2.6.6 the same author [119]. In the investigation of nonlinear systems by vector Liapunov functions the notion of multistability of comparison system was used by Lakshmikantham, Leela and Rao [95].

2.7. In subsection 2.7.1 the solution of the Lefschetz [100] problem is presented according to Martynyuk [111]. Moreover, the results by Djordjević [29] are used. The results of subsections 2.7.2 - 2.7.3 are taken from Martynyuk and Miladzhanov [125]. The results of subsection 2.7.4 are taken from Freedman and Martynyuk [37].

3

STABILITY OF SINGULARLY-PERTURBED SYSTEMS

3.1 Introduction

The physical system can consist of subsystems that react differently to the external impacts. Moreover, each of the subsystems has its own scale of natural time. In the case when the subsystems are not interconnected, the dynamical properties of each subsystem are examined in terms of the corresponding time scale. It turned out that it is reasonable to use such information when the additional conditions on the subsystems are formulated in the investigation of large scale systems. The existence of various time scales related to the separated subsystems is mathematically expressed by arbitrarily small positive parameters μ_i present at the part of the higher derivatives in differential equation. If the parameters μ_i vanish, the number of differential equations of the large scale system is diminished and, hence the appearance of algebraic equations.

This is just the singular case allowing the consideration of various peculiarities of the system with different time scales.

The chapter is arranged as follows.

Section 3.2 provides mathematical description of the system with quick and slow variables and states the problem of investigation.

Section 3.3 deals with asymptotic stability conditions for singularly perturbed system in terms of scalar Liapunov function.

Section 3.4 deals with Lur'e-Postnikov systems in terms of scalar Liapunov function.

In Section 3.5 the notion of the property of having a fixed sign is formulated for matrix-valued function for singularly perturbed system.

In Section 3.6 the matrix-valued Liapunov function is introduced and the structure of estimation of this function total derivative along solution of the system under consideration is determined.

In the Section 3.7 and 3.8 general results of the direct Liapunov method are stated for singularly perturbed system via matrix-valued function.

In Section 3.9 the method of constructing elements of the matrix-valued function is concretized and linear singularly perturbed systems are investigated using this method.

Section 3.10 contains some applications of general results to systems modeling mechanics problems such as oscillating system of solid bodies and Lur'e-Postnikov system.

The final Section 3.11 is supplied with detailed bibliography comments to the sections of the chapter.

3.2 Description of Systems

The singularly perturbed system S being considered below, is described by two systems of nonlinear differential equations

$$\frac{dx}{dt} = f(t, x, y, \mu), \tag{3.2.1}$$

$$\mu\frac{dy}{dt} = g(t, x, y, \mu), \tag{3.2.2}$$

where $(x^{\mathrm{T}}, y^{\mathrm{T}})^{\mathrm{T}}$ is a vector of state of the whole system, $x \in R^n$, $y \in R^m$, $f \in C(R \times R^n \times R^m \times \mathcal{M}, R^n)$, $g \in C(R \times R^n \times R^m \times \mathcal{M}, R^m)$. The parameter μ is positive and is supposed to be arbitrarily small. We set $\mu \in (0, 1] = \mathcal{M}$.

The states $x = 0$ and $y = 0$ have open connected neighborhoods $\mathcal{N}_x \subseteq R^n$ and $\mathcal{N}_y \subseteq R^m$ respectively. The vector-function f and g are such that for $(x^{\mathrm{T}}, y^{\mathrm{T}})^{\mathrm{T}} = 0$ system (3.2.1), (3.2.2) has the only equilibrium state in the Cartesian product $\mathcal{N}_x \times \mathcal{N}_y$ of the sets $\mathcal{N}_x$ and $\mathcal{N}_y$ for any $\mu \in (0, 1]$. If μ takes zero value, system (3.2.1), (3.2.2) degenerates into system S_0, which is described by the differential and algebraic equation

$$\frac{dx}{dt} = f(t, x, y, 0), \tag{3.2.3}$$

$$0 = g(t, x, y, 0). \tag{3.2.4}$$

It is supposed that $g(t, x, y, 0)$ vanishes for any $t \in R$ and $x \in \mathcal{N}_x$, iff $y = 0$. This requirement is motivated by an effective application of the Liapunov's coordinates transformation by Hoppensteadt [74] in the investigation of singularly-perturbed systems. The system of lower order

$$\frac{dx}{dt} = f(t, x, 0, 0) \tag{3.2.5}$$

obtained in result, is of importance in the stability investigation of system (3.2.1), (3.2.2). If $\mu > 0$ is a sufficiently small value of the parameter, then system (3.2.1), (3.2.2) consists of the parts which accomplish slow and quick motions. The quick system S_τ (or the boundary layer) is obtained from system (3.2.1), (3.2.2) after the change of the time scale by introducing the variable

$$\tau = (t - t_0)\mu^{-1}.$$

Then, the quick system corresponding to system (3.2.2) becomes

$$\frac{dy}{d\tau} = g(\alpha, b, y, 0). \tag{3.2.6}$$

In this system α and b, $b = (\beta_1, \dots, \beta_n)$, are scalar and vector parameters, introduced instead of $t \in R$ and $x \in \mathcal{N}_x$ respectively. We suppose as earlier, that g vanishes for any $t \in R$, $x \in \mathcal{N}_x$, $\mu \in (0,1]$ iff $y = 0$. The separation of the time-scales in the investigation of stability of system (3.2.1), (3.2.2) is essential due to the fact that the analysis of the degenerate system S_0 (3.2.5) and the quick system S_τ (3.2.6) is a more simple problem in comparison with the general problem of stability of system (3.2.1), (3.2.2). The next problem to be considered is to establish conditions for the vector-function f and g under which the property of uniform asymptotic stability in the product $\mathcal{N}_x \times \mathcal{N}_y$ of system (3.2.1), (3.2.2) can be obtained from the same property of solutions of system (3.2.5) and (3.2.6).

3.3 Asymptotic Stability Conditions

Let

$$\mathcal{N}_{x0} = \{x : x \in \mathcal{N}_x,\ x \neq 0\}, \qquad \mathcal{N}_{y0} = \{y : y \in \mathcal{N}_y,\ y \neq 0\}.$$

The function $V(\alpha, b, y) \in C^{(1,1,1)}(R \times R^n \times R^m, R)$ and

$$V_\alpha = \frac{\partial V}{\partial \alpha}, \qquad V_b = \left(\frac{\partial V}{\partial \beta_1}, \frac{\partial V}{\partial \beta_2}, \dots, \frac{\partial V}{\partial \beta_n}\right)^{\mathrm{T}}.$$

We introduce two assumptions on systems (3.2.5) and (3.2.6) connected with positive definite functions θ and V.

ASSUMPTION 3.3.1. There exist

(1) a decreasing positive definite on $\mathcal{N}_x$ and radially unbounded for $\mathcal{N}_x = R^n$ function $\theta \in C^{(1,1)}(R \times \mathcal{N}_{xo}, R_+)$;
(2) positive definite function $\varphi \in C(R^n, R_+)$ and $\psi \in C(R^m, R_+)$ on $\mathcal{N}_x$ and $\mathcal{N}_y$, respectively;
(3) non-negative numbers ζ_1 and ζ_2, $\zeta_1 < 1$, and the conditions are satisfied:
   (a) $\theta_t(t,x) + \theta_x^{\mathrm{T}}(t,x)f(t,x,0) \le -\varphi(x)$ $\forall (t,x) \in R \times \mathcal{N}_{x0}$;
   (b) $\theta_x^{\mathrm{T}}(t,x)[f(t,x,y,\mu) - f(t,x,y,0)] \le \zeta_1\varphi(x) + \zeta_2\psi(y)$, $\forall (t,x,y,\mu) \in R \times \mathcal{N}_{xo} \times \mathcal{N}_{yo} \times \mathcal{M}$.

Conditions (1)–(3)(a) of Assumption 3.2.1 ensure uniform asymptotic stability of $x = 0$ of system (3.2.5) in the whole, when $\mathcal{N}_x = R^n$. Condition (3)(b) is a requirement to the qualitative properties of the vector-function f on $\mathcal{N}_x \times \mathcal{N}_y$.

ASSUMPTION 3.3.2. There exist

(1) a decreasing positive definite on $\mathcal{N}_x \times \mathcal{N}_y$ and radially unbounded in y uniformly relatively $x \in \mathcal{N}_x$ for $\mathcal{N}_y = R^m$ function $V(t,x,y) \in C^{(1,1,1)}(R \times \mathcal{N}_x \times \mathcal{N}_y, R_+)$ (or $V(t,y) \in C^{(1,1)}(R \times \mathcal{N}_{yo}, R_+)$ decreasing and positive definite on $\mathcal{N}_y$ and radially unbounded for $\mathcal{N}_y = R^m$);
(2) non-negative numbers ξ_1, ξ_2, ξ_3, ξ_4 ($\xi_1 < 1$, $\xi_2 < 1$) and an integer $\pi > 1$;
(3) positive definite functions $\varphi \in C(R^n, R_+)$, $\psi \in C(R^m, R_+)$ on $\mathcal{N}_x$ and $\mathcal{N}_y$ respectively and the following conditions are satisfied
   (a) $V_y^{\mathrm{T}} g(\alpha,b,y,0) \le -\psi(y)$ $\forall(\alpha,b,y) \in R \times \mathcal{N}_x \times \mathcal{N}_y$ or ($\forall(\alpha,b,y) \in R \times \mathcal{N}_x \times \mathcal{N}_{yo}$) respectively;
   (b) $V_y^{\mathrm{T}}[g(\alpha,b,y,\mu) - g(\alpha,b,y,0)] \le \xi_1\mu^\pi\varphi(b) + \xi_2\psi(y)$ $\forall(\alpha,b,y,\mu) \in R \times \mathcal{N}_x \times \mathcal{N}_y \times \mathcal{M}$ or ($\forall(\alpha,b,y,\mu) \in R \times \mathcal{N}_x \times \mathcal{N}_{yo} \times \mathcal{M}$) respectively;
   (c) $V_\alpha + V_b^{\mathrm{T}} f(\alpha,b,y,\mu) \le \xi_3\varphi(b) + \xi_4\psi(y)$ $\forall(\alpha,b,y,\mu) \in R \times \mathcal{N}_x \times \mathcal{N}_y \times \mathcal{M}$ or ($\forall(\alpha,b,y,\mu) \in R \times \mathcal{N}_x \times \mathcal{N}_{yo} \times \mathcal{M}$) respectively.

The constants ζ_1, ζ_2, ξ_1, ξ_2 and ξ_3, ξ_4 mentioned in Assumption 3.3.1, 3.3.2 must be taken as small as possible. If the function V does not depend on x, then it is to be positive definite on $\mathcal{N}_y$ only. If, in addition $\mathcal{N}_y$ is time-invariant, then condition (c) in Assumption 3.3.2 is omitted.

Let

$$\widetilde{\mu} = \frac{1 - \xi_2}{\zeta_2 + \xi_4}.$$

This value is a lower estimate of the upper boundary of the admissible change of μ.

THEOREM 3.3.1. *In order that the equilibrium state* $(x^{\mathrm{T}}, y^{\mathrm{T}})^{\mathrm{T}} = 0$ *of system (3.2.1), (3.2.2) to be uniformly asymptotically stable, it is sufficient that conditions of Assumptions 3.3.1 and 3.3.2 be satisfied for every* $\mu \in (0, \widetilde{\mu})$ *and for* $\mu \to 0$ *as soon as the inequality*

$$1 > \zeta_1 + \xi_1 \widetilde{\mu}^{\pi - 1} + \xi_3$$

holds.

If moreover $\mathcal{N}_y \times \mathcal{N}_y = R^{m+n}$, *then the equilibrium state is uniformly asymptotically stable in the whole for every* $\mu \in (0, \widetilde{\mu})$ *and for* $\mu \to 0$.

PROOF. Let the function ν be defined by the formula $\nu = \theta + V$. Then $\nu(t, x, y) \in C^{(1,1,1)}(R \times \mathcal{N}_{xo} \times \mathcal{N}_{yo})$ and, since the conditions of Assumptions 3.3.1 and 3.3.2 are satisfied, it is decreasing and positive on $\mathcal{N}_x \times \mathcal{N}_y$. The Euler derivative $\dfrac{d\nu(t, x(t), y(t), \mu)}{dt}$ of it along the motion of system (3.2.1), (3.2.2) $z(t) = (x^{\mathrm{T}}(t), y^{\mathrm{T}}(t))^{\mathrm{T}} \neq 0$ ($z(t) = 0,\ t \in [t_0, +\infty[$) means that the equilibrium state is reachable and therefore is not considered, due to system (3.2.1), (3.2.2) is

$$\frac{d\nu}{dt} = \theta_t + \theta_x^{\mathrm{T}} f + V_t + V_x^{\mathrm{T}} f + \frac{1}{\mu} V_y^{\mathrm{T}} g.$$

The right-side part of this expression is transformed to the form

$$\begin{aligned}\frac{d\nu}{dt} &= \theta_t + \theta_x^{\mathrm{T}} f(t, x, 0, 0) + \theta_x^{\mathrm{T}} [f(t, x, y, \mu) - f(t, x, 0, 0)] + V_x^{\mathrm{T}} f(t, x, y, 0) \\ &\quad + \frac{1}{\mu} V_y^{\mathrm{T}} g(t, x, y, 0) + \frac{1}{\mu} V_y^{\mathrm{T}} [g(t, x, y, \mu) - g(t, x, y, 0)].\end{aligned}$$

Conditions (3)(a) and (3)(b) of Assumption 3.3.1 and (3)(a)–(3)(c) of Assumption 3.3.2 lead to the estimate

$$\begin{aligned}\frac{d\nu}{dt} &\leq -(1 - \zeta_1 - \xi_1 \mu^{\pi - 1} - \xi_3)\varphi(x) - \frac{1}{\mu}[1 - \xi_2 - \mu(\zeta_2 + \xi_4)]\psi(y), \\ &\forall \mu \in (0, \widetilde{\mu}) \quad \mu \to 0 \quad \forall\, (t, x, y) \in R \times \mathcal{N}_{x0} \times \mathcal{N}_{y0}.\end{aligned} \tag{3.3.1}$$

Let

$$\mathcal{N}_{0x} = \{z \colon x = 0, y \in N_{y0}\}, \qquad \mathcal{N}_{0y} = \{z \colon x \in \mathcal{N}_{x0}, y = 0\},$$
$$\mathcal{N}_0 = \mathcal{N}_{0x} \times \mathcal{N}_{0y}.$$

It is clear that

$$\mathcal{N}_x \times \mathcal{N}_y = \mathcal{N}_{x0} \times \mathcal{N}_{y0} \times \mathcal{N}_0 \times \{z \colon z = 0\}.$$

Let ν_M be a maximal positive number, for which the largest connected neighborhood $U_{\nu_M}(t)$ of point $z = 0$ is such, that

$$\nu(t, x, y) \in [0, \nu_M), \quad \forall\, (x, y) \in U_{\nu_M}(t) \quad \forall\, t \in R,$$

is a subset of the product $\mathcal{N} = \mathcal{N}_x \times \mathcal{N}_y$ for every $t \in R$. The existence of the value $\nu_M > 0$ is implied by the positive definiteness of function ν on $\mathcal{N}$ and the time-invariance of the neighborhood of point $z = 0$.

Let τ_i, τ_i^*, $t_0 \le \tau_i < \tau_i^* \le +\infty$ denote the times when $z(t) \in U_{\nu_M}(t) \setminus \mathcal{N}_0$ $\forall\, t \in (\tau_i, \tau_i^*)$, $\tau_i > t_0$ and $z(t) \in \mathcal{N}_0$ $\forall\, t \in [\tau_{i-1}^*, \tau_i]$. If $z(t_0) \in U_{\nu_M}(t_0) \setminus \mathcal{N}_0$ then $i = 0$, $\tau_0 = t_0$, $[\tau_0, \tau^*) = [t_0, \tau^*)$ is the first interval to be considered and the next is $[\tau_0^*, \tau_1]$. If $z(t_0) \in \mathcal{N}_0$, then $i = 1$, $\tau_0^* = t_0$ and $[\tau_0^*, \tau_1]$ is the first interval to be considered, and the next is (τ_1, τ_1^*). In what follows, $i \ge 0$ is an integer.

Let

$$\zeta(t; t_0, z_0, \mu) = (\chi^{\mathrm{T}}(t; t_0, z_0, \mu), \eta^{\mathrm{T}}(t; t_0, z_0, \mu))^{\mathrm{T}},$$
$$\zeta(t_0; t_0, z_0, \mu) \equiv z_0,$$

is a motion of system (3.2.1), (3.2.2) for the initial values z_0 and $t = t_0$ when $\mu > 0$.

PROPOSITION 3.3.1. *The function ν is strictly decreasing in $t \in [\tau_{i-1}^*, \tau_i]$ along motions $\zeta(t; t_0, z_0, \mu)$ of system (3.2.1), (3.2.2) for every $\mu \in (0, \tilde{\mu})$ and for $\mu \to 0$.*

PROOF.

Part 1. Let there exist a time $\hat{t} \in [\tau_{i-1}^*, \tau_i[$ when $\nu(t, x(t), y(t)) \le \nu(\hat{t}, x(\hat{t}), y(\hat{t}))$ for some $t \in (\tau_{i-1}, \tau_{i-1}^*)$. If $\hat{t} = \tau_{i-1}^*$, then there exist $\overline{\tau}_1$, $\overline{\tau}_2 \in (\tau_{i-1}, \tau_{i-1}^*)$, $\overline{\tau}_1 < \overline{\tau}_2$ such that $\nu(\tau_1, x(\tau_1), y(\tau_1)) \le \nu(\tau_2, x(\tau_2), y(\tau_2))$ due to the continuity of function V and ζ at $t \in \mathcal{T}_0$, $\forall\, t \in R$ which ensures the continuity of functions f and g. Therefore, there exists a $\tau_3 \in [\overline{\tau}_1, \overline{\tau}_2]$, when

$$\left.\frac{d\nu}{dt}\right|_{t=\tau_3} \ge 0.$$

However, this contradicts estimate (3.3.1) because of the positive definiteness of functions φ and ψ and the fact that

$$(1 - \zeta_1 - \xi_1 \mu^{\pi - 1} - \xi_3) > 0, \quad \frac{1}{\mu}[1 - \xi_2 - \mu(\zeta_2 + \xi_4)] > 0 \quad \forall\, \mu \in (0, \tilde{\mu}).$$

Hence, the equality $\hat{t} = \tau_{i-1}^*$ is impossible and a value $\hat{t} \in (\tau_{i-1}^*, \tau_i]$ is to be considered. Let $T_1 \subseteq [\tau_{i-1}^*, \tau_i)$ be a set of all times t such that $x(t) = 0$, $T_2 \subseteq [\tau_{i-1}^*, \tau_i)$ be a set of all times t, such that $y(t) = 0$. Since $z(t) = 0$ is excluded $\forall\, t \in [t_0, +\infty[$, then, by virtue of the continuity of the system motion it should be $T_1 = [\tau_{i-1}^*, \tau_i)$ or $T_2 = [\tau_{i-1}^*, \tau_i)$. To be specific, we suppose that $T_1 = [\tau_{i-1}^*, \tau_i)$. Then $\theta(t, x(t)) = \theta(t, 0)$ $\forall\, t \in T_1$ and $\nu(t, x(t), y(t)) = \nu(t, 0, y(t))$. Moreover,

$$\frac{d}{dt}\nu(t,0,y(t)) = \frac{d}{dt}V(t,0,y(t)) \leq -\frac{1}{\mu}(1 - \xi_2 - \xi_4)\psi(y(t)) \tag{3.3.2}$$
$$\forall t \in T_1, \quad \forall \mu \in (0, \widetilde{\mu}), \quad \mu \to 0.$$

This contradicts the assumption that $\hat{t} \in T_1$. Now let $T_2 = [\tau_{i-1}^*, \tau_i)$. Then $y(t) = 0$ $\forall\, t \in T_2$. Therefore

$$\nu(t, x(t), y(t)) = \nu(t, x(t), 0) \qquad \forall\, t \in T_2,$$
$$\frac{d}{dt}\nu(t, x(t), 0) \leq -(1 - \zeta_1 - \zeta_3)\varphi(x(t)) \qquad \forall\, t \in T_2,$$

that contradicts the assumption that $\hat{t} \in T_2$. In general, there exists no value $\hat{t} \in [\tau_{i-1}^*, \tau_i[$ mentioned above.

Part 2. Inequalities (3.3.1), (3.3.2), estimates of $\widetilde{\mu}$ and conditions $1 > \zeta_1 + \xi_1\widetilde{\mu}^{\pi-1} + \xi_3$, $\zeta_2 > 0$, $\xi_3 > 0$ together with the positive definiteness of functions φ and ψ prove that the function ν strictly decreases on interval $[\tau_{i-1}^*, \tau_i)$, $\tau_{i-1}^* \geq t_0$, $\forall\, i \geq 1$.

Part 3. Let there exist $\hat{t} \in [\tau_{i-1}^*, \tau_i]$ such that $\nu(t, x(t), y(t)) \geq \nu(\hat{t}, x(\hat{t}), y(\hat{t}))$ for some $t \in (\tau_i, \tau_i^*)$. Hence, there exist $\overline{\tau}_1, \overline{\tau}_2 \in (\tau_i, \tau_i^*)$, $\overline{\tau}_1 < \overline{\tau}_2$ such that $\nu(\overline{\tau}_1, x(\overline{\tau}_1), y(\overline{\tau}_1)) \leq \nu(\overline{\tau}_2, x(\overline{\tau}_2), y(\overline{\tau}_2))$ due to the continuity of $\nu(t, x(t), y(t))$ and ζ in t and because of description of Section 3.2.

Therefore, $\exists\, \overline{\tau}_3 \in [\overline{\tau}_1, \overline{\tau}_2]$ is such that $\frac{d}{dt}\nu(t, x(t), y(t))|_{t=\overline{\tau}_3} \geq 0$ and this contradicts condition (3.3.1).

The combination of assertions of Parts 1–3 proves Proposition 3.3.1.

In view of the positive definiteness of ν we establish according to the results Part 1 the uniform stability of state $z = 0$ of system (3.2.1), (3.2.2) for $\forall\, \mu \in (0, \widetilde{\mu})$ and for $\mu \to 0$. Further on, because of the positive definiteness of functions φ and ψ and the fact that $(1 - \zeta_1 - \xi_1\mu^{\pi-1} - \xi_3) > 0$ and $(1 - \zeta_1 - \xi_1\mu^{\pi-1}) > 0$ $\forall\, \mu \in (]0, \widetilde{\mu})$ as $\mu \to 0$ and due to the estimate of $\widetilde{\mu}$, $\frac{d}{dt}\nu$ is proved to be smaller than a negative definite function on $\mathcal{N}_{xo} \times \mathcal{N}_{yo}$, on $\mathcal{N}_{ox}$ and on $\mathcal{N}_{oy}$. This result together with the conditions of positive definiteness and decrease of function ν proves uniform attraction in the whole of the state $z = 0$ of system (3.2.1), (3.2.2) and completes the prove of the first assertion of the theorem. In the case when $\mathcal{N}_x \times \mathcal{N}_y = R^{n+m}$, the function ν will be radially unbounded and this together with the other conditions proves the second assertion of the theorem. This theorem is applied in the absolute stability analysis of singularly perturbed Lur'e-Postnikov systems.

3.4 Singularly Perturbed Lur'e-Postnikov Systems

Let system (3.2.1), (3.2.2) be the Lur'e-Postnikov type system (see Grujić [54])

$$\begin{aligned} \frac{dx}{dt} &= A_{11}x + A_{12}y + q_1\Phi_1(\sigma_1), \\ \sigma_1 &= c_{11}^{\mathrm{T}}x + c_{12}^{\mathrm{T}}y; \end{aligned} \tag{3.4.1}$$

$$\begin{aligned} \mu\frac{dy}{dt} &= \mu A_{21}x + A_{22}y + q_2\Phi_2(\sigma), \\ \sigma_2 &= \mu c_{21}^{\mathrm{T}}x + c_{22}^{\mathrm{T}}y. \end{aligned} \tag{3.4.2}$$

The matrices $A_{(\cdot)}$ and vectors $c_{(\cdot)}$ and $q_{(\cdot)}$ are of the appropriate dimensions. The nonlinearities Φ_i, $i = 1, 2$ are continuous, $\Phi_i(0) = 0$, and in Lur'e sectors $[0, k_i]$, $k_i \in (0, +\infty)$ satisfy the conditions

$$\frac{\Phi_i(\sigma_i)}{\sigma_i} \in [0, k_i], \quad i = 1, 2; \quad \forall\, \sigma_i \in (-\infty, +\infty)\,.$$

The nonlinearities Φ_i are considered incidentally, for which the state $x = 0$, $y = 0$ is the only equilibrium state of the degenerate system

$$\frac{dx}{dt} = A_{11}x + q_1\varphi_1(\sigma_1^0), \qquad \sigma_1^0 = c_{11}^{\mathrm{T}}x \tag{3.4.3}$$

and the system, describing the boundary layer respectively

$$\frac{dy}{dt} = A_{22}y + q_2\Phi_2(\sigma_2^0), \qquad \sigma_2^0 = c_{22}^{\mathrm{T}}y. \tag{3.4.4}$$

This assumption is valid if

$$c_{ii}^{\mathrm{T}} A_{ii}^{-1} q_i > 0, \qquad i = 1, 2.$$

We suppose the matrix A_{11} is stable, the pair (A_{11}, q_1) is controlled and there exist numbers $\psi_1 \in [0, +\infty[$ and $\varepsilon_1 \in (0, +\infty)$ such that

$$\begin{aligned} k_1^{-1} &+ \mathrm{Re}(1 + j\psi_1\omega) c_{11}^{\mathrm{T}} (A_{11} - j\omega I_1)^{-1} q_1 \\ &- \varepsilon_1 q_1^{\mathrm{T}} (A_{11}^{\mathrm{T}} + j\omega I_n)^{-1} (A_{11} - j\omega I_n)^{-1} q_1 \geq 0 \qquad \forall \omega \in [0, +\infty]. \end{aligned}$$

Then

$$\Theta(x) = \left(x^{\mathrm{T}} H_1 x + \psi_1 \int_0^{\sigma_1^0} \Phi_1(\sigma_1^0)\, d\sigma_1^0 \right)^{1/2}$$

is the Liapunov function for degenerate system (3.4.3) for any Φ_i taking the values in $[0, K_1]$, where H_1 is a solution of the equations

$$(3.4.5) \qquad A_{11}^{\mathrm{T}} H_1 + H_1 A_{11} + q_1 q_1^{\mathrm{T}} = -\varepsilon_1 I_1, \qquad h_1 + H_1 q_1 = -\sqrt{\gamma}\, q_1$$

for

$$(3.4.6) \qquad \gamma = k_1^{-1} - \zeta_1 c_{11}^{\mathrm{T}} q_1, \qquad h_1 = \frac{1}{2}(\psi_1 A_{11}^{\mathrm{T}} c_{11} + c_{11}).$$

Now we shall verify the conditions of Assumptions 3.3.1 and 3.3.2.

The verification of conditions of Assumption 3.3.1: Let H_1 and $\theta(x)$ be defined as above. Hence, the function $\theta(x)$ is decreasing positive definite on R^n and radially unbounded. We shall check up the condition (3)(a) first

(a) in this case $\theta_t = 0$ and

$$\Theta_x^{\mathrm{T}}(x) f(x, 0, 0) \leq -\frac{1}{2}\varepsilon_1 \eta_2^{-1} \|x\| \qquad \forall\, (x \neq 0) \in R^n,$$

where $\eta_2 = \Lambda^{1/2}(H_1 + \frac{1}{2}\psi_1 k_1 c_{11} c_{11}^{\mathrm{T}})$ and $\Lambda(\cdot)$ is a maximal eigenvalue of matrix $(\cdot)$. Hence

$$\varphi(x) = \eta_3 \|x\|, \qquad \eta_3 = \frac{1}{2}\varepsilon_1 \eta_2^{-1}$$

and

$$\theta_t + \theta_x^{\mathrm{T}} f(x, 0, 0) \leq -\varphi(x) \qquad \forall\, (x \neq 0) \in R^n$$

and, besides, $\mathcal{N}_x = R^n$, $\mathcal{N}_{x0} = \{x\colon x \neq 0, x \in R^n\}$;

(b) for the function $\theta(x)$ we have

$$\theta_x^{\mathrm{T}}[f(x,y,\mu) - f(x,0,0)] = \frac{1}{2}\theta(x)x^{\mathrm{T}}\left(2H_1 + \zeta\frac{\Phi_1(\sigma_1^0)}{\sigma_1^0}c_{11}c_{11}^{\mathrm{T}}\right)$$
$$\times\{A_{12}y + q_1[\Phi_1(\sigma) - \Phi_1(\sigma_1^0)]\} \le \zeta_1\varphi(x) + \zeta_2\psi(y),$$
$$\forall\, x \in \mathcal{N}_{x0}, \quad \forall\, y \in R^m \quad \forall\, \mu \in (0,1].$$

Incidentally

$$\psi(y) = \rho_3\|y\|, \qquad \zeta_1 = k_1(\eta_1\eta_3)^{-1}\eta_2\|q_1\|\,\|c_{11}\|$$

and

$$\zeta_2 = (\eta_1\rho_3)^{-1}\eta_2(k_1\,\|c_{12}\|\,\|q_1\| + \|A_{12}\|), \qquad \eta_1 = \lambda^{1/2}(H_1),$$

where $\lambda(\cdot)$ is a minimal eigenvalue of matrix $(\cdot)$. The value $\rho_3 > 0$ will be defined below. The numbers ζ_1 and ζ_2 and the functions θ, φ and ψ satisfy the conditions of Assumption 3.3.1.

The verification of the conditions of Assumption 3.3.2: We take the function $V(y) = \|y\|$ as the auxiliary function. This choice shows the alternative to the choice of the Liapunov functions. The function V is decreasing positive definite in R^n and radially unbounded. In order to verify condition (3)(a) of Assumption 3.3.2, we present the system of the boundary layer in the form suggested by Rosenbrok

$$\frac{dy}{d\tau} = D_{22}(\alpha_2)y,$$

where

$$D_{22}(\alpha_2) = A_{22} + \alpha_2(\sigma_2^0)q_2c_{22}^{\mathrm{T}}, \qquad \alpha_2(\sigma_2^0) = \frac{\Phi_2(\sigma_2^0)}{\sigma_2^0}.$$

The matrix $\hat{D}_2(\alpha_2) = D_{22}^{\mathrm{T}}(\alpha_2) + D_{22}(\alpha_2)$ is negative definite for each $(\sigma,\varphi_2) \in R \times \mathcal{N}_0([0,K_2])$ iff $D_{22}(0)$ and $D_{22}(K)$ are negative definite. In the case under consideration this assumption is fulfilled. At last $\psi(y) = \rho_3(y)$ and $V_y^{\mathrm{T}}\,g(\alpha,b,y,0) \le -\psi(y)\ \forall\,(y \neq 0) \in R^m$ ensure the satisfaction of condition (3)(a).

For condition (3)(a) we have

$$V_y^{\mathrm{T}}[g(\alpha,b,y,\mu) - g(\alpha,b,y,0)] = \frac{1}{V}y^{\mathrm{T}}\{\mu A_{21}b + q_2[\Phi_2(\sigma_2) - \Phi_2(\sigma_2^0)]\}$$
$$\forall\,(y \neq 0) \in R^m.$$

Let

$$\xi_1 = 2\varepsilon_1\eta_2 \sup_{\alpha\in[0,k_2]} ||A_{21} + \alpha q_2 c_{21}^{\mathrm{T}}||,$$
$$\xi_2 = k_2||q_2 c_{21}^{\mathrm{T}}||\rho_3^{-1}.$$

We assume that $\xi_2 < 1$, then

$$V_y^{\mathrm{T}}[g(\alpha, b, y, \mu) - g(\alpha, b, 0, 0)] \leq \xi_1\mu\varphi(b) + \xi_2\psi(y)$$
$$\forall\,(\alpha, b, y, \mu) \in R \times R^n \times R^m \times (0, \infty).$$

This corresponds to condition (3) in Assumption 3.3.2 for $\pi = 1$. Checking up condition (3)(c) we take into account that $V_\alpha \equiv 0$ and $V_b \equiv 0$ and, therefore, $\xi_3 = 0$ and $\xi_4 = 0$. The lower estimate of the upper bound of the parameter μ changes and has the form

$$\tilde{\mu} = \frac{1 - \xi_2}{\zeta_2}.$$

Now the inequality $1 > \zeta_1 + \xi_1$ ensures absolute stability of the state $z = (x^{\mathrm{T}}, y^{\mathrm{T}})^{\mathrm{T}} = 0$ of system (3.4.1), (3.4.2).

EXAMPLE 3.4.1. Let

$$A_{11} = \begin{pmatrix} 0 & 1 \\ -1 & -2 \end{pmatrix}, \qquad q_1 = \begin{pmatrix} 0 \\ 10^{-1} \end{pmatrix}, \qquad c_{11} = \begin{pmatrix} -10^{-2} \\ 0 \end{pmatrix},$$
$$A_{12} = I, \qquad c_{12} = \begin{pmatrix} 1 \\ 1 \end{pmatrix}, \qquad k_1 = 2$$

and

$$A_{21} = 10^{-3} I_2, \qquad c_{21} = \begin{pmatrix} 10^{-3} \\ 0 \end{pmatrix}, \qquad k_2 = 1,$$
$$A_{22} = \begin{pmatrix} -4 & 1 \\ 1 & -4 \end{pmatrix}, \qquad q_2 = \begin{pmatrix} 1 \\ 1 \end{pmatrix}, \qquad c_{22} = \begin{pmatrix} 1 \\ 0 \end{pmatrix}.$$

In this example we take $\psi_1 = 1$, $\varepsilon_1 = 10^{-1}$ so that

$$\frac{1}{k_1} + \operatorname{Re}(1 + j\psi_1\omega)c_{11}^{\mathrm{T}}(A_{11} - j\omega I_2)^{-1}q_1$$
$$- \varepsilon_1 q_1^{\mathrm{T}}(A_{11}^{\mathrm{T}} + j\omega I_2)^{-1}(A_{11} - j\omega I_2)^{-1}q_1 \equiv \frac{1}{k_1} > 2.$$

Further

$$g_1 = \begin{pmatrix} 0 \\ 0 \end{pmatrix}, \qquad H_1 = \begin{pmatrix} h_{11} & h_{12} \\ h_{21} & h_{22} \end{pmatrix}$$

is defined from the equation

$$\begin{pmatrix} 0 & -1 \\ -1 & 2 \end{pmatrix} \begin{pmatrix} h_{11} & h_{12} \\ h_{21} & h_{22} \end{pmatrix} + \begin{pmatrix} h_{11} & h_{21} \\ h_{12} & h_{22} \end{pmatrix} \begin{pmatrix} 0 & 1 \\ -1 & -2 \end{pmatrix} = -\frac{1}{10} \begin{pmatrix} 1 & 0 \\ 0 & 1 \end{pmatrix}$$

in the form

$$H_1 = \frac{1}{20} \begin{pmatrix} 3 & 1 \\ 1 & 1 \end{pmatrix}.$$

Hence $\eta_1 = 0.16$ and $\eta_2 = 0.45$. The matrix $\hat{D}_{22}(\alpha_2)$ reads

$$\hat{D}_{22}(\alpha_2) = \begin{pmatrix} -8 + 2\alpha_{22} & 2 + \alpha_{22} \\ 2 + \alpha_{22} & -8 \end{pmatrix}.$$

The matrices $\hat{D}_{22}$ and $\hat{D}_{22}(1)$ are negative definite. Finally, $\zeta_1 = 0.05$, $\zeta_2 = 1.88$, $\xi_1 = 0.02$ and $\xi_2 = 0.002$. Therefore $\widetilde{\mu} = 0.52$. Since $\zeta_1 + \xi_1 = 0.53$ is smaller then 1, the state $z = (x^{\mathrm{T}}, y^{\mathrm{T}})^{\mathrm{T}} = 0$ of the system defined in this example is absolutely stable for each $\mu \in (0; \widetilde{\mu})$, i.e. $\mu \in (0;\ 0.52)$ on $\mathcal{N}_0(L)$, $L = [0, K]$, $K = \operatorname{diag}(2, 1)$. The advantage of the separation of the time-scales in this example is that the order of the system in question is diminished. Namely, instead of the system of the fourth order one investigates two systems of the second order and verifies the inequality $1 > \zeta_1 + \xi_1$. Moreover, the lowering of the order of the systems simplifies the construction of the Liapunov functions.

However, the dimensions m and n of the reduced systems (3.2.5) and (3.4.3) and the systems of the boundary layer (3.2.6) and (3.4.4) are high enough so that one faces the problem of the lowering their order again.

3.5 The Property of Having a Fixed Sign of Matrix-Valued Function

Alongside the system (3.2.1)–(3.2.2) we shall consider first a more simple case.

3.5.1 Case A.

Let perturbed motion equations be given in the form

$$\frac{dx}{dt} = f(t, x, y), \tag{3.5.1}$$

$$\mu \frac{dy}{dt} = g(t, x, y), \tag{3.5.2}$$

where $x \in R^n$, $y \in R^m$, $f \in C(R \times R^n \times R^m, R^n)$ and $g \in C(R \times R^n \times R^m, R^m)$, $\mu \in \mathcal{M}$. For $\mu = 0$ we obtain from (3.5.1) and (3.5.2)

$$\frac{dx}{dt} = f(t, x, y), \tag{3.5.3}$$

$$0 = g(t, x, y), \tag{3.5.4}$$

Assume that $g(t, x, y)$ vanishes if and only if $y = 0$. Then we get from system (3.5.3)–(3.5.4) the system

$$\frac{dx}{dt} = f(t, x, 0), \tag{3.5.5}$$

which describes slow motions in system (3.5.1)–(3.5.2). The quick system (boundary layer) corresponding to system (3.5.2) has the form

$$\frac{dy}{dt} = g(\alpha, b, y), \tag{3.5.6}$$

where $\tau = (t - t_0)\mu^{-1}$, α and b are the same as in system (3.2.6).

We define the functions

$$f^*(t, x, y) = f(t, x, y) - f(t, x, 0);$$
$$g^*(t, x, y) = g(t, x, y) - g(\alpha, b, y).$$

and represent system (3.5.1)–(3.5.2) as

$$\begin{aligned} \frac{dx}{dt} &= f(t, x, 0) + f^*(t, x, y), \\ \mu \frac{dy}{dt} &= g(\alpha, b, y) + g^*(t, x, y). \end{aligned} \tag{3.5.7}$$

In order to investigate systems (3.5.1) and (3.5.2) with subsystems (3.5.5) and (3.5.6) we shall consider the matrix-valued function

$$U(t,x,y,\mu) = \begin{pmatrix} v_{11}(t,x) & v_{12}(t,x,y,\mu) \\ v_{21}(t,x,y,\mu) & v_{22}(t,y,\mu) \end{pmatrix}. \tag{3.5.8}$$

The elements v_{11} and v_{22} of matrix U corresponds to the subsystems (3.5.5) and (3.5.6) and functions $v_{12} = v_{21}$ are responsible for the interconnections of the subsystems. Using the matrix-valued function $U(t,x,y,\mu)$ we introduce the scalar function

$$V(t,x,y,\mu) = w^{\mathrm{T}} U(t,x,y,\mu) w, \tag{3.5.9}$$

where $w \in R^2$.

DEFINITION 3.5.1. The matrix-valued function $U\colon R_+ \times R^m \times R^n \times \mathcal{M} \to R^{2\times 2}$ is referred to as

(i) *positive definite*, iff there exist connected neighborhoods $\mathcal{N}_x$ and $\mathcal{N}_y$ of points $x = 0$ and $y = 0$ $\mathcal{N}_x \subseteq R^m$, $\mathcal{N}_y \subseteq R^n$ such that
  (a) $U \in C(R_+ \times \mathcal{N}_x \times \mathcal{N}_y \times \mathcal{M}, R^{2\times 2})$
  (b) $U(t,0,0,\mu) = 0 \ \ \forall\, t \in R_+, \ \forall\, \mu \in \mathcal{M}$;
  (c) $w^{\mathrm{T}} U(t,x,y,\mu) w > u(x,y) \ \ \forall\, (t, x \neq 0, y \neq 0, w \neq 0) \in R_+ \times \mathcal{N}_x \times \mathcal{N}_y \times \mathcal{M} \times R^2$;

(ii) *positive definite on S* iff the conditions of Definition 3.5.1, (i) are satisfied on $\mathcal{N}_x \times \mathcal{N}_y = S$;

(iii) *positive definite in the whole*, iff all conditions of Definition 3.5.1, (i) are satisfied for $\mathcal{N}_x \times \mathcal{N}_y = R^m \times R^n$.

REMARK 3.5.1. It can be easily seen that this definition of the property of having a fixed sign of matrix-valued function U agrees with the well-known notions such as

(i) positive definiteness of the numerical matrix;
(ii) positive definiteness of the scalar Liapunov function;
(iii) conceptual applicability of function (3.5.9) in the construction of the direct Liapunov's method of motion stability investigation.

In many problems of stability it is sufficient to use a fixed vector $\eta \in R^2$ (or $\eta \in R^2_+$) instead of the vector in formula (3.5.9).

Let $\eta = (\eta_1, \eta_2)^{\mathrm{T}}$, $\eta_i > 0$, $i = 1, 2$ then

$$V(t,x,y,\mu) = \eta^{\mathrm{T}} U(t,x,y,\mu) \eta. \tag{3.5.10}$$

DEFINITION 3.5.2. The matrix-valued function $U \in C(R_+ \times R^m \times R^n \times \mathcal{M}, R^{2\times 2})$ is called

(i) *η-positive definite*, iff there exist connected neighborhoods $\mathcal{N}_x$ and $\mathcal{N}_y$ of points $x = 0$ and $y = 0$, $\mathcal{N}_x \subseteq R^m$, $\mathcal{N}_y \subseteq R^n$ such that

  (a) $U \in C(R_+ \times \mathcal{N}_x \times \mathcal{N}_y \times \mathcal{M}, R^{2\times 2})$;

  (b) $U(t, 0, 0, \mu) = 0 \ \forall\, t \in R_+, \ \forall\, \mu \in \mathcal{M}$;

  (c) $\eta^{\mathrm{T}} U(t, x, y, \mu)\eta > u(x, y) \ \forall\,(x \neq 0, y \neq 0) \in \mathcal{N}_x \times \mathcal{N}_y$, $\forall\,(t, \mu) \in R_+ \times \mathcal{M}$;

(ii) *η-positive definite on S*, iff all conditions of Definition 3.5.2 (i) are satisfied for $\mathcal{N}_x \times \mathcal{N}_y = S$;

(iii) *η-positive definite in the whole*, iff all conditions of Definition 3.5.2 (i) are satisfied for $\mathcal{N}_x \times \mathcal{N}_y = R^m \times R^n$.

Definition 3.5.2 agrees with points (i)–(iii) of Remark 3.5.1. In particular, the vector η can be unique, i.e. $\eta_i = 1$ and $i = 1, 2$.

REMARK 3.5.2. The definitions of positive semi-definiteness and η-positive semi-definiteness of matrix-valued function U are introduced on the basis of Definitions 3.5.1 and 3.5.2, in conditions (c) of which the $u(x, y)$ should be replaced by ≥ 0.

REMARK 3.5.3. Functions (3.5.9) and (3.5.10) can be also constructed in the form

$$V(t, x, y, \mu, w) = w^{\mathrm{T}} U^{\mathrm{T}} U w, \qquad w \in R^2$$

or

$$V(t, x, y, \mu) = \eta^{\mathrm{T}} U^{\mathrm{T}} U \eta, \qquad \eta \in R^2_+.$$

In addition, the requirements to the elements of matrix-valued function U satisfying the conditions of Definitions 3.5.1 and 3.5.2 can be weakened.

The algebraic conditions of the property of having a fixed sign of function (3.5.10) are formulated in terms of the assumptions on elements $v_{ij}(t, \cdot)$ of the matrix-valued function U.

ASSUMPTION 3.5.1. There exist functions $v_{11}(t, x)$, $v_{22}(t, y, \mu)$, $v_{12}(t, x, y, \mu)$, functions φ_i and ψ_i of class $K(KR)$, $i = 1, 2$ and constants $\underline{\alpha}_{ii} > 0$, $\overline{\alpha}_{ii} > 0$, $i = 1, 2$ and $\underline{\alpha}_{12}$, $\overline{\alpha}_{12}$ such that

(1) $\underline{\alpha}_{11}\varphi_1^2(\|x\|) \leq v_{11}(t, x) \leq \overline{\alpha}_{11}\varphi_2^2(\|x\|)$
$\forall (t, x) \in R_+ \times \mathcal{N}_x$ $(\forall (t, x) \in R_+ \times R^m)$;

(2) $\mu\underline{\alpha}_{22}\psi_1^2(\|y\|) \leq v_{22}(t, y, \mu) \leq \mu\overline{\alpha}_{22}\psi_2^2(\|y\|)$

$\forall (t, y, \mu) \in R_+ \times \mathcal{N}_y \times \mathcal{M}\ (R_+ \times R^n \times \mathcal{M})$;

(3) $\mu \underline{\alpha}_{12} \varphi_1(\|x\|)\psi_1(\|y\|) \le v_{12}(t, x, y, \mu) \le \mu \overline{\alpha}_{12} \varphi_2(\|x\|)\psi_2(\|y\|)$
$\forall (t, x, y, \mu) \in R_+ \times \mathcal{N}_x \times \mathcal{N}_y \times \mathcal{M} \quad (R_+ \times R^m \times R^n \times \mathcal{M})$;

(4) $v_{12}(t, x, y, \mu) = v_{21}(t, x, y, \mu)$
$\forall (t, x, y, \mu) \in R_+ \times \mathcal{N}_x \times \mathcal{N}_y \times \mathcal{M}\ (R_+ \times R^m \times R^n \times \mathcal{M})$.

The following assertion is valid.

PROPOSITION 3.5.1. *If for the elements $v_{ij}(t, \cdot)$, $i, j = 1, 2$, of matrix-valued function (3.5.8) the conditions of Assumption 3.5.1 are satisfied, then function (3.5.10) satisfies the bilateral estimate*

$$\begin{gathered} u_1^{\mathrm{T}} A(\mu) u_1 \le V(t, x, y, \mu) \le u_2^{\mathrm{T}} B(\mu) u_2 \\ \forall (t, x, y, \mu) \in R_+ \times \mathcal{N}_x \times \mathcal{N}_y \times \mathcal{M}(R_+ \times R^m \times R^n \times \mathcal{M}), \end{gathered} \tag{3.5.11}$$

where $u_1^{\mathrm{T}} = (\varphi_1, \psi_1)$, $u_2^{\mathrm{T}} = (\varphi_2, \psi_2)$,

$$A(\mu) = H^{\mathrm{T}} A_1(\mu) H, \quad B(\mu) = H^{\mathrm{T}} A_2(\mu) H, \quad H = \mathrm{diag}(\eta_1, \eta_2); \quad \eta_1, \eta_2 > 0,$$

$$A_1(\mu) = \begin{pmatrix} \underline{\alpha}_{11} & \mu \underline{\alpha}_{12} \\ \mu \underline{\alpha}_{21} & \mu \underline{\alpha}_{22} \end{pmatrix}, \qquad A_2(\mu) = \begin{pmatrix} \overline{\alpha}_{11} & \mu \overline{\alpha}_{12} \\ \mu \overline{\alpha}_{21} & \mu \overline{\alpha}_{22} \end{pmatrix},$$

$$\underline{\alpha}_{12} = \underline{\alpha}_{21}; \quad \overline{\alpha}_{12} = \overline{\alpha}_{21}.$$

PROOF. We get the estimate from above in inequality (3.5.11). In view of expression (3.5.10) and inequalities (1)–(4) of Assumption 3.5.1 we have

$$V(t, x, y, \mu) \le \begin{pmatrix} \eta_1 & \varphi_2 \\ \eta_2 & \psi_2 \end{pmatrix}^{\mathrm{T}} \begin{pmatrix} \overline{\alpha}_{11} & \mu \overline{\alpha}_{12} \\ \mu \overline{\alpha}_{21} & \mu \overline{\alpha}_{22} \end{pmatrix} \begin{pmatrix} \eta_1 & \varphi_2 \\ \eta_2 & \psi_2 \end{pmatrix}$$

or

$$V(t, x, y, \mu) \le \begin{pmatrix} \varphi_2 \\ \psi_2 \end{pmatrix}^{\mathrm{T}} \begin{pmatrix} \eta_1 & 0 \\ 0 & \eta_2 \end{pmatrix}^{\mathrm{T}} \begin{pmatrix} \overline{\alpha}_{11} & \mu \overline{\alpha}_{12} \\ \mu \overline{\alpha}_{21} & \mu \overline{\alpha}_{22} \end{pmatrix} \begin{pmatrix} \eta_1 & 0 \\ 0 & \eta_2 \end{pmatrix} \begin{pmatrix} \varphi_2 \\ \psi_2 \end{pmatrix}.$$

Hence, in view of the designations adopted in Proposition 3.5.1 we get the estimate from above in inequality (3.5.11). The estimate from below is obtained in the same way.

3.6 Matrix-Valued Liapunov Function

The conception of the property of having a fixed sign of matrix-valued function admitted in Definitions 3.5.1 and 3.5.2 allows us to introduce the matrix-valued Liapunov function in the following way. We introduce the designations

$$D^*V(t,x,y,\mu,w) = w^{\mathrm{T}} D^*U(t,x,y,\mu)w,$$
$$D^*U(t,x,y,\mu) = [D^*v_{ij}(t,\cdot)]; \qquad i,j = 1,2.$$

The sign D^*U shows that both derivatives D^+U and D_+U can be used, where

$$\begin{aligned} D^+U(t,x,y,\mu) &= \limsup\{[U(t+\theta, x(t+\theta,\cdot), y(t+\theta,\cdot),\mu) \\ &\quad - U(t,x,y,\mu)]\theta^{-1}: \theta \to 0^+\}; \\ D_+U(t,x,y,\mu) &= \liminf\{[U(t+\theta, x(t+\theta,\cdot), y(t+\theta,\cdot),\mu) \\ &\quad - U(t,x,y,\mu)]\theta^{-1}: \theta \to 0^+\}. \end{aligned}$$

In this notation D^+U (D_+U) is the upper (lover) right-side Dini derivative of matrix-valued function U relatively (t,x,y).

DEFINITION 3.6.1. Matrix-valued function $U\colon R_+ \times \mathcal{N}_x \times \mathcal{N}_y \times \mathcal{M} \to R^{2\times 2}$ is referred to as

(i) *matrix-valued Liapunov function of the $\mathcal{S}(w)$ type*, if
 (a) the matrix-valued function $U(t,x,y,\mu)$ is positive definite and decreasing on $R_+ \times \mathcal{N}_x \times \mathcal{N}_y \times \mathcal{M} \to R^{2\times 2}$;
 (b) the matrix-valued function $D^*U(t,x,y,\mu)$ is nonpositive on $R_+ \times \mathcal{N}_x \times \mathcal{N}_y$ for $\mu \in (0,\mu_0)$ and as $\mu \to 0$ and $D^*U(t,0,0,\mu) = 0$ for all $t \in R_+$;
(ii) *matrix-valued Liapunov function of $AS(w)$ type*, if
 (a) the matrix-valued function $U(t,x,y,\mu)$ is positive definite and decreasing on $R_+ \times \mathcal{N}_x \times \mathcal{N}_y \times \mathcal{M}$;
 (b) the matrix-valued function $D^*U(t,x,y,\mu)$ is strictly negative on $R_+ \times \mathcal{N}_{x0} \times \mathcal{N}_{y0}$ for $\mu \in (0,\mu_0)$ and for $\mu \to 0$ and $D^*(t,0,0,\mu) = 0$ for $t \in R_+$, $\mathcal{N}_{x0} = \{(x \neq 0) \in \mathcal{N}_x\}$, $\mathcal{N}_{y0} = \{(y \neq 0) \in \mathcal{N}_y\}$.
(iii) *matrix-valued Liapunov-Chetayev function of $NS(w)$ type*, if there exist a $t_0 \in (\tau,\infty)$, $\tau \in R$, some value $\varepsilon > 0$ $(\overline{B}_\varepsilon \subset \mathcal{N}_x \times \mathcal{N}_y)$

and an open set $D \in B_\varepsilon$ such that on $[t_0, \infty) \times D$ the following conditions are satisfied

(a) $0 < U(t,x,y,\mu) \le Q < \infty$ component wise, there Q is a 2×2 matrix;

(b) $w^{\mathrm{T}} D^+ U(t,x,y,\mu) w \ge a(V(t,x,y,\mu,w))$ for $\mu \in (0, \mu_0)$ and for $\mu \to 0$, where a is of class K, and moreover

(c) $(x = 0, y = 0) \in \partial\mathcal{D}$;

(d) $U(t,x,y,\mu) = 0$ on $[t_0, \infty[\times (\partial\mathcal{D} \cap B_\varepsilon)$.

The definitions of η – matrix-valued function are formulated in a similar manner, using the definitions of η-positive definiteness of matrix-valued function $U(t,x,y,\mu)$.

ASSUMPTION 3.6.1. There exist

(1) functions φ_i, ψ_i, $i = 1,2$ and v_{sk}; s, $k = 1,2$ mentioned in Assumption 3.5.1 and, moreover

(a) function $v_{11}(t,x) \in C(R_+ \times \mathcal{N}_{x0}, R_+)$;

(b) function $v_{22}(t,y,\mu) \in C(R_+ \times \mathcal{N}_{y0} \times \mathcal{M}, R_+)$;

(c) function $v_{12}(t,x,y,\mu) \in C(R_+ \times \mathcal{N}_{x0} \times \mathcal{N}_{y0} \times \mathcal{M}, R)$,

(2) constants ρ_{ij} $(i = 1,2;\ j = 1,\dots,8)$ and the following conditions are satisfied

(a) $D_t^+ v_{11} + (D_x^+ v_{11})^{\mathrm{T}} f(t,x,0) \le \rho_{11}\varphi_2^2(\|x\|)$
$\forall\,(t,x) \in R_+ \times \mathcal{N}_x$;

(b) $D_t^+ v_{22} + (D_y^+ v_{22})^{\mathrm{T}} g(\alpha,b,y,0) \le \mu\rho_{21}\psi_2^2(\|y\|)$
$\forall(\alpha,b,y,\mu) \in R_+ \times \mathcal{N}_x \times \mathcal{N}_y \times \mathcal{M}$;

(c) $(D_x^+ v_{11})^{\mathrm{T}}[f(t,x,y) - f(t,x,0)] \le \rho_{12}\varphi_2^2(\|x\|)$
$+\rho_{13}\varphi_2(\|x\|)\psi_2(\|y\|)$ $\forall(t,x,y) \in R_+ \times \mathcal{N}_x \times \mathcal{N}_y$;

(d) $(D_y^+ v_{22})^{\mathrm{T}}[g(\alpha,b,y,\mu) - g(\alpha,b,y,0)] \le \rho_{22}\psi_2^2(\|y\|)$
$+\mu\rho_{23}\varphi_2(\|x\|)\psi_2(\|y\|)$ $\forall(t,b,y,\mu) \in R_+ \times \mathcal{N}_x \times \mathcal{N}_y \times \mathcal{M}$;

(e) $D_t^+ v_{12} + (D_x^+ v_{12})^{\mathrm{T}} f(t,x,0) \le \mu\rho_{14}\varphi_2^2(\|x\|)$
$+\mu\rho_{15}\varphi_2(\|x\|)\psi_2(\|y\|)$ $\forall(t,x,y,\mu) \in R_+ \times \mathcal{N}_x \times \mathcal{N}_y \times \mathcal{M}$;

(f) $(D_x^+ v_{12})^{\mathrm{T}}[f(t,x,y) - f(t,x,0)] \le \mu\rho_{16}\varphi_2^2(\|x\|)$
$+\mu\rho_{17}\varphi_2(\|x\|)\psi_2(\|y\|) + \mu\rho_{18}\psi_2^2(\|y\|)$
$\forall(t,x,y,\mu) \in R_+ \times \mathcal{N}_x \times \mathcal{N}_y \times \mathcal{M}$;

(g) $(D_y^+ v_{12})^{\mathrm{T}} g(\alpha,b,y,0) \le \mu\rho_{24}\psi_2^2(\|y\|) + \mu\rho_{25}\varphi_2(\|x\|)\psi_2(\|y\|)$
$\forall(t,x,y,\mu) \in R_+ \times \mathcal{N}_x \times \mathcal{N}_y \times \mathcal{M}$;

(h) $(D_y^+ v_{12})^{\mathrm{T}}[g(\alpha, b, y, \mu) - g(\alpha, b, y, 0)] \le \mu\rho_{26}\varphi_2^2(\|x\|)$
$+\mu\rho_{27}\varphi_2(\|x\|)\psi_2(\|y\|) + \mu\rho_{28}\psi_2^2(\|y\|)$
$\forall (t, b, y, \mu) \in R_+ \times \mathcal{N}_x \times \mathcal{N}_y \times \mathcal{M}$.

PROPOSITION 3.6.1. *If all conditions of Assumption 3.6.1 are satisfied, then for the upper right Dini derivative of function (3.5.10) the upper estimate*

$$(3.6.1)\quad D^+V(t, x, y, \mu) \le u_2^{\mathrm{T}} C(\mu) u_2 \qquad \forall (t, x, y, \mu) \in R_+ \times \mathcal{N}_x \times \mathcal{N}_y \times \mathcal{M}$$

is satisfied, where

$$C(\mu) = [c_{ij}(\mu)], \quad c_{12}(\mu) = c_{21}(\mu); \quad i, j = 1, 2,$$

and

$$c_{11}(\mu) = \eta_1^2(\rho_{11} + \rho_{12}) + 2\eta_1\eta_2(\mu\rho_{14} + \mu\rho_{16} + \rho_{26});$$
$$c_{22}(\mu) = \eta_2^2(\rho_{22} + \rho_{21}) + 2\eta_1\eta_2(\mu\rho_{18} + \mu\rho_{24} + \rho_{28});$$
$$c_{12}(\mu) = \frac{1}{2}(\eta_1^2\rho_{13} + \eta_2^2\rho_{23}) + \eta_1\eta_2(\mu\rho_{15} + \rho_{25} + \mu\rho_{17} + \rho_{27}).$$

PROOF. In view of the fact that

$$D^+V(t, \cdot) = \eta^{\mathrm{T}} D^+[v_{ij}(t, \cdot)]\eta, \qquad i, j = 1, 2$$

the estimates (a)–(h) for the elements of matrix $U(t, x, , \mu)$ lead to inequality (3.6.1).

We introduce the values μ_j, $j = 1, \dots, 4$, μ_0, μ^* by the formulas

$$\mu_1 = -\frac{2\eta_2\rho_{26} + \eta_1(\rho_{11} + \rho_{12})}{2\eta_2(\rho_{14} + \rho_{16})};$$
$$\mu_2 = -\frac{\eta_2(\rho_{21} + \rho_{22}) + 2\eta_1(\rho_{24} + \rho_{28})}{2\eta_1\rho_{18}};$$
$$\mu_3 = \frac{-b + \sqrt{b^2 - 4ac}}{2a}, \qquad \mu_4 = \frac{\underline{\alpha}_{11}\underline{\alpha}_{22}}{\underline{\alpha}_{12}^2},$$
$$\mu_0 = \min(\mu_1, \mu_2, \mu_3), \qquad \mu^* = \min(\mu_0, \mu_4).$$

Besides,

$$
\begin{aligned}
a &= 4\eta_1^2\eta_2^2[(\rho_{15}+\rho_{17})^2 - \rho_{18}(\rho_{14}+\rho_{16})];\\
b &= \left[\frac{1}{2}\eta_1^2\rho_{13} + \frac{1}{2}\eta_2^2\rho_{23} + \eta_1\eta_2(\rho_{25}+\rho_{27})\right] 2\eta_1\eta_2(\rho_{15}+\rho_{17})\\
&\quad - 2\eta_1\eta_2\rho_{18}[\eta_1^2(\rho_{11}+\rho_{12}) + 2\eta_1\eta_2\rho_{26}]\\
&\quad - 2\eta_1\eta_2(\rho_{14}+\rho_{16})[\eta_2^2(\rho_{21}+\rho_{22}) + 2\eta_1\eta_2(\rho_{24}+\rho_{28})];\\
c &= \left[\frac{1}{2}\eta_1^2\rho_{13} + \frac{1}{2}\eta_2^2\rho_{23} + \eta_1\eta_2(\rho_{25}+\rho_{27})\right]^2\\
&\quad - [\eta_1^2(\rho_{11}+\rho_{12}) + 2\eta_1\eta_2\rho_{26}][\eta_2^2(\rho_{21}+\rho_{22}) + 2\eta_1\eta_2(\rho_{24}+\rho_{28})].
\end{aligned}
$$

Should $\mu_0 > 1$, we shall consider $\mu \in (0, 1]$.

PROPOSITION 3.6.2. *The matrix $C(\mu)$ is negative definite for every $\mu \in (0, \mu_0)$ and for $\mu \to 0$, provided that*

(a) $\rho_{14} + \rho_{16} > 0$;
(b) $2\eta_2\rho_{26} + \eta_1(\rho_{11} + \rho_{12}) < 0$;
(c) $\eta_2(\rho_{26} + \rho_{22}) + 2\eta_1(\rho_{24} + \rho_{28}) < 0$;
(d) $\rho_{18} > 0$;
(e) $a > 0$;
(f) $c < 0$.

PROOF. Conditions (a) and (b) imply that $c_{11} < 0$ for every $\mu \subset (0, \mu_1)$ and $\mu \to 0$; conditions (c) and (d) imply that $c_{22} < 0$ for every $\mu \in (0, \mu_2)$ and for $\mu \to 0$; and conditions (e) and (f) imply that $c_{11}c_{22} - c_{12}^2 > 0$ for every $\mu \in (0, \mu_3)$ and for $\mu \to 0$.

All these conditions hold for every $\mu \in (0, \mu_0)$ and for $\mu \to 0$, where $\mu_0 = \min(\mu_1, \mu_2, \mu_3)$. The conditions are sufficient for the matrix $C(\mu)$ negative definite.

REMARK 3.6.1. If for conditions (a)–(c), (e) and (f) Proposition 3.6.2 is satisfied and $\rho_{18} \leq 0$, then its assertion is true for $\mu_0 = \min(\mu_1, \mu_3)$.

REMARK 3.6.2. If for conditions (b)–(f) Proposition 3.6.2 is satisfied and $\rho_{14} + \rho_{16} \leq 0$, then its assertion is true for $\mu_0 = \min(\mu_2, \mu_3)$.

REMARK 3.6.3. If for conditions (b), (c), (e) and (f) Proposition 3.6.2 is satisfied and $\rho_{18} \leq 0$, $\rho_{14} + \rho_{16} \leq 0$, then its assertion is true for $\mu_0 = \mu_3$.

We note that the quadratic form $u_2^{\mathrm{T}} C(\mu) u_2$ is given in the cone R_+^2 formed by the functions (φ_2, ψ_2). Therefore the following result is valid.

PROPOSITION 3.6.3. *The matrix $C(\mu)$ is conditionally negative definite, i.e. $u_2^T C(\mu) u_2 < 0$ for $u_2 \in R_+^2 \setminus 0$, for every $\mu \in (0, \mu_0)$ and for $\mu \to 0$, if*

$$\max(\det[-C(\mu)], c_{12}(\mu)) > 0$$

for every $\mu \in (0, \mu_0)$ and for $\mu \to 0$.

Estimates (3.5.11) and (3.6.1) allow us to formulate the generalizations of the classical results on stability and instability of unperturbed motion of system (3.5.1), (3.5.2) as follows.

3.7 General Theorems on Stability and Instability in Case A

The equilibrium state $(x^T, y^T)^T = 0$ of system (3.5.1), (3.5.2) is investigated by means of function (3.5.10) being a special case of function (3.5.9). Estimates (3.5.11) and (3.6.1) allows us to formulate algebraic condi tions ensuring the presence of some properties of the equilibrium state $(x^T, y^T)^T = 0$.

THEOREM 3.7.1. *Let the motion $(x^T(t; t_0, x_0, \mu), y^T(t; t_0, y_0, \mu))^T$ of system (3.5.1), (3.5.2) be continuous for $(t_0, x_0, y_0) \in R_+ \times \mathcal{N}_x \times \mathcal{N}_y$ and $\mu \in \mathcal{M}^0 \subset \mathcal{M}$. In order that the equilibrium state $(x^T, y^T)^T = 0$ of system (3.5.1), (3.5.2) be uniformly stable for every $\mu \in (0, \mu_0)$ and for $\mu \to 0$ it is necessary that all conditions of Assumptions 3.5.1 and 3.6.1 be satisfied and it is sufficient that*

1. *the matrices $A_1(\mu) + A_1^T(\mu)$ and $A_2(\mu) + A_2^T(\mu)$ be conditionally positive;*
2. *the matrix $C(\mu)$ be non-positive for every $\mu \in (0, \mu_0)$ and for $\mu \to 0$.*

If in addition, $\mathcal{N}_x \times \mathcal{N}_y = R^{m+n}$, then the equilibrium state $(x^T, y^T)^T = 0$ is uniformly stable in the whole for every $\mu \in (0, \mu_0)$ and for $\mu \to 0$.

PROOF. Estimate (3.5.11) implies that if Assumption 3.5.1 and condition (1) of Theorem 3.7.1 hold, the function $V(t, x, y, \mu)$ is definite positive and decreasing. The conditions of Assumption 3.6.1 and condition (2) of Theorem 3.7.1 ensure nonpositiveness of function $D^+V(t, x, y, \mu)$ on $R_+ \times \mathcal{N}_x \times \mathcal{N}_y$ for every $\mu \in (0, \mu_0)$ and for $\mu \to 0$. The combination of this conditions is equivalent to the conditions of Liapunov's theorem on stability of the equilibrium state $(x^{\mathrm{T}}, y^{\mathrm{T}})^{\mathrm{T}} = 0$ (see Liapunov [101], and Grujić, Martynyuk and Ribbens-Pavella [57]).

If $\mathcal{N}_x \times \mathcal{N}_y = R^{m+n}$, then functions (φ_i, ψ_i) belong to class KR and estimates (3.5.11) and (3.6.1) are satisfied for all $(x, y) \in R^{m+n}$. Together with conditions (1) and (2) of Theorem 3.7.1 this ensures stability in the whole of the state $(x^{\mathrm{T}}, y^{\mathrm{T}})^{\mathrm{T}} = 0$.

The theorem is proved.

THEOREM 3.7.2. *Let the motion $(x^{\mathrm{T}}(t; t_0, x_0, \mu), y^{\mathrm{T}}(t; t_0, y_0, \mu))^{\mathrm{T}}$ of system (3.5.1) and (3.5.2) be continuous for $(t_0, x_0, y_0) \in R_+ \times \mathcal{N}_x \times \mathcal{N}_y$ and $\mu \in \mathcal{M}^0 \subset \mathcal{M}$. For the equilibrium state $(x^{\mathrm{T}}, y^{\mathrm{T}})^{\mathrm{T}} = 0$ of system (3.5.1), (3.5.2) be uniform asymptotically stable for every $\mu \in (0, \mu_0)$ and for $\mu \to 0$ it is necessary that all conditions of Assumptions 3.3.1 and 3.3.2 be satisfied and it is sufficient that*

(1) *the matrices $A_1(\mu) + A_1^{\mathrm{T}}(\mu)$ and $A_2(\mu) + A_2^{\mathrm{T}}(\mu)$ be conditionally positive;*
(2) *the matrix $C(\mu)$ be conditionally negative for every $\mu \in (0, \mu_0)$ and for $\mu \to 0$.*

If, in addition, $\mathcal{N}_x \times \mathcal{N}_y = R^{m+n}$, then the equilibrium state $(x^{\mathrm{T}}, y^{\mathrm{T}})^{\mathrm{T}}$ is uniformly asymptotically stable in the whole for every $\mu \in (0, \mu_0)$ and for $\mu \to 0$.

PROOF. The proof of Theorem 3.7.2 is similar to that of Theorem 3.7.1, taking into account that its conditions are equivalent to the conditions of the theorem on uniform asymptotic stability (Grujić, Martynyuk and Ribbens-Pavella [57]).

The theorem is proved.

PROPOSITION 3.7.1. *Let in Assumption 3.6.1 in conditions (a)–(h) the inequality sign "$\leq$" be replaced by "$\geq$", the constants ρ_{ij} $(i = 1, 2;$ $j = 1, \ldots, 8)$ be replaced by $\widetilde{\rho}_{ij}$ $(i = 1, 2;\ j = 1, \ldots, 8)$ and the pair of functions (φ_2, ψ_2) be replaced by the pair of function (φ_1, ψ_1).*

Then for the upper right-side Dini derivative of function (3.5.10) the estimate from below

$$(3.7.1)\quad D^+V(t, x, y, \mu) \geq u_1^{\mathrm{T}} \widetilde{C}(\mu) u_1 \qquad \forall\, (t, x, y, \mu) \in R_+ \times \mathcal{N}_x \times \mathcal{N}_y \times \mathcal{M}$$

is satisfied, where the matrix $\widetilde{C}(\mu)$ has the same structure as the matrix $C(\mu)$.

THE PROOF of Proposition 3.7.1 is similar to that of Proposition 3.6.1.

THEOREM 3.7.3. *Let the motion* $(x^{\mathrm{T}}(t;t_0,x_0,\mu),y^{\mathrm{T}}(t;t_0,y_0,\mu))^{\mathrm{T}}$ *of system (3.5.1) and (3.5.2) be continuous for* $(t_0,x_0,y_0)\in R_+\times\mathcal{N}_x\times\mathcal{N}_y$. *For the equilibrium state* $(x^{\mathrm{T}},y^{\mathrm{T}})^{\mathrm{T}}=0$ *of system (3.5.1), (3.5.2) is unstable for every* $\mu\in(0,\mu_0)$ *and for* $\mu\to 0$ *it is necessary that the conditions of Assumption 3.5.1 and Proposition 3.7.1 be satisfied, and it is sufficient that*

(1) *the matrices* $A_1(\mu)+A_1^{\mathrm{T}}(\mu)$ *and* $A_2(\mu)+A_2^{\mathrm{T}}(\mu)$ *be conditionally positive;*
(2) *the matrix* $\widetilde{C}(\mu)$ *be conditionally positive for every* $\mu\in(0,\mu_0)$ *and for* $\mu\to 0$.

PROOF. Due to inequality (3.5.11) and condition (1) of Theorem 3.7.3 the function $V(t,x,y,\mu)$ is positive definite and bounded for every $\mu\in(0,\mu_0)$ and for $\mu\to 0$. Inequality (3.7.1) and condition (2) of Theorem 3.7.3 together with the above condition are equivalent to the conditions of the second Liapunov's theorem on instability (see Liapunov [101]).

This completes the proof.

3.8 General Theorems on Stability and Instability in Case B

We consider the general system (3.2.1)–(3.2.2) and matrix-valued function (3.5.8). Systems of (3.2.1)–(3.2.2) type are attributed to Case B of inclusion of a small parameter. Functions

$$
\begin{aligned}
f^0(t,x,y,\mu)&=f(t,x,y,\mu)-f(t,x,y,0),\\
g^0(t,x,y,\mu)&=g(t,x,y,\mu)-g(\alpha,b,y,0).
\end{aligned}
$$

are considered as perturbed systems describing slow motions and as a boundary layer of systems (3.2.5) and (3.2.6) respectively.

ASSUMPTION 3.8.1. For the systems of equations (3.2.1) and (3.2.2) all conditions of Assumption 3.5.1 are satisfied, and for function (3.5.8) estimates (3.5.11) are valid.

ASSUMPTION 3.8.2. There exist

(1) the functions φ_i, $\psi_i\in K$, $i=1,2$, v_{sk}, $s,k=1,2$ mentioned in Assumption 3.6.1;
(2) a constants ρ_{ij} $(i=1,2,\ j=1,2,\dots,8)$ such that

(a) $D_t^+ v_{11} + (D_x^+ v_{11})^{\mathrm{T}} f(t,x,0,0) \le \rho_{11}\varphi_2^2(\|x\|)$
$\forall (t,x) \in R_+ \times \mathcal{N}_x$;

(b) $D_t^+ v_{22} + (D_y^+ v_{22})^{\mathrm{T}} g(\alpha,b,y,0) \le \mu\rho_{21}\psi_2^2(\|y\|)$
$\forall (\alpha,b,y,\mu) \in R_+ \times \mathcal{N}_x \times \mathcal{N}_y \times \mathcal{M}$;

(c) $(D_x^+ v_{11})^{\mathrm{T}}[f(t,x,y,0) - f(t,x,0,0)] \le \rho_{12}\varphi_2^2(\|x\|)$
$+\rho_{13}\varphi_2(\|x\|)\psi_2(\|y\|) \quad \forall\,(t,x,y) \in R_+ \times \mathcal{N}_x \times \mathcal{N}_y$;

(d) $(D_y^+ v_{22})^{\mathrm{T}}[g(\alpha,b,y,\mu) - g(\alpha,b,y,0)] \le \rho_{22}\psi_2^2(\|y\|)$
$+\mu\rho_{23}\varphi_2(\|x\|)\psi_2(\|y\|) \quad \forall\,(\alpha,b,y,\mu) \in R_+ \times \mathcal{N}_x \times \mathcal{N}_y \times \mathcal{M}$;

(e) $D_t^+ v_{12} + (D_x^+ v_{12})^{\mathrm{T}} f(t,x,0,0) \le \mu\rho_{14}\varphi_2^2(\|x\|)$
$+\mu\rho_{15}\varphi_2(\|x\|)\psi_2(\|y\|) \quad \forall\,(t,x,y,\mu) \in R_+ \times \mathcal{N}_x \times \mathcal{N}_y \times \mathcal{M}$;

(f) $(D_x^+ v_{12})^{\mathrm{T}}[f(t,x,y,0) - f(t,x,0,0)] \le \mu\rho_{16}\varphi_2^2(\|x\|)$
$+\mu\rho_{17}\varphi_2(\|x\|)\psi_2(\|y\|) + \mu\rho_{18}\psi_2^2(\|y\|)$
$\forall\,(t,x,y,\mu) \in R_+ \times \mathcal{N}_x \times \mathcal{N}_y \times \mathcal{M}$;

(g) $(D_y^+ v_{12})^{\mathrm{T}} g(\alpha,b,y,0) \le \mu\rho_{24}\psi_2^2(\|y\|) + \mu\rho_{25}\varphi_2(\|x\|)\psi_2(\|y\|)$
$\forall\,(\alpha,b,y,\mu) \in R_+ \times \mathcal{N}_x \times \mathcal{N}_y \times \mathcal{M}$;

(h) $(D_y^+ v_{12})^{\mathrm{T}}[g(\alpha,b,y,\mu) - g(\alpha,b,y,0)] \le \mu\rho_{26}\varphi_2^2(\|x\|)$
$+\mu\rho_{27}\varphi_2(\|x\|)\psi_2(\|y\|) + \mu\rho_{28}\psi_2^2(\|y\|)$
$\forall\,(\alpha,b,y,\mu) \in R_+ \times \mathcal{N}_x \times \mathcal{N}_y \times \mathcal{M}$.

PROPOSITION 3.8.1. *If all conditions of Assumption 3.8.2 are satisfied, then for the upper right Dini derivative of function (3.5.8) along a solution of (3.2.1)–(3.2.2) the upper estimate*

$$D^+V(t,x,y,\mu) \le u_2^{\mathrm{T}} C^0(\mu) u_2 \qquad \forall\,(t,x,y,\mu) \in R_+ \times \mathcal{N}_x \times \mathcal{N}_y \times \mathcal{M}$$

is satisfied, where

$$C^0(\mu) = [s_{ij}(\mu)], \quad s_{12}(\mu) = s_{21}(\mu), \quad i,\, j = 1,2,$$

and

$$s_{11} = \eta_1^2(\rho_{11} + \rho_{12}) + 2\eta_1\eta_2\mu(\rho_{14} + \rho_{16} + \rho_{26});$$
$$s_{22} = \eta_2^2(\rho_{21} + \rho_{22}) + 2\eta_1\eta_2(\mu\rho_{18} + \rho_{24} + \rho_{28});$$
$$s_{12} = s_{21} = \frac{1}{2}\eta_1^2\rho_{13} + \frac{1}{2}\eta_2^2\mu\rho_{23} + \eta_1\eta_2(\mu\rho_{15} + \mu\rho_{17} + \rho_{25} + \rho_{27}).$$

THE PROOF is similar to the proof of Proposition 3.6.1.

We introduce the values $\widetilde{\mu}_i$, $\widetilde{\mu}_0$, $\widetilde{\mu}^*$, $i = 1, 2, 3, 4$ by the formulas

$$\widetilde{\mu}_1 = -\frac{\eta_1(\rho_{11} + \rho_{12})}{2\eta_2(\rho_{14} + \rho_{16} + \rho_{26})},$$

$$\widetilde{\mu}_2 = -\frac{\eta_2(\rho_{21} + \rho_{22}) + 2\eta_1(\rho_{24} + \rho_{28})}{2\eta_1\rho_{18}},$$

$$\widetilde{\mu}_3 = \frac{-\widetilde{b} + \sqrt{\widetilde{b}^2 - 4\widetilde{a}\widetilde{c}}}{2\widetilde{a}},$$

$$\widetilde{\mu}_4 = \mu_4, \quad \widetilde{\mu}_0 = \min(\widetilde{\mu}_1, \widetilde{\mu}_2, \widetilde{\mu}_3), \quad \widetilde{\mu}^* = \min(\widetilde{\mu}_0, \widetilde{\mu}_4).$$

Moreover,

$$\widetilde{a} = \left[\frac{1}{2}\eta_2^2\rho_{23} + \eta_1\eta_2(\rho_{15} + \rho_{17})\right]^2 - 4\eta_1\eta_2\rho_{18}(\rho_{14} + \rho_{16} + \rho_{26});$$

$$\widetilde{b} = \left[\frac{1}{2}\eta_1^2\rho_{13} + \eta_1\eta_2(\rho_{25} + \rho_{27})\right]\left[\frac{1}{2}\eta_2^2\rho_{23} + \eta_1\eta_2(\rho_{15} + \rho_{17})\right]$$
$$- 2\eta_1\eta_2\rho_{18}(\rho_{11} + \rho_{12}) - 2\eta_1\eta_2(\rho_{14} + \rho_{16} + \rho_{26})$$
$$\times [\eta_2^2(\rho_{21} + \rho_{22}) + 2\eta_1\eta_2(\rho_{24} + \rho_{28})],$$

$$\widetilde{c} = \left[\frac{1}{2}\eta_1^2\rho_{13} + \eta_1\eta_2(\rho_{25} + \rho_{27})\right]^2$$
$$- \eta_1^2(\rho_{11} + \rho_{12})[\eta_2^2(\rho_{21} + \rho_{22}) + 2\eta_1\eta_2(\rho_{24} + \rho_{26})].$$

PROPOSITION 3.8.2. *The matrix $C^0(\mu)$ is negative definite for every $\mu \in (0, \widetilde{\mu}_0)$ and for $\mu \to 0$, provided that*

(a) $\rho_{11} + \rho_{12} < 0$;
(b) $\rho_{14} + \rho_{16} + \rho_{26} > 0$;
(c) $\eta_2(\rho_{21} + \rho_{22}) + 2\eta_1(\rho_{24} + \rho_{28}) < 0$;
(d) $\rho_{18} > 0$;
(e) $\widetilde{a} > 0$;
(f) $\widetilde{c} < 0$.

THE PROOF is similar to that of Proposition 3.6.2.

REMARK 3.8.1. If conditions (a), (b), (c), (e) and (f) of Proposition 3.8.2 are satisfied and $\rho_{18} \leq 0$, then its assertion is true for $\widetilde{\mu}_0 = \min(\widetilde{\mu}_1, \widetilde{\mu}_3)$.

REMARK 3.8.2. If conditions (a), (c), (e) and (f) of Proposition 3.8.2 are satisfied and $\rho_{14} + \rho_{16} + \rho_{26} \leq 0$, then its assertion is true for $\widetilde{\mu}_0 = \min(\widetilde{\mu}_2, \widetilde{\mu}_3)$.

REMARK 3.8.3. If conditions (a), (c), (e) and (f) of Proposition 3.8.2 are satisfied and $\rho_{18} \leq 0$, $\rho_{14} + \rho_{16} + \rho_{26} \leq 0$, then its assertion is true for $\widetilde{\mu}_0 = \widetilde{\mu}_3$.

THEOREM 3.8.1. *Let motion* $(x^{\mathrm{T}}(t; t_0, x_0, \mu); y^{\mathrm{T}}(t; t_0, y_0, \mu))^{\mathrm{T}}$ *of the system (3.2.1)–(3.2.2) be continuous for* $(t_0, x_0, y_0) \in R_+ \times \mathcal{N}_x \times \mathcal{N}_y$ *and* $\mu \in M^0 \subset M$. *In order that the equilibrium state* $(x^{\mathrm{T}}, y^{\mathrm{T}})^{\mathrm{T}} = 0$ *of system (3.2.1)–(3.2.2) be uniformly asymptotically stable for every* $\mu \in (0, \widetilde{\mu}_0)$ *and for* $\mu \to 0$ *it is sufficient that*

(1) *conditions of Assumptions 3.8.1 and 3.8.2 be satisfied;*
(1) *matrices* $A_1(\mu) + A_1^{\mathrm{T}}(\mu)$ *and* $A_2(\mu) + A_2^{\mathrm{T}}(\mu)$ *be conditionally positive definite;*
(2) *matrix* $C^0(\mu)$ *be negative definite for every* $\mu \in (0, \mu_0)$ *and for* $\mu \to 0$.

If, moreover, $\mathcal{N}_x \times \mathcal{N}_y = R^{m+n}$, *functions* φ_1, $\psi_1 \in KR$, $i = 1, 2$, *then the equilibrium state* $(x^{\mathrm{T}}, y^{\mathrm{T}})^{\mathrm{T}} = 0$ *of the system (3.2.1)–(3.2.2) is uniformly asymptotically stable in the whole.*

THE PROOF is similar to that of Theorem 3.7.1.

Sufficient instability conditions for state $(x^{\mathrm{T}}, y^{\mathrm{T}})^{\mathrm{T}} = 0$ of the system (3.2.1)–(3.2.2) are established in the same way as in Theorem 3.7.2.

3.9 Asymptotic Stability of Linear Autonomous Systems

For the mentioned class of systems two cases of singular perturbation are considered.

3.9.1 Case A

Consider the system

$$\frac{dx}{dt} = A_{11}x + A_{12}y, \qquad \mu\frac{dy}{dt} = A_{21}x + A_{22}y, \tag{3.9.1}$$

where $x \in R^n$, $y \in R^m$, A_{11}, A_{12}, A_{21}, A_{22} are constant matrices with corresponding dimensions, $\mu \in [0,1]$ is a small parameter.

We construct the matrix-valued function (3.5.1) of elements v_{ij} $(i,j) \in [1,2])$ in the form

$$\begin{aligned} v_{11}(x) &= x^{\mathrm{T}} B_1 x, \\ v_{22}(y,\mu) &= \mu y^{\mathrm{T}} B_2 y; \\ v_{12}(x,y,\mu) &= v_{21}(x,y,\mu) = \mu x^{\mathrm{T}} B_3 y. \end{aligned} \tag{3.9.2}$$

Besides, matrices B_1 and B_2 are symmetric and positive definite, and B_3 is a constant matrix.

Further we need the following estimate (see Djordjević [28])

PROPOSITION 3.9.1. *Let an $A \in R^{n\times m}$ and $B \in R^{m\times r}$, $x \in R^n$, $y \in R^r$. Then the bilinear form $x^{\mathrm{T}}ABy$ satisfies the bilateral estimate*

$$-\lambda_M^{1/2}(AA^{\mathrm{T}})\lambda_M^{1/2}(B^{\mathrm{T}}B)\|x\|\,\|y\| \le x^{\mathrm{T}}ABy \le \lambda_M^{1/2}(AA^{\mathrm{T}})\lambda_M^{1/2}(B^{\mathrm{T}}B)\|x\|\,\|y\|,$$

where $\lambda_M(AA^{\mathrm{T}})$ and $\lambda_M(B^{\mathrm{T}}B)$ are maximal eigenvalues of the matrices AA^{T} and $B^{\mathrm{T}}B$ respectively.

PROOF. Let $\alpha \in R$. We construct the vector

$$w = \alpha A^{\mathrm{T}} x + By,$$

and consider the inequality

$$w^{\mathrm{T}} w \ge 0 \tag{3.9.3}$$

Since $w^{\mathrm{T}} = \alpha x^{\mathrm{T}} A + y^{\mathrm{T}} B^{\mathrm{T}}$, then (3.9.3) is equal to

$$\alpha^2 x^{\mathrm{T}} AA^{\mathrm{T}} x + 2\alpha x^{\mathrm{T}} ABy + y^{\mathrm{T}} B^{\mathrm{T}} By \ge 0. \tag{3.9.4}$$

In order that the polynomial (3.9.4) be non-negative it is sufficient that its discriminant be non-positive. Hence, we get

$$(x^{\mathrm{T}}ABy)^2 \le (x^{\mathrm{T}}AA^{\mathrm{T}}x)(y^{\mathrm{T}}B^{\mathrm{T}}By)$$

and

$$|x^{\mathrm{T}}ABy| \le (x^{\mathrm{T}}AA^{\mathrm{T}}x)^{1/2}(y^{\mathrm{T}}B^{\mathrm{T}}By)^{1/2}. \tag{3.9.5}$$

Hence, it follows the estimate from Proposition 3.9.1.

COROLLARY 3.9.1. If in Proposition 3.9.1 $B = I$ (I is an identity matrix) and $r = n$, then bilateral estimate becomes

$$-\lambda_M^{1/2}(AA^{\mathrm{T}})\|x\|\,\|y\| \le x^{\mathrm{T}}Ay \le \lambda_M^{1/2}(AA^{\mathrm{T}})\|x\|\,\|y\| \tag{3.9.6}$$

for all $A \in R^n$, $x \in R^n$, $y \in R^n$.

In view of estimates typical for the quadratic forms and with regard to Corollary 3.9.1 it is easily seen that for functions (3.9.2) the following inequalities are valid:
(3.9.7)

$$\begin{gathered}
v_{11}(x) \ge \lambda_m(B_1)\|x\|^2 \qquad \forall\, x \in \mathcal{N}_{x0};\\
v_{22}(y,\mu) \ge \mu\lambda_m(B_2)\|y\|^2 \qquad \forall\,(y,\mu) \in \mathcal{N}_{y0o} \times \mathcal{M};\\
v_{12}(x,y,\mu) = v_{21}(x,y,\mu) \ge -\mu\lambda_M^{1/2}(B_3B_3^{\mathrm{T}})\|x\|\,\|y\|\\
\forall\,(x,y,\mu) \in \mathcal{N}_{x0} \times \mathcal{N}_{y0} \times \mathcal{M}.
\end{gathered}$$

For the function $V(x,y,\mu) = \eta^{\mathrm{T}}U(x,y,\mu)\eta$, $\eta \in R_+^2$, the matrix $A_1(\mu)$ from estimate (3.5.11) has the form

$$A_1(\mu) = \begin{pmatrix} \lambda_m(B_1) & -\mu\lambda_M^{1/2}(B_3B_3^{\mathrm{T}}) \\ -\mu\lambda_M^{1/2}(B_3B_3^{\mathrm{T}}) & \mu\lambda_m(B_2) \end{pmatrix}.$$

Since by assumption on matrix B_1 have $\lambda_m(B_1) > 0$, and then for the function $V(x,y,\mu)$ to be positive definite it is sufficient that

$$\lambda_m(B_1)\lambda_m(B_2) > \mu\lambda_M(B_3B_3^{\mathrm{T}}) \tag{3.9.8}$$

for every $\mu \in (0,\mu_0^*)$ and for $\mu \to 0$.

The fact that

$$\frac{dV(x,y,\mu)}{dt} = \eta^{\mathrm{T}}\frac{dU(x,y,\mu)}{dt}\eta$$

yields

$$\frac{1}{2}\frac{dV(x,y,\mu)}{dt} = z^{\mathrm{T}}\begin{pmatrix} c_{11} & c_{12}+\mu\sigma_{12} \\ c_{12}+\mu\sigma_{12} & c_{22}+\mu\sigma_{22} \end{pmatrix} z, \tag{3.9.9}$$

where $z = (x^{\mathrm{T}}, y^{\mathrm{T}})^{\mathrm{T}}$ and

$$\begin{aligned}
c_{11} &= \frac{1}{2}\eta_1^2(B_1A_{11} + A_{11}^{\mathrm{T}}B_1) + \frac{1}{2}\eta_1\eta_2(B_2A_{21} + A_{21}^{\mathrm{T}}B_3^{\mathrm{T}});\\
c_{22} &= \frac{1}{2}\eta_2^2(B_2A_{22} + A_{22}^{\mathrm{T}}B_2),\\
\sigma_{22} &= \frac{1}{2}\eta_1\eta_2(B_3^{\mathrm{T}}A_{12} + A_{12}^{\mathrm{T}}B_3),\\
c_{12} &= \frac{1}{2}\eta_1^2B_1A_{12} + \frac{1}{2}\eta_1\eta_2B_3A_{22} + \frac{1}{2}\eta_2^2A_{21}^{\mathrm{T}}B_2;\\
\sigma_{12} &= \frac{1}{2}\eta_1\eta_2A_{11}^{\mathrm{T}}B_3, \quad \eta_1 > 0, \quad \eta_2 > 0.
\end{aligned}$$

Let $\frac{d}{dt}V_M(x, y, \mu)$ be an upper bound of the expression (3.9.6). It is easy to verify that

$$\frac{d}{dt}V_M(x, y, \mu) \le 2u^{\mathrm{T}}C(\mu)u, \tag{3.9.10}$$

where $u = (\|x\|, \|y\|)^{\mathrm{T}}$ and

$$C(\mu) = \begin{pmatrix} \lambda_M(c_{11}) & \lambda_M^{1/2}(c_{12}c_{12}^{\mathrm{T}}) + \mu\lambda_M^{1/2}(\sigma_{12}\sigma_{12}^{\mathrm{T}}) \\ \lambda_M^{1/2}(c_{12}c_{12}^{\mathrm{T}}) + \mu\lambda_M^{1/2}(\sigma_{12}\sigma_{12}^{\mathrm{T}}) & \lambda_M(c_{22}) + \mu\lambda_M(\sigma_{22}) \end{pmatrix}.$$

Here $\lambda_M(c_{ii})$ and $\lambda_M(\sigma_{22})$ are maximal eigenvalues of matrices c_{ii}, $i = 1, 2$ and σ_{22} respectively; and $\lambda_M^{1/2}(c_{12}c_{12}^{\mathrm{T}})$ and $\lambda_M^{1/2}(\sigma_{12}\sigma_{12}^{\mathrm{T}})$ are norms of matrices c_{12} and σ_{12} respectively.

In this case, the values μ_2, μ_3 and μ_0 are expressed as follows

$$\mu_2 = -\lambda_M(c_{22})/\lambda_M(\sigma_{22}), \quad \mu_3 = (-b+\sqrt{b^2 - 4ac})/2a, \quad \mu_0 = \min(\mu_2, \mu_3),$$

where

$$\begin{aligned}
a &= \lambda_M(\sigma_{12}\sigma_{12}^{\mathrm{T}}),\\
b &= \lambda_M^{1/2}(c_{12}c_{12}^{\mathrm{T}})\lambda_M^{1/2}(\sigma_{12}\sigma_{12}^{\mathrm{T}}) - \lambda_M(c_{11})\lambda_M(\sigma_{22}),\\
c &= \lambda_M(c_{12}c_{12}^{\mathrm{T}}) - \lambda_M(c_{11})\lambda_M(c_{22}).
\end{aligned}$$

Sufficient conditions for uniform asymptotic stability of the state $(x^{\mathrm{T}}, y^{\mathrm{T}})^{\mathrm{T}} = 0$ of (3.9.1) are established in terms of Theorem 3.7.1. Namely, the equilibrium state $(x^{\mathrm{T}}, y^{\mathrm{T}})^{\mathrm{T}} = 0$ of system (3.9.1) is uniformly asymptotically stable in the whole if

(1) inequality (3.9.8) is satisfied;
(2) the following inequalities are satisfied
   (a) $\lambda_M(c_{11}) < 0$;
   (b) $\lambda_M(c_{22}) < 0$;
   (c) $\lambda_M(\sigma_{22}) > 0$;
   (d) $\lambda_M(c_{12}c_{12}^{\mathrm{T}}) - \lambda_M(c_{11})\lambda_M(c_{22}) < 0$.

This assertion follows from the fact that for functions (3.9.2) under condition (3.9.8) the function $V(x, y, \mu) = \eta^{\mathrm{T}}U(x, y)\eta$ is positive definite and radially unbounded, and under condition (2) $D^+V(x, y, \mu)$ along solutions of system (3.9.1) is negative definite. Therefore, all conditions of Theorem 3.7.1 are satisfied.

3.9.2 Case B

Consider the system

$$
\begin{aligned}
\frac{dx}{dt} &= A_{11}x + A_{12}y, \\
\mu\frac{dy}{dt} &= \mu A_{21}x + A_{22}y,
\end{aligned}
\tag{3.9.11}
$$

where, $x \in R^n$, $y \in R^m$, $\mu \in (0, 1]$ and matrices $A_{11}, \dots, A_{22}$ are the same as in system (3.9.1).

In order to establish conditions for uniform asymptotic stability of equilibrium state $(x^{\mathrm{T}}, y^{\mathrm{T}})^{\mathrm{T}} = 0$ of system (3.9.11) we incorporate the Theorem 3.7.3. To this end we take the elements of a matrix-valued function in the form of (3.9.2) and assume that the estimate (3.5.11) is satisfied for the function $V(x, y, \mu)$.

We have for the total derivative of function $V(x, y, \mu)$ along a solutions of system (3.9.11)

$$
\frac{1}{2}\frac{d}{dt}V(x, y, \mu) = z^{\mathrm{T}} \begin{pmatrix} c_{11}^0 + \mu\sigma_{11}^0 & c_{12}^0 + \mu\sigma_{12}^0 \\ c_{12}^0 + \mu\sigma_{12}^0 & c_{22}^0 + \mu\sigma_{22}^0 \end{pmatrix} z,
\tag{3.9.12}
$$

where

$$\begin{aligned}
c_{11}^0 &= \frac{1}{2}\eta_1^2(B_1A_{11} + A_{11}^{\mathrm{T}}B_1);\\
\sigma_{11}^0 &= \frac{1}{2}\eta_1\eta_2(B_3A_{21} + A_{21}^{\mathrm{T}}B_3^{\mathrm{T}});\\
c_{22}^0 &= \frac{1}{2}\eta_2^2(B_2A_{22} + A_{22}^{\mathrm{T}}B_2);\\
\sigma_{22}^0 &= \frac{1}{2}\eta_1\eta_2(B_3^{\mathrm{T}}A_{12} + A_{12}^{\mathrm{T}}B_3);\\
c_{12}^0 &= \frac{1}{2}\eta_1^2B_1A_{12} + \frac{1}{2}\eta_1\eta_2B_3A_{22};\\
\sigma_{12}^0 &= \frac{1}{2}\eta_2^2A_{21}^{\mathrm{T}}B_2 + \frac{1}{2}\eta_1\eta_2A_{11}^{\mathrm{T}}B_3, \quad \eta_1 > 0, \quad \eta_2 > 0.
\end{aligned}$$

For the upper bound $\frac{d}{dt}V_M(x,y,\mu)$ of expression (3.9.12) we have the estimate

$$\frac{d}{dt}V_M(x,y,\mu) \le 2u^{\mathrm{T}}C^0(\mu)u, \tag{3.9.13}$$

where

$$C^0(\mu) = \begin{pmatrix} \lambda_M(c_{11}^0) + \mu\lambda_M(\sigma_{11}^0) & \lambda_M^{1/2}(c_{12}^0c_{12}^{0T}) + \mu\lambda_M^{1/2}(\sigma_{12}^0\sigma_{12}^{0T}) \\ \lambda_M^{1/2}(c_{12}^0c_{12}^{0T}) + \mu\lambda_M^{1/2}(\sigma_{12}^0\sigma_{12}^{0T}) & \lambda_M(c_{22}^0) + \mu\lambda_M(\sigma_{22}^0) \end{pmatrix}.$$

In this case, the values $\widetilde{\mu}_i$, $i = \overline{1,4}$, $\widetilde{\mu}_0$ and $\widetilde{\mu}^*$ are defined as

$$\begin{aligned}
\widetilde{\mu}_1 &= -\frac{\lambda_M(c_{11}^0)}{\lambda_M(\sigma_{11}^0)} & \widetilde{\mu}_2 &= -\frac{\lambda_M(c_{22}^0)}{\lambda_M(\sigma_{22}^0)},\\
\widetilde{\mu}_3 &= \frac{-b_1 + \sqrt{b_1^2 - 4a_1c_1}}{2a_1}, & \widetilde{\mu}_4 &= \mu_4,\\
\widetilde{\mu}_0 &= \min(\widetilde{\mu}_1, \widetilde{\mu}_2, \widetilde{\mu}_3), & \widetilde{\mu}^* &= \min(\widetilde{\mu}_0, \widetilde{\mu}_4);
\end{aligned}$$

where

$$\begin{aligned}
a_1 &= \lambda_M(\sigma_{12}^0\sigma_{12}^{0T}) - \lambda_M(\sigma_{11}^0)\lambda_M(\sigma_{22}^0);\\
b_1 &= \lambda_M^{1/2}(c_{12}^0c_{12}^{0T})\lambda_M^{1/2}(\sigma_{12}^0\sigma_{12}^{0T}) - \lambda_M(c_{11}^0)\lambda_M(\sigma_{22}^0) - \lambda_M(\sigma_{11}^0)\lambda_M(\sigma_{22}^0);\\
c_1 &= \lambda_M(c_{12}^0c_{12}^{0T}) - \lambda_M(c_{11}^0)\lambda_M(c_{22}^0).
\end{aligned}$$

According to estimates (3.5.11) and (3.9.13) for functions $V(x, y, \mu)$ and $DV_M(x, y, \mu)$ the sufficient conditions for uniform asymptotic stability in the whole of state $(x^{\mathrm{T}}, y^{\mathrm{T}})^{\mathrm{T}} = 0$ of the system (3.9.11) are

(a) $\lambda_m(B_1)\lambda_m(B_2) > \mu\lambda_M^{1/2}(B_3 B_3^{\mathrm{T}})$;

(b) $\lambda_M(c_{11}^0) < 0$;

(c) $\lambda_M(\sigma_{11}^0) > 0$;

(d) $\lambda_M(c_{22}^0) < 0$;

(e) $\lambda_M(\sigma_{22}^0) > 0$;

(f) $\lambda_M(\sigma_{12}^0\sigma_{12}^{0T}) - \lambda_M(\sigma_{11}^0)\lambda_M(\sigma_{22}^0) > 0$;

(g) $\lambda_M(c_{12}^0 c_{12}^{0T}) - \lambda_M(c_{11}^0)\lambda_M(c_{22}^0) < 0$.

3.9.3 Example

Let the system (3.9.1) be

$$\begin{aligned} \frac{dx}{dt} &= \begin{pmatrix} 0.5 & 0.1 \\ -0.5 & 0.6 \end{pmatrix} x + \begin{pmatrix} 3.6 & 0.3 \\ -0.2 & 5 \end{pmatrix} y, \\ \mu\frac{dy}{dt} &= \begin{pmatrix} -7 & 0.5 \\ -1 & -8 \end{pmatrix} x + \begin{pmatrix} -3 & 0.5 \\ 1 & -8 \end{pmatrix} y, \end{aligned} \tag{3.9.14}$$

where $x \in R^2$, $\mu \in (0, 1]$.

We take for the system (3.9.14) the matrix valued function $U(x, y, \mu)$ with the elements

$$\begin{aligned} v_{11}(x) &= x^{\mathrm{T}}\operatorname{diag}[2, 2]x; \\ v_{22}(y, \mu) &= \mu y^{\mathrm{T}}\operatorname{diag}[1, 1]y; \\ v_{12}(x, y, \mu) = v_{21}(x, y, \mu) &= \mu x^{\mathrm{T}}\operatorname{diag}[0.4; 0.4]y. \end{aligned} \tag{3.9.15}$$

It is easy to see that $v_{ij}(\cdot)$, $i, j = 1, 2$ satisfy the estimates

$$\begin{aligned} v_{11}(x) &\geq 2\|x\|^2 && \forall\, (x) \in \mathcal{N}_{x0}, \\ v_{22}(y, \mu) &\geq \mu\|y\|^2 && \forall\, (y, \mu) \in \mathcal{N}_{y0} \times \mathcal{M}, \\ v_{12}(x, y, \mu) &\geq -0.4\mu\|x\|\,\|y\| && \forall\, (x, y, \mu) \in \mathcal{N}_{x0} \times \mathcal{N}_{y0} \times \mathcal{M}. \end{aligned} \tag{3.9.16}$$

Let $\eta = (1, 1)$. Then matrix $A_1(\mu)$ in estimate (3.5.11) for the function

$$V(x, y, \mu) = \eta^{\mathrm{T}} U(x, y, \mu)\eta, \qquad \eta \in R_+^2$$

with elements (3.9.15) and the estimates (3.9.16) has the form

$$A_1(\mu) = \begin{pmatrix} 2 & -0.4\mu \\ -0.4\mu & \mu \end{pmatrix}.$$

It can be easily verified that matrix $A_1(\mu)$ is positive definite for every $\mu \in (0, 1]$ and for $\mu \to 0$.

The elements of matrix $C(\mu)$ from the estimate

$$\frac{d}{dt} V_M(x, y, \mu) \leq 2u^{\mathrm{T}} C(\mu) u,$$

where $u = (\|x\|, \|y\|)^{\mathrm{T}}$, have the values

$$\begin{aligned} \lambda_M(c_{11}) &= -1.291723; \\ \lambda_M(c_{22}) &= -2.89; \\ \lambda_M(\sigma_{22}) &= 2.000713; \\ \lambda_M^{1/2}(c_{12}c_{12}^{\mathrm{T}}) &= 0.784953; \\ \lambda_M^{1/2}(\sigma_{12}\sigma_{12}^{\mathrm{T}}) &= 0.165452; \end{aligned} \tag{3.9.17}$$

The values of parameters μ_2, μ_3 and μ_0 are

$$\mu_2 = 1.444485; \qquad \mu_3 = 1.779742;$$
$$\mu_0 = \min(\mu_2, \mu_3) = 1.444485.$$

With regard to (3.9.17) we find that

(a) $\lambda_M(c_{11}) < 0$;
(b) $\lambda_M(c_{22}) < 0$;
(c) $\lambda_M(\sigma_{22}) > 0$;
(d) $\lambda_M(c_{12}c_{12}^{\mathrm{T}}) - \lambda_M(c_{11})\lambda_M(c_{22}) = -3.117332 < 0$,

and $\mu_0 = 1.444485$.

By Theorem 3.7.3 the equilibrium state $(x^{\mathrm{T}}, y^{\mathrm{T}})^{\mathrm{T}} = 0$ of the system (3.9.14) is uniformly asymptotically stable in the whole for every $\mu \in (0, 1]$ and for $\mu \to 0$.

3.10 Applications

Consider some applications of general results to the problems of mechanics.

3.10.1 Plane two-component pendulum

Let two absolutely solid bodies form a pendulum as shown on Figure 3.10.1.

Body I is rotating around hinge O_1 and contains a sphere cavity. A round body II is placed into this cavity and is freely connected with body I at point O_2. For the sake of simplicity we assume that the center of mass of body II coincides with point O_2.

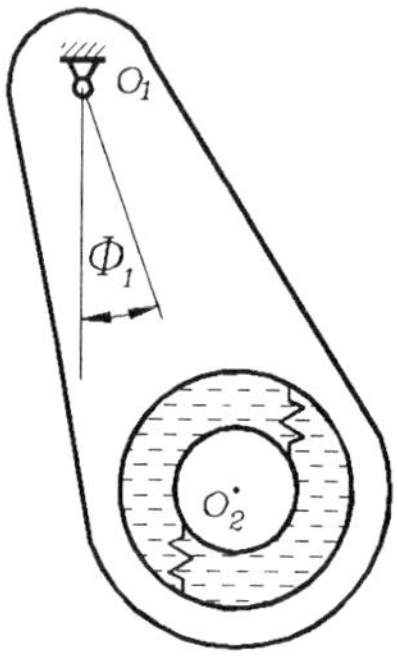

FIGURE 3.10.1 Plane two-component pendulum

The bodies forming such a pendulum are subjected to the weight force and moments of elasticity force and friction with a large coefficient of proportionality to relative rotation angulars and relative angular velocities of the links. Body I moves in the medium with viscous friction. The motion equations of this system in the form of moment of momentum equations for the total system relative to point O_1 and for body II relative to point O_2 are

$$\frac{d}{d\tau}(I_1\Omega_1 + I_2\Omega_2) = -Pl\sin\Phi_1 - N_1\Omega_1,$$

$$\frac{d}{d\tau}I_2\Omega_2 = -K_2(\Phi_2 - \Phi_1) - N_2(\Omega_2 - \Omega_1), \tag{3.10.1}$$

$$\frac{d\Phi_1}{d\tau} = \Omega_1, \qquad \frac{d\Phi_2}{d\tau} = \Omega_2.$$

Here we designate by Φ_1 and Φ_2 the rotation angulars of the system elements, by Ω_1 and Ω_2 its angular velocities, by I_1 and I_2 the moments of inertia, by τ the natural time, by P the total weight of the system, by l the distance from point O_1 to the center of masses, by N_1 the coefficient of moment of friction of outer forces for the system, by K_2 and N_2 the coefficients of stiffness and friction of moments of interaction forces between the bodies.

In system (3.10.1) we get over from variables Φ_1, Φ_2, Ω_1, Ω_2 to the set φ_1, Ω_1, Δ, U containing the variables $\Delta = \Phi_2 - \Phi_1$, $U = \Omega_2 - \Omega_1$ with respect to which tight co-actions take place. Then we obtain the following equations

$$
\begin{aligned}
I_1 \frac{d}{d\tau} &= -Pl \sin \Phi_1 - N_1 \Omega_1 + K_2 \Delta + N_2 U, \\
I_2 \frac{d}{d\tau} &= \frac{I_2}{I_1} Pl \sin \Phi_1 - (1 + \frac{I_2}{I_1})(K_2 \Delta + N_2 U), \\
\frac{d\Phi_1}{d\tau} &= \Omega_1, \\
\frac{d\Delta}{d\tau} &= U.
\end{aligned}
\tag{3.10.2}
$$

In the system (3.10.2) we get over to the pure normalized values

$$
\begin{aligned}
t &= \frac{\tau}{\tau_*}, \quad i_1 = \frac{I_1}{I_*}, \quad i_2 = \frac{I_2}{I_*}, \quad \varphi_1 = \frac{\Phi_1}{\Phi_*}, \\
\delta &= \frac{\Delta}{\Delta_*}, \quad \omega_1 = \frac{\Omega_1}{\Omega_*}, \quad u = \frac{U}{U_*},
\end{aligned}
\tag{3.10.3}
$$

Let us consider a class of motions for which

(a) the oscillations of body I are large ($\Phi_* = I$);

(b) the moments of inertia are of the same order ($I_* = I_1$);

(c) the stiffness of elastic forces is essentially larger than the coefficient of regeneration $K_1 = Pl$ due to the condition $K_1 \ll K_2$.

We estimate partial time constants of the system. Time constants τ_i of slow oscillations due to condition (c) are estimated by $\tau_1^2 = I_1/K_1$, the time constant τ_2 of quick oscillations of body II due to elasticity is estimated by the correlation $\tau_2 = I_2/k_2$. For $K_2 \geq K_1$ we have $\mu = \tau_2/\tau_1 \ll 1$. We estimate characteristic angular velocities of the system with respect to variables Ω_1, U by the correlations $\Omega_* = \Phi_*/\tau_1$ and $U_* = \Delta_*/\tau_2$.

Assume that the oscillation moments and moments of forces of elastic interaction are the values of the same order ($K_2\Delta_* = K_1$). We take the value of the order of slow partial oscillations $(\tau_* = \tau_1)$ as the characteristic time. In result of the normalization of (3.10.3) equations (3.10.2) become

$$
\begin{aligned}
\frac{d\omega}{dt} &= -\sin\varphi_1 - 2\xi_1\omega_1 + \delta + 2\xi_2 u,\\
\frac{d\varphi_1}{dt} &= \omega_1;\\
\mu\frac{du}{dt} &= i_2\sin\varphi_1 - (1+i_2)(\delta + 2\xi_2 u),\\
\mu\frac{d\delta}{dt} &= u.
\end{aligned}
\tag{3.10.4}
$$

Here all variables φ_1, ω_1, δ and u have the values of the order of one, $i_2 = I_2/I_1$ and ξ_1, ξ_2 are dimensionless coefficients of damping of the first and second partial oscillating links. In system (3.10.4) we make the change of variables

$$\sin\varphi_1 = x_1, \quad \omega_1 = x_2, \quad u = y_1, \quad (1+i_2)\delta - i_2\sin\varphi_1 = y_2$$

and linearize the system. In result we get

$$
\begin{aligned}
\frac{dx}{dt} &= A_{11}x + A_{12}y,\\
\mu\frac{dy}{dt} &= A_{21}x + A_{22}y,
\end{aligned}
\tag{3.10.5}
$$

where

$$
A_{11} = \begin{pmatrix} 0 & 1\\ \dfrac{1}{1+i_2} & -2\xi_1 \end{pmatrix}, \qquad
A_{12} = \begin{pmatrix} 0 & 0\\ 2\xi_2 & \dfrac{1}{1+i_2} \end{pmatrix},
$$

$$
A_{21} = \begin{pmatrix} 0 & 0\\ 0 & -i_2 \end{pmatrix}, \qquad
A_{22} = \begin{pmatrix} -2(1+i_2)\xi_2 & -1\\ 1+i_2 & 0 \end{pmatrix},
$$

$x = (x_1, x_2)^{\mathrm T}$, $y = (y_1, y_2)^{\mathrm T}$, μ is a small parameter.

For system (3.10.5) we construct matrix-valued function with elements

$$
\begin{aligned}
v_{11}(x) &= x^{\mathrm T}\begin{pmatrix} 2(\xi_1\gamma_1 + 1) & \gamma_1\\ \gamma_1 & 2(1+i_2) \end{pmatrix} x;\\
v_{12}(x, y, \mu) &= \mu x^{\mathrm T}\begin{pmatrix} 0 & 0\\ 2 & 0.01 \end{pmatrix} y,\\
v_{22}(y, \mu) &= \mu y^{\mathrm T}\begin{pmatrix} 2(1+i_2) & \gamma_2\\ \gamma_2 & 2(\xi_2\gamma_2 + 1) \end{pmatrix} y,
\end{aligned}
\tag{3.10.6}
$$

where γ_1 and γ_2 are constants satisfying the conditions

$$\gamma_1 < 4\xi_1(1+i_2), \qquad \gamma_2 < 4\xi_2(1+i_2) \tag{3.10.7}$$

Functions (3.10.6) satisfy the estimates

$$\begin{aligned} v_{11}(x) &\geq k_1\|x\|^2 && \forall\, x \in R^2 \\ v_{12}(x,y,\mu) &\geq -2\mu\|x\|\,\|y\| && \forall\,(x,y) \in R^2 \times R^2 \\ v_{22}(y,\mu) &\geq \mu k_2\|y\|^2 && \forall\, y \in R^2, \end{aligned}$$

where

$$k_1 = \xi_1\gamma_1 + i_2 + 2 - \sqrt{(\xi_1\gamma_1 - i_2)^2 + \gamma_1^2};$$
$$k_2 = \xi_2\gamma_2 + i_2 + 2 - \sqrt{(\xi_2\gamma_2 - i_2)^2 + \gamma_2^2}.$$

It can be easily verified that when inequalities (3.10.7) are satisfied, then $k_1 > 0$ and $k_2 > 0$.

Matrix $A_1(\mu)$ in estimate (3.5.11) for matrix-valued function with elements (3.10.6) has the form

$$A_1(\mu) = \begin{pmatrix} k_1 & -2\mu \\ -2\mu & k_2\mu \end{pmatrix} \tag{3.10.8}$$

and is positive definite for any $\mu \in (0, \widetilde{\mu}_4)$, where

$$\widetilde{\mu}_4 = \frac{1}{4}k_1k_2.$$

If $\eta^{\mathrm{T}} = (1,1)$, then the elements of matrix $C^0(\mu)$ are

$$\lambda_M(c_{11}^0) = \max\left(-\frac{\gamma_1}{1+i_2}; -4\xi_1(1+i_2) + \gamma_1\right);$$
$$\lambda_M(\sigma_{11}^0) = 0;$$
$$\lambda_M(c_{22}^0) = \max\left(-4(1+i_2)^2\xi_2 + (1+i_2)\gamma_2; -\gamma_2\right);$$
$$\lambda_M(\sigma_{22}^0) = 2\xi_2 + \frac{1}{400(1+i_2)}$$
$$+ \left[\left(2\xi_2 + \frac{1}{400(1+i_2)}\right)^2 + \left(\frac{\xi_2}{100(1+i_2)}\right)^2 + \left(\frac{1}{1+i_2}\right)^2\right]^{1/2}.$$

$$\lambda_M^{1/2}(c_{12}^0 c_{12}^{0T}) = \left[\frac{1}{2}P_1 + \left(\frac{1}{4}P_1^2 - \left(\frac{\gamma_1}{200}\right)^2\right)^{1/2}\right]^{1/2};$$

$$P_1 = (\xi_1\gamma_1)^2 + \left(\frac{\gamma_1}{2(1+i_2)}\right)^2 + \left(\frac{1+i_2}{100}\right)^2;$$

$$\lambda_M^{1/2}(\sigma_{12}^0 \sigma_{12}^{0T}) = \left[\frac{1}{2}P_2 + \left(\frac{1}{4}P_2^2 - \left(i_2(\xi_2\gamma_2+1) - \frac{i_2\gamma_2}{100}\right)^2\right)^{1/2}\right]^{1/2};$$

$$P_2 = 1 + \left(\frac{1}{200}\right)^2 + \left(2\xi_1 + \frac{i_2\gamma_2}{2}\right)^2 + \left(\frac{\xi_1}{100} + i_2(\xi_2\gamma_2+1)\right)^2.$$

Matrix $C^0(\mu)$ is negative definite for every $\mu \in (0, \widetilde{\mu}_0)$ and $\mu \to 0$, where $\widetilde{\mu}_0 = \min(\widetilde{\mu}_2, \widetilde{\mu}_3)$ and

$$\widetilde{\mu}_2 = -\frac{\lambda_M(c_{22}^0)}{\lambda_M(\sigma_{22}^0)}, \qquad \widetilde{\mu}_3 = \frac{-b_1 + \sqrt{b_1^2 - 4a_1c_1}}{2a_1},$$

where

$$a_1 = \lambda_M(\sigma_{12}^0\sigma_{12}^{0T}),$$
$$b_1 = \lambda_M^{1/2}(c_{12}^0c_{12}^{0T})\lambda_M^{1/2}(\sigma_{12}^0\sigma_{12}^{0T}) - \lambda_M(c_{11}^0)\lambda_M(\sigma_{22}^0);$$
$$c_1 = \lambda_M(c_{12}^0c_{12}^{0T}) - \lambda_M(c_{11}^0)\lambda_M(c_{22}^0)$$

if one of the following conditions (i)–(iv) is satisfied

(i) $\dfrac{\gamma_1\gamma_2}{1+i_2} > \lambda_M\left(c_{12}^0c_{12}^{0T}\right)$ for $\gamma_1 < \dfrac{4\xi_1(1+i_2)^2}{2+i_2}$ and $\gamma_2 < \dfrac{4\xi_2(1+i_2)^2}{2+i_2}$;

(ii) $(4\xi_2(1+i_2) - \gamma_2)\gamma_1 > \lambda_M\left(c_{12}^0c_{12}^{0T}\right)$ for $\gamma_1 < \dfrac{4\xi_1(1+i_2)^2}{2+i_2}$ and $\dfrac{4\xi_2(1+i_2)^2}{2+i_2} < \gamma_2 < 4\xi_2(1+i_2)$;

(iii) $(4\xi_1(1+i_2) - \gamma_1)\gamma_2 > \lambda_M\left(c_{12}^0c_{12}^{0T}\right)$ for $\dfrac{4\xi_1(1+i_2)^2}{2+i_2} < \gamma_1 < 4\xi_1(1+i_2)$ and $\gamma_2 < \dfrac{4\xi_2(1+i_2)^2}{2+i_2}$;

(iv) $(1+i_2)(4\xi_1(1+i_2) - \gamma_1)(4\xi_2(1+i_2) - \gamma_2) > \lambda_M\left(c_{12}^0c_{12}^{0T}\right)$ for $\dfrac{4\xi_1(1+i_2)^2}{2+i_2} < \gamma_1 < 4\xi_1(1+i_2)$ and $\dfrac{4\xi_2(1+i_2)^2}{2+i_2} < \gamma_2 < 4\xi_2(1+i_2)$.

By Theorem 3.7.1 the equilibrium state $(x^{\mathrm{T}}, y^{\mathrm{T}})^{\mathrm{T}} = 0$ of system (3.10.5) is uniformly asymptotically stable for every $\mu \in (0, \widetilde{\mu}^*)$ and for $\mu \to 0$, where $\widetilde{\mu}_* = \min(\widetilde{\mu}_4, \widetilde{\mu}_0)$.

3.10.2 Singularly perturbed Lur'e systems

In this section, the stability of a singularly perturbed system of the Lur'e form is analyzed on the basis of the Liapunov matrix-valued function. We obtain sufficient conditions for the absolute stability of a system of the Lur'e form and we indicate the bounds of the variation of the small parameter.

3.10.2.1 Singularly Perturbed Lur'e System. Case A. We consider the autonomous singularly perturbed system of Lur'e type

$$\begin{aligned} \frac{dx}{dt} &= A_{11}x + A_{12}y + q_1 f_1(\sigma_1), \quad \sigma_1 = c_{11}^{\mathrm{T}} + c_{12}^{\mathrm{T}} y; \\ \mu\frac{dy}{dt} &= A_{21}x + A_{22}y + q_2 f_2(\sigma_2), \quad \sigma_2 = c_{21}^{\mathrm{T}} + c_{22}^{\mathrm{T}} y, \end{aligned} \tag{3.10.9}$$

where $x \in N_x \subseteq R^n$, $y \in N_y \subseteq R^m$, $\mu \in (0, 1]$ is a small parameter, the matrices $A(\cdot)$ and the vectors $c(\cdot)$, $q(\cdot)$ having appropriate dimensions. The nonlinearities f_i, $i = 1, 2$, are continuous, $f_i(0) = 0$ and in the Lur'e sectors $[0, k_i]$, $k_i \in (0, +\infty)$ satisfy the conditions $f_i(\sigma_i)/\sigma_i \in (0, k_i]$, $i = 1, 2$; $\forall\, \sigma_i \in (-\infty, +\infty)$.

Moreover, we consider only those nonlinearities f_i for which the state $(x^{\mathrm{T}}, y^{\mathrm{T}})^{\mathrm{T}} = 0$ is the unique equilibrium state of the degenerate system

$$\frac{dx}{dt} = A_{11}x + q_1 f_1(\sigma_1^0); \qquad \sigma_1^0 = c_{11}^{\mathrm{T}} x \tag{3.10.10}$$

and of the system, describing the boundary layer,

$$\mu\frac{dy}{dt} = A_{22}y + q_2 f_2(\sigma_2^0); \qquad \sigma_2^0 = c_{22}^{\mathrm{T}} y \tag{3.10.11}$$

This assumption holds if

$$c_{ii}^{\mathrm{T}} A_{ii}^{-1} q_i > 0.$$

We introduce the following notations:

$$\begin{aligned} f(x, 0) &= A_{11}x + q_1 f_1(\sigma_1^0); \\ f^*(x, y) &= A_{12}y + q_1[f_1(\sigma_1) - f_1(\sigma_1^0)]; \\ g(0, y) &= A_{22}y + q_2 f_2(\sigma_2^0); \\ g^*(x, y) &= A_{21}x + q_2[f_2(\sigma_2) - f_2(\sigma_2^0)]. \end{aligned}$$

Then the system (3.10.9) takes the form

$$\frac{dx}{dt} = f(x,0) + f^*(x,y);$$
$$\mu\frac{dy}{dt} = g(0,y) + g^*(x,y).$$

Together with system (3.10.9) and subsystems (3.10.10), (3.10.11) we shall consider the matrix-valued function

$$U(x,y,\mu) = \begin{pmatrix} v_{11}(x) & v_{12}(x,y,\mu) \\ v_{21}(x,y,\mu) & v_{22}(y,\mu) \end{pmatrix}; \qquad v_{12} = v_{21}, \tag{3.10.12}$$

where

$$v_{11} = x^{\mathrm{T}}B_1x; \quad v_{12} = \mu y^{\mathrm{T}}B_2y; \quad v_{12} = \mu x^{\mathrm{T}}B_3y;$$

where B_1 and B_2 are symmetric, positive-definite matrices; B_3 is a constant matrix. With the aid of the matrix-valued function (3.10.12) we introduce the scalar function

$$V(x,y,\mu) = \eta^{\mathrm{T}}U(x,y,\mu)\eta, \tag{3.10.13}$$

where $\eta^{\mathrm{T}} = (\eta_1,\eta_2)$; $\eta \in R^2_+$; $\eta_i > 0$, $i = 1,2$.

We assume that the elements of the matrix-valued function (3.10.12) satisfy the estimates

(3.10.14)
$$\begin{aligned} v_{11}(x) &\geq \lambda_m(B_1)\|x\|^2 && \forall\, x \in \mathcal{N}_{x0} = \{x : x \in \mathcal{N}_x; x \neq 0\}; \\ v_{22}(y,\mu) &\geq \mu\lambda_m(B_2)\|y\|^2 && \forall\,(y,\mu) \in \mathcal{N}_{y0} \times \mathcal{M}; \\ v_{12}(x,y,\mu) &\geq -\mu\lambda_M^{1/2}(B_3B_3^{\mathrm{T}})\|x\|\,\|y\| && \forall\,(x,y,\mu) \in \mathcal{N}_{x0} \times \mathcal{N}_{y0} \times \mathcal{M}, \end{aligned}$$

where $\lambda_m(B_i)$ are the minimal eigenvalues of the matrices B_i, $i = 1,2$; $\lambda_M^{1/2}(B_3B_3^{\mathrm{T}})$ is the norm of the matrix $(B_3B_3^{\mathrm{T}})$; $\lambda_M(B_3B_3^{\mathrm{T}})$ is the maximal eigenvalue of the matrix $B_3B_3^{\mathrm{T}}$; $\mathcal{N}_{y0} = \{y : y \in \mathcal{N}_y,\ y \neq 0\}$; $\mathcal{M} = (0,1]$.

Under the estimates (3.10.14), for the function (3.10.13) we have the estimate

$$v(x,y,\mu) \geq u^{\mathrm{T}}H^{\mathrm{T}}AHu \qquad \forall\,(x,y,\mu) \in \mathcal{N}_x \times \mathcal{N}_y \times \mathcal{M},$$

where $u^{\mathrm{T}} = (\|x\|,\|y\|)$, $H = \operatorname{diag}(\eta_1,\eta_2)$;

$$A(\mu) = \begin{pmatrix} \lambda_m(B_1) & -\mu\lambda_M^{1/2}(B_3B_3^{\mathrm{T}}) \\ -\mu\lambda_M^{1/2}(B_3B_3^{\mathrm{T}}) & \mu\lambda_m(B_2) \end{pmatrix}.$$

For the derivatives of the elements of the matrix-valued function (3.10.12) along the solutions of the system (3.10.9) we have the following estimates;

$$\begin{aligned}
&\text{(a)} \quad (\nabla_x v_{11})^{\mathrm{T}} f(x,0) \le \rho_{11}\|x\|^2 \quad \forall x \in \mathcal{N}_{x0};\\
&\text{(b)} \quad (\nabla_x v_{11})^{\mathrm{T}} f^*(x,y) \le \rho_{12}\|x\|^2 + 2\rho_{13}^{1/2}\|x\|\,\|y\|\\
&\qquad \forall (x,y) \in \mathcal{N}_{x0} \times \mathcal{N}_{y0};\\
&\text{(c)} \quad (\nabla_y v_{22}) T g(0,y) \le \mu\rho_{21}\|y\|^2 \quad \forall (y,\mu) \in \mathcal{N}_{y0} \times \mathcal{M};\\
&\text{(d)} \quad (\nabla_y v_{22}) T g^*(x,y) \le \mu\rho_{22}\|y\|^2 + \mu\rho_{23}^{1/2}\|x\|\,\|y\|\\
&\qquad \forall (x,y,\mu) \in \mathcal{N}_{x0} \times \mathcal{N}_{y0} \times \mathcal{M};\\
(3.10.15)\quad &\text{(e)} \quad (\nabla_x v_{12}) T f(x,0) \le \mu\rho_{15}^{1/2}\|x\|\,\|y\|\\
&\qquad \forall (x,y,\mu) \in \mathcal{N}_{x0} \times \mathcal{N}_{y0} \times \mathcal{M};\\
&\text{(f)} \quad (\nabla_x v_{12}) T f^*(x,y) \le \mu\rho_{17}^{1/2}\|x\|\,\|y\| + \mu\rho_{18}\|y\|^2\\
&\qquad \forall (x,y,\mu) \in \mathcal{N}_{x0} \times \mathcal{N}_{y0} \times \mathcal{M};\\
&\text{(g)} \quad (\nabla_y v_{12})^{\mathrm{T}} g(0,y) \le \mu\rho_{25}^{1/2}\|x\|\,\|y\|\\
&\qquad \forall (x,y,\mu) \in \mathcal{N}_{x0} \times \mathcal{N}_{y0} \times \mathcal{M};\\
&\text{(h)} \quad (\nabla_y v_{12})^{\mathrm{T}} g^*(x,y) \le \mu\rho_{26}\|x\|^2 + \mu\rho_{27}^{1/2}\|x\|\,\|y\|\\
&\qquad \forall (x,y,\mu) \in \mathcal{N}_x \times \mathcal{N}_y \times \mathcal{M},
\end{aligned}$$

where ρ_{11}, ρ_{12}, ρ_{21}, ρ_{22}, ρ_{18}, ρ_{26} are the maximal eigenvalues of the matrices

$$\begin{gathered}
B_1A_{11} + A_{11}^{\mathrm{T}}B_1 + B_1q_1k_1^*c_{11}^{\mathrm{T}} + (q_1k_1^*c_{11}^{\mathrm{T}})^{\mathrm{T}}B_1,\\
B_1q_1k_1^*c_{11}^{\mathrm{T}} + (q_1k_1^*c_{11}^{\mathrm{T}})^{\mathrm{T}}B_1,\\
B_2A_{22} + A_{22}^{\mathrm{T}}B_2 + B_2q_2k_2^*c_{22}^{\mathrm{T}} + (q_2k_2^*c_{22}^{\mathrm{T}})^{\mathrm{T}}B_2,\\
B_2q_2k_2^*c_{22}^{\mathrm{T}} + (q_2k_2^*c_{22}^{\mathrm{T}})^{\mathrm{T}}B_2,\\
A_{12}^{\mathrm{T}}B_3 + (q_1k_1^*c_{12}^{\mathrm{T}})^{\mathrm{T}}B_3,\\
B_3A_{21} + B_3q_2k_2^*c_{21}^{\mathrm{T}},
\end{gathered}$$

respectively; $\rho_{13}^{1/2}$, $\rho_{23}^{1/2}$, $\rho_{15}^{1/2}$, $\rho_{17}^{1/2}$, $\rho_{25}^{1/2}$, $\rho_{27}^{1/2}$ are the norms of the matrices

$$B_1A_{12} + B_1q_1k_1^*c_{12}^{\mathrm{T}},$$
$$B_2A_{21} + B_2q_2k_2^*c_{21}^{\mathrm{T}},$$
$$A_{11}^{\mathrm{T}}B_3 + (q_1k_1^*c_{11}^{\mathrm{T}})^{\mathrm{T}}B_3,$$
$$(q_1k_1^*c_{11}^{\mathrm{T}})^{\mathrm{T}}B_3,$$
$$B_3A_{22} + B_3q_2k_2^*c_{22}^{\mathrm{T}},$$
$$B_3q_2k_2^*c_{22}^{\mathrm{T}},$$

respectively,

$$k_i^* = \begin{cases} k_i & \text{for } \sigma_iq_iB_jx > 0 \text{ (or } \sigma_iq_iB_jy > 0); \\ 0 & \text{for } \sigma_iq_iB_jx \le 0 \text{ (or } \sigma_iq_iB_jy \le 0); \end{cases} \qquad (i = 1,2; \quad j = 1,2,3).$$

Denoting the upper bound of the derivative of the function (3.10.13) by $\dfrac{d}{dt}V_M(x,y,\mu)$, we find the estimate

$$\frac{d}{dt}V_M(x,y,\mu) \le u^{\mathrm{T}}C(\mu)u, \tag{3.10.16}$$

where

$$C(\mu) = \begin{pmatrix} \sigma_{11} & \sigma_{12} \\ \sigma_{21} & \sigma_{22} \end{pmatrix}, \qquad \sigma_{12} = \sigma_{21};$$
$$\sigma_{11} = \eta_1^2(\rho_{11} + \rho_{12}) + 2\eta_1\eta_2\rho_{26};$$
$$\sigma_{22} = \eta_2^2(\rho_{21} + \rho_{22}) + 2\mu\eta_1\eta_2\rho_{18};$$
$$\sigma_{12} = \eta_1^2\rho_{13}^{1/2} + \eta_2^2\rho_{23}^{1/2} + \eta_1\eta_2(\mu\rho_{15}^{1/2} + \mu\rho_{17}^{1/2} + \rho_{25}^{1/2} + \rho_{27}^{1/2}).$$

We introduce the quantities

$$\mu_1 = -\frac{\eta_2(\rho_{21} + \rho_{22})}{2\eta_1\rho_{18}}; \qquad \mu_2 = \frac{-b + \sqrt{b^2 - 4ac}}{2a}; \qquad \mu_0 = \min(\mu_1, \mu_2),$$

where

$$a = \eta_1^2\eta_2^2(\rho_{15}^{1/2} + \rho_{17}^{1/2})^2;$$
$$b = \eta_1\eta_2(\rho_{15}^{1/2} + \rho_{17}^{1/2})[\eta_1^2\rho_{13}^{1/2} + \eta_2^2\rho_{23}^{1/2} + \eta_1\eta_2(\rho_{25}^{1/2} + \rho_{27}^{1/2})] - 2\eta_1\eta_2\rho_{18}\sigma_{11};$$
$$c = [\eta_1^2\rho_{13}^{1/2} + \eta_2^2\rho_{23}^{1/2} + \eta_1\eta_2(\rho_{25}^{1/2} + \rho_{27}^{1/2})]^2 - \eta_2^2(\rho_{21} + \rho_{22})\sigma_{11}.$$

If it turns out that $\mu_0 > 1$, then we consider $\mu \in (0, 1]$.

PROPOSITION 3.10.1. *The matrix $C(\mu)$ is negative-definite for every $\mu \in (0,1]$ and for $\mu \to 0$ if the following conditions hold:*

(a) $\sigma_{11} < 0$,
(b) $\eta_1 \rho_{18} > 0$,
(c) $\eta_2(\rho_{21} + \rho_{22}) < 0$,
(d) $c < 0$.

REMARK 3.10.1. If $\eta_1 \rho_{18} \leq 0$ and the conditions (a), (b), (d) of Proposition 3.10.1 are satisfied, then its assertion remains valid for $\mu_0 = \mu_2$.

THEOREM 3.10.1. *Assume that the singularly perturbed Lur'e system (3.10.9) is such that the matrix-valued function (3.10.12) has been constructed for it, the elements of which satisfy the estimates (3.10.14), and for the upper bound of the derivative of the function (3.10.13) the estimate (3.10.15) holds.*

In this case, if

(a) *the matrix A is positive-definite;*
(b) *the matrix $C(\mu)$ is negative-definite for every $\mu \in (0, \mu_0)$ and for $\mu \to 0$,*

then the equilibrium state $(x^{\mathrm{T}}, y^{\mathrm{T}}) = 0$ of the system (3.10.9) is uniformly asymptotically stable for every $\mu \in (0, \mu_0)$ and for $\mu \to 0$.

If, furthermore, $\mathcal{N}_x \times \mathcal{N}_y = R^{n+m}$ then the equilibrium state of the system (3.10.9) is uniformly asymptotically stable on the whole for every $\mu \in (0, \mu_0)$ and for $\mu \to 0$.

PROOF. On the basis of the matrix-valued function (3.10.12), with the aid of the vector $\eta \in R^2_+$, $\eta > 0$, we construct the scalar function (3.10.13). Under the estimates (3.10.14) one can show that

$$v(x, y, \mu) \geq u^{\mathrm{T}} H^{\mathrm{T}} A H u, \qquad \forall\, (x, y, \mu) \in \mathcal{N}_x \times \mathcal{N}_y \times \mathcal{M}.$$

Then from condition (a) of Theorem 3.10.1 there follows that the function $V(x, y, \mu)$ is positive-definite.

For the derivative $\dfrac{d}{dt}V(x, y, \mu)$ the estimate (3.10.15) holds. From here and from condition (b) of Theorem 3.10.1 there follows that the derivative $\dfrac{d}{dt}V(x, y, \mu)$ of the function (3.10.13) is negative-definite for every

$\mu \in (0, \mu_0)$ and for $\mu \to 0$. As is known (see Grujić, Martynyuk and Ribbens-Pavella [1]), these conditions are sufficient for the uniform asymptotic stability of the equilibrium state of the system (3.10.9).

In the case $\mathcal{N}_x \times \mathcal{N}_y = R^{n+m}$ the function $V(x, y, \mu)$ is radially unbounded which, together with the other conditions, proves the second assertion of this theorem. This is the absolute stability of the system (3.10.9), μ_0 being an estimate of the upper bound of the variation of the parameter μ.

3.10.2.2 Singularly Perturbed Lur'e System. Case B. Assume that the singularly perturbed system is the Lur'e-type system:

$$
\begin{aligned}
\frac{dx}{dt} &= A_{11}x + A_{12}y + q_1 f_1(\sigma_1), \qquad \sigma_1 = c_{11}^{\mathrm{T}}x + c_{12}^{\mathrm{T}}y; \\
\mu\frac{dy}{dt} &= \mu A_{21}x + A_{22}y + q_2 f_2(\sigma_2), \qquad \sigma_2 = c_{21}^{\mathrm{T}}x + c_{22}^{\mathrm{T}}y.
\end{aligned}
\tag{3.10.17}
$$

Here we preserve all the assumptions made regarding the system (3.10.9), including the assumption on the equilibrium state, i.e., the conditions on the system (3.10.10), (3.10.11).

We assume that for the system (3.10.17) we have constructed the matrix-valued function (3.10.12) for the elements of which the estimated (3.10.14) are satisfied. We introduce the following notations:

$$
\begin{aligned}
f(x,0) &= A_{11}x + q_1 f_1(\sigma_1^0); \\
f^*(x,y) &= A_{12}y + q_1[f_1(\sigma_1) - f_1(\sigma_1^0)]; \\
g(0,y) &= A_{22}y + q_2 f_2(\sigma_2^0); \\
g^*(x,y,\mu) &= \mu A_{21}x + q_2[f_2(\sigma_2) - f_2(\sigma_2^0)].
\end{aligned}
$$

The system (3.10.17) takes the form

$$
\begin{aligned}
\frac{dx}{dt} &= f(x,0) + f^*(x,y); \\
\mu\frac{dy}{dt} &= g(0,y) + g^*(x,y,\mu).
\end{aligned}
$$

By virtue of the system (3.10.17), for the derivatives of the elements v_{ij} of the matrix-valued function (3.10.12) we have the estimates:

(3.10.18)

$$
\begin{aligned}
&\text{(a)}\quad (\nabla_x v_{11})^{\mathrm T} f(x,0) \le \rho_{11}\|x\|^2 \quad \forall\, x \in \mathcal{N}_{x0};\\
&\text{(b)}\quad (\nabla_x v_{11})^{\mathrm T} f^*(x,y) \le \rho_{12}\|x\|^2 + 2\rho_{13}^{1/2}\|x\|\,\|y\|\\
&\qquad \forall\,(x,y) \in \mathcal{N}_{x0}\times\mathcal{N}_{y0};\\
&\text{(c)}\quad (\nabla_y v_{22})^{\mathrm T} g(0,y) \le \mu\rho_{21}\|y\|^2 \quad \forall\,(y,\mu) \in \mathcal{N}_{y0}\times\mathcal{M};\\
&\text{(d)}\quad (\nabla_y v_{22})^{\mathrm T} g^*(x,y,\mu) \le \mu\rho_{22}\|y\|^2 + 2\mu^2\rho_{23}^{1/2}\|x\|\,\|y\|\\
&\qquad \forall\,(x,y,\mu) \in \mathcal{N}_{x0}\times\mathcal{N}_{y0}\times\mathcal{M};\\
&\text{(e)}\quad (\nabla_x v_{12})^{\mathrm T} f(x,0) \le \mu\rho_{15}^{1/2}\|x\|\,\|y\|\\
&\qquad \forall\,(x,y,\mu) \in \mathcal{N}_{x0}\times\mathcal{N}_{y0}\times\mathcal{M};\\
&\text{(f)}\quad (\nabla_x v_{12})^{\mathrm T} f^*(x,y) \le \mu\rho_{17}^{1/2}\|x\|\,\|y\| + \mu\rho_{18}\|y\|^2\\
&\qquad \forall\,(x,y,\mu) \in \mathcal{N}_{x0}\times\mathcal{N}_{y0}\times\mathcal{M};\\
&\text{(g)}\quad (\nabla_y v_{12})^{\mathrm T} g(0,y) \le \mu\rho_{25}^{1/2}\|x\|\,\|y\|\\
&\qquad \forall\,(x,y,\mu) \in \mathcal{N}_{x0}\times\mathcal{N}_{y0}\times\mathcal{M};\\
&\text{(h)}\quad (\nabla_y v_{12})^{\mathrm T} g^*(x,y,\mu) \le \mu^2\rho_{26}\|x\|^2 + \mu\rho_{27}^{1/2}\|x\|\,\|y\|\\
&\qquad \forall\,(x,y,\mu) \in \mathcal{N}_{x0}\times\mathcal{N}_{y0}\times\mathcal{M}.
\end{aligned}
$$

Here, ρ_{11}, ρ_{12}, ρ_{21}, ρ_{22}, ρ_{18}, ρ_{26} and $\rho_{13}^{1/2}$, $\rho_{23}^{1/2}$, $\rho_{15}^{1/2}$, $\rho_{17}^{1/2}$, $\rho_{25}^{1/2}$, $\rho_{27}^{1/2}$ are the same quantities as in the estimates (3.10.15). We note that the presence of the small parameter μ in the right-hand side of the system (3.10.18) leads only to the modification of the estimates (3.10.15)(d) and (3.10.15)(h) to the form (3.10.18)(d).

Denoting the upper bound of the derivative of the function (3.10.13) along the solution of the system (3.10.17) by $\dfrac{d}{dt}V(x,y,\mu)$, we find the estimate

$$
\frac{d}{dt}V_M(x,y,\mu) \le u^{\mathrm T}\widetilde{C}(\mu)u, \tag{3.10.19}
$$

where

$$
\widetilde{C}(\mu) = \begin{pmatrix} \widetilde{\sigma}_{11} & \widetilde{\sigma}_{12} \\ \widetilde{\sigma}_{21} & \widetilde{\sigma}_{22} \end{pmatrix}; \qquad \widetilde{\sigma}_{12} = \widetilde{\sigma}_{21};
$$

$$
\widetilde{\sigma}_{11} = \eta_1^2(\rho_{11}+\rho_{12}) + 2\mu\eta_1\eta_2\rho_{26};
$$

$$
\widetilde{\sigma}_{22} = \eta_2^2(\rho_{21}+\rho_{22}) + 2\mu\eta_1\eta_2\rho_{18};
$$

$$
\widetilde{\sigma}_{12} = \eta_1^2\rho_{13}^{1/2} + \mu\eta_2^2\rho_{23}^{1/2} + \eta_1\eta_2(\mu\rho_{15}^{1/2} + \mu\rho_{17}^{1/2} + \rho_{25}^{1/2} + \rho_{27}^{1/2}).
$$

We introduce the quantities

$$\widetilde{\mu}_1 = -\frac{\eta_1(\rho_{11}+\rho_{12})}{2\eta_2\rho_{26}}; \quad \widetilde{\mu}_2 = -\frac{\eta_2(\rho_{21}+\rho_{22})}{2\eta_1\rho_{18}}; \quad \widetilde{\mu}_3 = -\frac{\widetilde{b}+\sqrt{\widetilde{b}^2-4\widetilde{a}\widetilde{c}}}{2\widetilde{a}};$$

$$\widetilde{\mu}_0 = \min(\widetilde{\mu}_1, \widetilde{\mu}_2, \widetilde{\mu}_3);$$

$$\widetilde{a} = [\eta_2^2\rho_{23}^{1/2} + \eta_1\eta_2(\rho_{15}^{1/2} + \rho_{17}^{1/2})]^2 - 4\eta_1^2\eta_2^2\rho_{18}\rho_{26};$$

$$\begin{aligned}\widetilde{b} = &[\eta_1^2\rho_{13}^{1/2} + \eta_1\eta_2(\rho_{25}^{1/2} + \rho_{27}^{1/2})][\eta_2^2\rho_{23}^{1/2} + \eta_1\eta_2(\rho_{15}^{1/2} + \rho_{17}^{1/2})] \\ &- 2\eta_1\eta_2[\eta_1^2\rho_{18}(\rho_{11}+\rho_{12}) + \eta_2^2\rho_{26}(\rho_{21}+\rho_{22})];\end{aligned}$$

$$\widetilde{c} = [\eta_1^2\rho_{13}^{1/2} + \eta_1\eta_2(\rho_{25}^{1/2} + \rho_{27}^{1/2})]^2 - \eta_1^2\eta_2^2(\rho_{11}+\rho_{12})(\rho_{21}+\rho_{22}).$$

If it turns out that $\mu_0 > 1$, then we consider $\mu \in (0,1)$.

PROPOSITION 3.10.2. *The matrix $\widetilde{C}(\mu)$ is negative-definite for every $\mu \in (0,\mu_0)$ and for $\mu \to 0$ the following conditions hold:*

(a) $\eta_1(\rho_{11}+\rho_{12}) < 0$;
(b) $\eta_2\rho_{26} > 0$;
(c) $\eta_2(\rho_{21}+\rho_{22}) < 0$;
(d) $\eta_1\rho_{18} > 0$;
(e) $\widetilde{a} > 0$;
(f) $\widetilde{c} < 0$.

REMARK 3.10.2. If $\eta_2\rho_{26} \leq 0$ and conditions (a), (c)–(f) of Proposition 3.10.2 are satisfied, then its assertion remains valid for $\widetilde{\mu}_0 = \min(\widetilde{\mu}_2, \widetilde{\mu}_3)$.

REMARK 3.10.3. If $\eta_1\rho_{18} \leq 0$ and conditions (a)–(c), (e), (f) of Proposition 3.10.2 are satisfied, then its assertion remains valid for $\widetilde{\mu}_0 = \min(\widetilde{\mu}_1, \widetilde{\mu}_3)$.

REMARK 3.10.4. If $\eta_2\rho_{26} \leq 0$, $\eta_1\rho_{18} \leq 0$ and conditions (a), (b), (e), (f) of Proposition 3.10.2 are satisfied, then its assertion remains valid for $\widetilde{\mu}_0 = \widetilde{\mu}_3$.

THEOREM 3.10.2. *Assume that the singularly perturbed Lur'e system (3.10.17) is such that the matrix-valued function (3.10.12) has been constructed for it, the elements of which satisfy the estimates (3.10.14), and for the upper bound of the derivative of the function (3.10.13) the estimate (3.10.19) holds.*

In this case, if

(a) *the matrix A is positive-definite;*

(b) *the matrix* $\widetilde{C}(\mu)$ *is negative-definite for every* $\widetilde{\mu} \in (0, \widetilde{\mu}_0)$ *and for* $\mu \to 0$,

then the equilibrium state $(x^{\mathrm{T}}, y^{\mathrm{T}})^{\mathrm{T}} = 0$ *of the system (3.10.17) is uniformly asymptotically stable for every* $\widetilde{\mu} \in (0, \widetilde{\mu}_0)$ *and for* $\mu \to 0$.

If, furthermore, $\mathcal{N}_x \times \mathcal{N}_y = R^{n+m}$ *then the equilibrium state of the system (3.10.17) is uniformly asymptotically stable on the whole for every* $\widetilde{\mu} \in (0, \widetilde{\mu}_0)$ *and for* $\mu \to 0$.

3.10.2.3 Example. We consider a system of the form (3.10.17) in which

$$A_{11} = \begin{pmatrix} 0 & 1 \\ -1 & -2 \end{pmatrix}; \qquad q_1 = \begin{pmatrix} 0 \\ 0.1 \end{pmatrix}; \qquad c_{11} = \begin{pmatrix} -0.01 \\ 0 \end{pmatrix};$$

$$A_{12} = \begin{pmatrix} 1 & 0 \\ 0 & 1 \end{pmatrix}; \qquad c_{12} = \begin{pmatrix} 1 \\ 1 \end{pmatrix} \qquad (k_1 = 2);$$

$$A_{21} = \begin{pmatrix} 0.001 & 0 \\ 0 & 0.001 \end{pmatrix}; \qquad c_{21} = \begin{pmatrix} 0.001 \\ 0 \end{pmatrix}; \qquad q_2 = \begin{pmatrix} 1 \\ 1 \end{pmatrix};$$

$$A_{22} = \begin{pmatrix} -4 & 1 \\ 1 & -4 \end{pmatrix}; \qquad c_{22} = \begin{pmatrix} 1 \\ 0 \end{pmatrix} \qquad (k_2 = 1).$$

The matrix-valued function (3.10.12) has the elements

$$v_{11}(x) = x^{\mathrm{T}} \begin{pmatrix} 0.3 & 0.1 \\ 0.1 & 0.3 \end{pmatrix} x; \qquad v_{22}(y, \mu) = \mu y^{\mathrm{T}} \begin{pmatrix} 2 & 0 \\ 0 & 2 \end{pmatrix} y;$$

$$v_{12}(x, y, \mu) = v_{21}(x, y, \mu) = \mu x^{\mathrm{T}} \begin{pmatrix} 0.01 & 0 \\ 0 & 0.01 \end{pmatrix} y,$$

for which we have the estimates

$$v_{11}(x) \ge 0.2\|x\|^2; \qquad v_{22}(y, \mu) \ge 2\mu\|y\|^2;$$
$$v_{12}(x, y, \mu) \ge -0.01\mu\|x\|\,\|y\|.$$

If $\eta_i = 1$, $i = 1, 2$, then the matrix

$$A = \begin{pmatrix} 0.2 & -0.01\mu \\ -0.01\mu & 2\mu \end{pmatrix}$$

is positive-definite for every $\mu \in (0, 1)$.

Moreover, for the elements of the matrix $\widetilde{C}(\mu)$ we have:

(1) for $k_i^* = k_i$: $\rho_{11} = -0.15290$, $\rho_{12} = 0.00043$, $\rho_{21} = -7.67545$, $\rho_{22} = 4.82843$, $\rho_{18} = 0.012$, $\rho_{26} = 0.00002$, $\rho_{23}^{1/2} = 0.00490$, $\rho_{13}^{1/2} = 0.46165$, $\rho_{15}^{1/2} = 0.02415$, $\rho_{17}^{1/2} = 0.00002$, $\rho_{25}^{1/2} = 0.05117$, $\rho_{27}^{1/2} = 0.01414$, and $\widetilde{a} = 0.00084$, $\widetilde{b} = 0.01909$, $\widetilde{c} = -0.15638$, $\widetilde{\mu}_{11} = 7623.205$, $\widetilde{\mu}_{12} = 118.6257$, $\widetilde{\mu}_{13} = 6.13236$;

(2) for $k_i^* = 0$: $\rho_{11} = -0.15279$, $\rho_{21} = -12$, $\rho_{18} = 0.01$, $\rho_{26} = 0.00001$, $\rho_{13}^{1/2} = 0.4$, $\rho_{23}^{1/2} = 0.002$, $\rho_{15}^{1/2} = 0.02414$, $\rho_{25}^{1/2} = 0.05$, $\rho_{12} = \rho_{22} = \rho_{17}^{1/2} = \rho_{27}^{1/2} = 0$, and $\widetilde{a} = 0.00068$, $\widetilde{b} = 0.01506$, $\widetilde{c} = -1.6398$, $\widetilde{\mu}_{21} = 7639.5$, $\widetilde{\mu}_{22} = 600$, $\widetilde{\mu}_{23} = 28.84152$.

It is easy to verify that in both cases the conditions of Proposition 3.10.2 are satisfied.

The quantity $\widetilde{\mu}_0 = \min(\mu_{ij},\ i = 1,2;\ j = 1,2,3) = 6.13236 > 1$.

Thus, on the basis Proposition 3.10.2, in the given example the matrix $\widetilde{C}(\mu)$ is negative-definite for every $\mu \in (0,1)$ and for $\mu \to 0$. On the basis of Theorem 3.10.2, the state $(x^{\mathrm{T}}, y^{\mathrm{T}}) = 0$ of the system, determined in this example, is absolutely stable for every $\mu \in (0,1)$ and for $\mu \to 0$.

We note that this example has been investigated in (Grujić, Martynyuk and Ribbens-Pavella [57]) by the vector function method. The obtained estimate has been $\widetilde{\mu} = 0.52$.

The use of the Liapunov matrix-valued function in the theory of absolute stability of a singularly perturbed system may turn out to be preferable to the method of the scalar or vector function because of two circumstances: the Liapunov matrix-valued function broadens the possibilities for the dynamical properties of the degenerate system (3.10.10) and of the boundary-layer system (3.10.11), and may give a more accurate estimate of the upper value of the parameter μ.

3.11 Notes

3.1. Singularly-perturbed systems are known to be rather widely used in the engineering and technology as models of real processes (see e.g. surveys by Vasiljeva and Butuzov [170]; Kokotović, O'Malley, and Sannuti [86, 87]; Grujić [50, 51]; and some others). Stability properties of SPS were studied by Gradshtein [43]; Tikhonov [169]; Klimushev and Krasovskii [84, 85]; Hoppensteadt [73–77]; Wilde and Kokotović [172]; Šiljak [168]; Zien [176];

Porter [156–158]; Habets [64, 65]; E. Geraschenko and M. Geraschenko [41]; Grujić [48, 52, 54]; Martynyuk and Gutowsky [123]; Martynyuk [114]; Martynyuk and Miladzhanov [126–128]. Monograph by Grujić, Martynyuk and Ribbens-Pavella [57] contains rather full list of bibliography on the SPS stability.

The present chapter describes a way of the Liapunov's direct method application basing on auxiliary matrix-valued function. This approach admits a weakening of some requirements to dynamical properties of the subsystems. In the chapter the investigation of the problem on absolute stability of singularly-perturbed Lur'e-Postnikov system is made minutely.

3.2. The description of system (3.2.1), (3.2.2) follows Grujić [48] and Grujić, Martynyuk and Ribbens-Pavella [57].

3.3–3.4. The contents of Section 3.3 and 3.4 may be found in Grujić, Martynyuk and Ribbens-Pavella [57].

3.5–3.8. The presentation of these sections is based on results by Martynyuk [114] and Martynyuk and Miladzhanov [128].

3.9. The results of this section are due to Martynyuk and Miladzhanov [126, 127].

3.10. The motion equations of the plane two-component pendulum are due to Novozhilov [149]. The investigation of these equations made in this section corresponds to Miladzhanov [145].

The problem of absolute stability plays a central role in stability theory as a consequence of its theoretical and applied importance. In 1944, Lur'e and Postnikov have shown that the mathematical model of hydraulic servosystems is described by a system of differential equations of a special form. These systems have been called Lur'e–Postnikov systems or Lur'e systems. The problem of absolute stability, closely related with these systems, has become classical in control theory. Since 1944, various approaches to the solution of the stability problem of Lur'e systems have been suggested. The majority of them are directed at the determination of sufficient conditions for absolute stability. The first results have been obtained by Lur'e. The conditions obtained by him are purely algebraic and the stability problem reduces to the verification of the existence of solutions for nonlinear algebraic equations. A sufficiently complete bibliography regarding this problem can be found in Grujić, Martynyuk and Ribbens-Pavella [57] and Gelig, Leonov and Yakubovich [39].

Popov's elegant frequency criteria (see Popov [155]) have been widely used and have stimulated further investigations in various directions of the theory of absolute stability. Grujić [53] has established necessary and sufficient conditions (Liapunov-type conditions) for absolute stability, from which follows that, in the family of functionals, one has to use more than one function, the form of which need not be affected by the form of the nonlinearities of the system. Likhtarnikov and Yakubovich [103] have presented a new approach for the analysis of the absolute stability of nonlinear systems. The essence of this approach consists in the fact that to a linear block in an automatic system one associates a linear manifold in some functional space; the non-linear blocks are described in an analogous manner. Moreover, the intersection of the sets of all possible processes on the input $\sigma(t)$ and the output $\xi(t)$ characterizes a closed system (the class of the corresponding systems). Then, on the basis of the theorem on the minimization of quadratic functionals in linear spaces under quadratic constraints, one constructs absolute stability criteria.

The investigation of the Lur'e-Postnikov system in subsection 3.10.2 is presented in accordance with Martynyuk and Miladzhanov [126].

4

STABILITY ANALYSIS OF STOCHASTIC SYSTEMS

4.1 Introduction

The impact estimation of perturbations, both determined and random ones, is of a great importance for the functioning of real physical systems. Therefore, it is reasonable to consider systems modeled by stochastic differential equations. The present chapter deals with the various types of probability stability for the above mentioned type of equations and develops the method of matrix-valued Liapunov functions with reference to the system of equations of Kats-Krasovskii's form [82] and Ito's form [78]. In the chapter sufficient conditions are formulated for stability and asymptotic stability with respect to probability, global stability with respect to probability, etc.

The notion of averaged derivative of matrix-valued Liapunov function along solutions of the system that has the meaning of infinitesimal operator [34] is crucial in the investigations of this chapter. In a large number of cases this operator defines unequivocally a random Markov process that models the perturbation in the system.

4.2 Stochastic Systems of Differential Equations in General

4.2.1 Notations

For the convenience of readers we collect the following additional nomenclature.

Let R^n be an n-dimensional Euclidean space with norm $\|\cdot\|$, $\nabla_u = \partial/\partial u$, $\nabla_{uv} = \partial^2/\partial u \partial v$, where u and v can be either scalars or vectors. For instance, if $x \in R^n$ and $v \in R^m \to R$, then $\nabla_x v$ denotes the gradient of vector v and $\nabla_{xx} v$ is a matrix with elements $\partial^2 v/\partial x_i \partial x_j$, $i, j \in [1, n]$. Let $\mathcal{T} = R_+ = [0, +\infty)$ and $(\Omega, \mathcal{A}, P)$ denote a probability space with probability measure P, defined on the σ-algebra $\mathcal{A}$ of ω-sets ($\omega \in \Omega$) in the sample space Ω. Every $\mathcal{A}$ measurable function on Ω is said to be random variable. A sequence of the random variables designated by $\{x(t),\, t \in \mathcal{T}\}$ is called a random process with parameter value t from $\mathcal{T}$. We designate by $R[\mathcal{T}, R[\Omega, R^n]]$ the class of random processes defined on $\mathcal{T}$ with the values in $R[\Omega, R^n]$. Random function $x \in R[[a,b]\colon R[\Omega, R^n]]$ is called measurable on the product, provided that $x(t,\omega)$ is a function measurable on $(\mathcal{A}' \times \mathcal{A})$ and defined on $[a,b] \times \Omega$ with the values in R^n, where A' designates the σ-algebra of measurable in the sense of Lebesque sets on $[a,b]$.

For the set $A \in \mathcal{A}$, $P(A)$ denotes the probability of event A and $P(A/B)$ means the conditional probability of event A under condition $B \in \mathcal{A}$. Function $x(t,\omega)$ is called continuous with respect to $t \in [a,b]$ if

$$P\left\{ \bigcup_{t\in[a,b]} \left\{ \lim_{\delta\to 0} [\|x(t+\delta) - x(t)\|] \neq 0 \right\} \right\} = 0,$$

where $\delta > (< 0)$ when $t = a(b)$.

We designate by $C[[a,b], R[\Omega, R^n]]$ the class of continuous functions defined on $[a,b]$.

Function $x(t)$ admits derivative $x'(t)$ for $t \in [a,b]$ provided

$$P\left\{ \bigcup_{t\in[a,b]} \left\{ \lim_{\delta\to 0} \left[\left\| \frac{x(t+\delta) - x(t)}{\delta} - x'(t) \right\| \right] \neq 0 \right\} \right\} = 0.$$

Let E denote the expectation operator and $\{x_t,\, t \in \mathcal{T}\}$ be a Markov process. Then $E_{x,s}x_t$ denotes the expected value of x_t at $t \in \mathcal{T}$ if it is known that $x_s = x$.

4.2.2 The Motion Equations of Random Parameter Systems

4.2.2.1 Equations of Kats-Krasovskii Form. We consider a system modeled by equations of the form

$$\frac{dx}{dt} = f(t, x, y(t)) \tag{4.2.1}$$

with determined initial conditions

$$x(t_0) = x_0, \tag{4.2.2}$$

$$y(t_0) = y_0. \tag{4.2.3}$$

Here $x \in R^n$, $t \in \mathcal{T}$ (or $t \in \mathcal{T}_\tau = [\tau, +\infty), \tau \geq 0$), $y(t)$ is a perturbation vector that can take the values from $Y \subset R^n$ for every $t \in \mathcal{T}$.

We assume that the vector function f is continuous with respect to every variable and satisfies Lipschitz condition in variable x, i.e.

$$\|f(t, x', y) - f(t, x'', y)\| \leq L\|x' - x''\|$$

in domain $B(\mathcal{T}, \rho, Y)$: $t \in \mathcal{T}$, $\|x\| < \rho$, $y \in Y$ ($\rho =$ const or $\rho = +\infty$) uniformly in $t \in \mathcal{T}$ and $y \in Y$, and is bounded for all $(t, y) \in \mathcal{T} \times Y$ in every bounded domain $\|x\| < \rho^*$ ($\rho^* =$ const > 0).

Moreover, we assume that

$$f(t, 0, y(t)) = 0 \qquad \forall\,(t, y) \in \mathcal{T} \times Y, \tag{4.2.4}$$

i.e. the unperturbed motion of system (4.2.1) corresponds to the solution $x(t) \equiv 0$.

In system (4.2.1) the random perturbation $y(t)$ is considered to be a random Markov process (see e.g. Doob [31] and Dynkin [34]). Further, two main types of random Markov functions are under consideration.

Case A. The vector $y(t)$ consists of components y_s, $s = 1, 2, \dots, r$ which are independent of each others pure discontinuous Markov processes, the transition functions $P\{y, \tau; \mathcal{A}, t\}$ of which admit the expansion

$$\begin{aligned} P\{y_s(t + \Delta t) \leq \beta,\ y_s(t + \Delta t) \neq \eta \mid y_s(t) = \eta\} \\ = q_s(t, \eta, \beta)\Delta t + o(\Delta t), \end{aligned} \tag{4.2.5}$$

$$\begin{aligned} P\{y_s(\tau) \equiv \eta,\ t < \tau \leq t + \Delta t \mid y_s(t) = \eta\} \\ = 1 - \tilde{q}_s(t, \eta)\Delta t + o(\Delta t). \end{aligned} \tag{4.2.6}$$

Here $o(\Delta t)$ is an infinitesimal value of the highest order of smallness relatively Δt, $q_s(t, \eta, \beta)$ and $\tilde{q}_s(t, \eta)$ are some known functions such that

$$q_s(t, \eta, \infty) = \tilde{q}_s(t, \eta), \qquad s = 1, 2, \dots, r.$$

In general we assume almost all realizations $y_s(t, \omega)$ of random process $y(t)$ to be piecewise constant functions continuous from the right.

It should be noted that if the set $Y = \{y_1, \dots, y_k\}$ is one-dimensional and finite, then the representation of functions $q(t, \eta, \beta)$ and $\tilde{q}(t, \eta)$ means the representation of transition matrix

$$p_{ij}(t + \Delta t) = q(t, i, j)\Delta t + o(\Delta t), \qquad i \neq j \tag{4.2.7}$$

where $p_{ij}(t, t + \Delta t)$ is a probability of transition $y_i \to y_j$ during the time from t to $t + \Delta t$.

The process $y(t)$ is called a homogeneous Markov chain with a finite number of states, if $q(t, i, j) = \tilde{q}(i, j)$.

Case B. Vector $y(t)$ is a solution of the generalized differential Ito equation (see e.g. Arnold [5] or Gikhman and Skorokhod [42]).

$$dy(t) = a(t, y(t))dt + b(t, y(t))d\omega(t) + \int c(t, y(t), u)\tilde{\nu}(dt, du) \tag{4.2.8}$$

Besides, $a(t, y)$ and $c(t, y, u)$ are r-component vectors with values in R^r, $y \in R^r$, $u \in R^r$, $b(t, y)$ is a $r \times m$-matrix, $\omega(t)$ is a standard m-dimensional Wienner process with independent coordinates, $\tilde{\gamma}(t, A) = \nu(t, A) - t\lambda(A)$, $\gamma(t, A)$ is a Poisson measure in R^r having a compact carrier, $E\nu(t, A) = t\lambda(A)$, the process $\omega(t)$ and the measure $\nu(t, A)$ are independent of each other.

For the existence conditions with only probability 1 and continuous from the right solution of the equation (4.2.8) see Gikhman and Skorokhod [42].

Following Kats and Krasovskii [82] we shall use the following descriptive interpretation of the solution of (4.2.1). Let almost every realization $y(t, \omega)$ of a random process $y(t)$ and the initial condition (4.2.2), (4.2.3) generate completely continuous realization $x(t, \omega)$ of solutions to the equation

$$\frac{dx}{dt} = f(t, x, y(t, \omega)) \tag{4.2.9}$$

lying in the domain $B(\mathcal{T}, \rho, Y)$ and continuable on $\mathcal{T}_\tau = [\tau, +\infty)$.

Then, the set of these realizations forms an $(n + r)$-dimensional random Markov process $\{x(t), y(t)\}$ that will be referred to as the solution of equations (4.2.1) satisfying conditions (4.2.2) and (4.2.3).

4.2.2.2 Equation of Ito Form. We consider the equation

$$dx = f(t, x)dt + \sigma(t, x)dy(t), \tag{4.2.10}$$

where $t \in \mathcal{T}$, $x_t \in R^n$, $f\colon \mathcal{T} \times R^n \to R^n$, $\sigma\colon \mathcal{T} \times R^n \to R^{n\times m}$ and $\{y(t),\, t \in \mathcal{T}\}$ is a Markov process with independent increments. The system of the equations (4.2.10) is perturbed by two specific types of stochastic processes.

Case C. $\{y(t),\, t \in \mathcal{T}\} \triangleq \{z_t, t \in \mathcal{T}\}$ is a normed m-dimensional Wienner process with independent components.

Case D. $\{y(t),\, t \in \mathcal{T}\} \triangleq \{q_t, t \in \mathcal{T}\}$ is a normed m-dimensional discontinuous Poisson process with independent components.

For the physical interpretation of equation (4.2.10) see e.g. Arnold [5], Kushner [90], et al. Functions f and σ are assumed to be smooth enough and there exists a separable and measurable Markov process $\{x_t,\, t \in \mathcal{T}\}$ satisfying system (4.2.10), that is completely continuous with probability 1.

4.2.3 The Concept of Probability Stability

The notions of probability stability are obtained in terms of Definitions 1.2.1–1.2.3 by replacement of ordinary convergence $x \to 0$, used there, by various types of the probability convergence (convergence with respect to probability, convergence in mean square or almost probable stability). Before we introduce the definitions let us pay attention to the following.

Let the process $y(t)$ be defined by Ito equation (4.2.8). Moreover, equations (4.2.1) and (4.2.8) and initial conditions (4.2.2) and (4.2.3) generate $(n+r)$-dimensional Markov process $\{x_t, y(t)\}$.

If $x(t_0) = 0$, then we have with probability 1 that $x(t) = 0$ for all $t \in \mathcal{T}$ and, therefore, the vector function $\{0, y(t)\}$ is a solution of this system. Let $y(t) \in Y$ for all $t \in \mathcal{T}$, and the set $D = \{0, Y\}$ is a time-invariant set for the process $\{x_t, y(t)\}$ in the sense that

$$P\{\,\{x(t), y(t)\} \in D \mid x(t_0) = x_0,\, y(t_0) = y_0\} = 1$$

for $\{x_0, y_0\} \in D$.

Similar equality is valid for the processes $\{x(t), y(t)\}$ generated by pure discontinuous Markov functions $y(t)$. Therefore, the notion of probability stability discussed herein is based on the stability of an invariant set, for instance $D = \{0, Y\}$.

DEFINITION 4.2.1. The state $x = 0$ of the system (4.2.1) is:

(i) *stable in probability with respect to* $\mathcal{T}_i$ if and only if for every $t_0 \in \mathcal{T}_i$ and every $\varepsilon > 0$, and $1 > p > 0$ there exists $\delta(t_0, \varepsilon) > 0$, such that

$$\|x_0\| < \delta(t_0, \varepsilon) \quad \text{and} \quad y_0 \in Y \tag{4.2.11}$$

implies

$$P\left\{\sup_{t \ge t_0} \|x(t; t_0, x_0, y_0\| < \varepsilon \mid x_0, y_0\right\} > 1 - p \tag{4.2.12}$$

for all $t \in \mathcal{T}_0$;

(ii) *uniformly stable in probability with respect to* $\mathcal{T}_i$ if and only if both (i) holds and for every $\varepsilon > 0$ the corresponding maximal δ_M obeying (i) satisfies

$$\inf\left[\delta_M(t_0, \varepsilon) \colon t_0 \in \mathcal{T}_i\right] > 0;$$

(iii) *stable in probability in the whole with respect to* $\mathcal{T}_i$ if and only if both (i) holds and

$$\delta_M(t_0, \varepsilon) \to +\infty \quad \text{as} \quad \varepsilon \to +\infty \quad \forall\, t_0 \in \mathcal{T}_i;$$

(iv) *uniformly stable in probability in the whole with respect to* $\mathcal{T}_i$ if and only if both (ii) and (iii) holds.

(v) *unstable in probability with respect to* $\mathcal{T}_i$ if and only if there are $t_0 \in \mathcal{T}_i$, $\varepsilon > 0$, $p > 0$ and $\tau \in \mathcal{T}_0$, $\tau > t_0$ such that for every $\delta > 0$ there is $x_0 \colon \|x_0\| < \delta$ and $y_0 \in Y$, for which

$$P\{\, \|x(\tau; t_0, x_0, y_0\| > \varepsilon \mid x_0, y_0\} > 1 - p.$$

The expression "with respect to $\mathcal{T}_i$" is omitted from (i)–(v) if and only if $\mathcal{T}_i = R$.

DEFINITION 4.2.2. The state $x = 0$ of the system (4.2.1) is:

(i) *attractive in probability with respect to* $\mathcal{T}_i$ if and only if for every $t_0 \in \mathcal{T}_i$ there exists $\Delta(t_0) > 0$ and for every $\varsigma > 0$ there exists $\tau(t_0, x_0, y_0, \varsigma) \in [0, +\infty)$ and $p > 0$ such that

$$\|x_0\| < \Delta(t_0) \quad \text{and} \quad y_0 \in Y$$

implies

$$P\left\{ \sup_{t \geq t_0 + \tau} \|x(t; t_0, x_0, y_0\| < \varsigma \mid x_0, y_0 \right\} > 1 - p;$$

(ii) (x_0, y_0)*-uniformly attractive in probability with respect to* $\mathcal{T}_i$ if and only if both (i) is true and for every $t_0 \in \mathcal{T}_i$ there exists $\Delta(t_0) > 0$ and for $\varsigma \in (0, +\infty)$ there exists $\tau_u[t_0, \Delta(t_0), Y, \varsigma] \in [0, +\infty)$ such that

$$\sup\left[\tau_m(t_0, x_0, y_0, \varsigma) \colon x_0 \in B_\Delta(t_0),\ y_0 \in Y\right] = \tau_u[t_0, \Delta(t_0), Y, \varsigma]$$

(iii) *t_0-uniformly attractive in probability with respect to $\mathcal{T}_i$* if and only if (i) is true, there is $\Delta > 0$ and for every $(x_0, y_0, \varsigma) \in B_\Delta \times Y \times (0, +\infty)$ there exists $\tau_u(\mathcal{T}_i, x_0, y_0, \varsigma) \in [0, +\infty)$ such that

$$\sup\left[\tau_m(t_0, x_0, y_0, \varsigma)\colon t_0 \in \mathcal{T}_i,\ y_0 \in Y\right] = \tau_u[\mathcal{T}_i, x_0, y_0, \varsigma];$$

(iv) *uniformly attractive in probability with respect to $\mathcal{T}_i$* if and only if both (ii) and (iii) hold, that is, that (i) is true, there exists $\Delta > 0$ and for every $\varsigma \in (0, +\infty)$ there is $\tau_u[\mathcal{T}_i, \Delta, Y, \varsigma) \in [0, +\infty)$ such that

$$\sup\left[\tau_m(t_0, x_0, y_0, \varsigma)\colon (t_0, x_0, y_0) \in \mathcal{T}_i \times B_\Delta \times Y\right] = \tau_u(\mathcal{T}_i, \Delta, Y, \varsigma).$$

(v) The properties (i)–(iv) hold "*in the whole*" if and only if (i) is true for every $\Delta(t_0) \in (0, +\infty)$ and every $t_0 \in \mathcal{T}_i$.

The expression "with respect to $\mathcal{T}_i$" is omitted if and only if $\mathcal{T}_i = R$.

DEFINITION 4.2.3. The state $x = 0$ of the system (4.2.1) is:

(i) *asymptotically stable in probability with respect to $\mathcal{T}_i$* if and only if it is both stable in probability with respect to $\mathcal{T}_i$ and attractive in probability with respect to $\mathcal{T}_i$;

(ii) *equi-asymptotically stable in probability with respect to $\mathcal{T}_i$* if and only if it is both stable in probability with respect to $\mathcal{T}_i$ and (x_0, y_0)-uniformly attractive in probability with respect to $\mathcal{T}_i$;

(iii) *quasi-uniformly asymptotically stable in probability with respect to $\mathcal{T}_i$* if and only if it is both uniformly stable in probability with respect to $\mathcal{T}_i$ and t_0-uniformly attractive in probability with respect to $\mathcal{T}_i$;

(iv) *uniformly asymptotically stable in probability with respect to $\mathcal{T}_i$* if it is both uniformly stable in probability with respect to $\mathcal{T}_i$ and uniformly attractive in probability with respect to $\mathcal{T}_i$;

(v) the properties (i)–(iv) hold "*in the whole*" if and only if both the corresponding stability in probability of $x = 0$ and the corresponding attraction in probability of $x = 0$ hold in the whole;

(vi) *exponentially stable in probability with respect to $\mathcal{T}_i$* if and only if there are $\Delta > 0$ and real numbers $\alpha \geq 1$, $\beta > 0$ and $0 < p < 1$ such that $\|x_0\| < \Delta$ and $y_0 \in Y$ implies

$$P\left\{\sup_{t \geq t_0} \|x(t; t_0, x_0, y_0\| < \alpha \|x_0\| \exp[-\beta(t - t_0)] \mid x_0, y_0\right\} > 1 - p.$$

This holds in the whole if and only if it is true for $\Delta = +\infty$.

The expression "with respect to $\mathcal{T}_i$" is omitted if and only if $\mathcal{T}_i = R$.

REMARK 4.2.1. The definitions of stability in probability based on the inequality

$$P\{\|x(t;t_0,x_0,y_0\| < \varepsilon \mid x(t_0) = x_0,\ y(t_0) = y_0\} > 1 - p \tag{4.2.13}$$

under the condition

$$\|x_0\| < \delta \quad \text{and} \quad y_0 \in Y$$

does not characterize separate realizations of the process $\{x(t), y(t)\}$. I.e. the solution can satisfy the condition (4.2.13), though at the same time almost all realizations may not leave the domain $\|x\| < \varepsilon$ (at various times). Therefore, following Kats and Krasovskii [82] we consider inequality (4.2.12) instead of (4.2.13).

REMARK 4.2.2. The probabilities mentioned in Definitions 4.2.1–4.2.3 are not specified in the general case by the finite dimensional distributions of the process $\{x(t), y(t)\}$ and may not exist. However, it is known (see Doob [31]) that a separable modification of the process $\{x(t), y(t)\}$ can be considered, having with probability 1 the realization continuous from the right. In this case all realizations in question have the meaning.

4.2.4 Stochastic Matrix-Valued Liapunov Function

We relate with the system (4.2.1) the stochastic matrix-valued function

$$\Pi(t, x, y(t)) = [v_{kl}(t, x, y(t))], \qquad k, l \in [1, s] \tag{4.2.14}$$

where $(t, x, y) \in B$ and $v_{kl}(t, 0, y(t)) \equiv 0\ \forall\, t \in \mathcal{T}$ and $y \in Y$, and, besides, $v_{kl}(t, \cdot) = v_{lk}(t, \cdot)\ \forall\,(k \neq l) \in [1, s]$, $v_{kl} \in C(\mathcal{T} \times R^n \times Y, R[Y, R])$.

Similar to the determined case (see Chapter 2) the property of having a fixed sign of matrix-valued stochastic function (4.2.14) is of importance in the stability investigation of a stochastic system (4.2.1).

The concept of the property of having a fixed sign must correspond to

(1) the property of having a fixed sign of stochastic matrix;
(2) the property of having a fixed sign of scalar stochastic Liapunov function;
(3) the construction of direct Liapunov method for stochastic systems.

To achieve this we act as follows.

Let $z \in R^s$ and function $V \in C(\mathcal{T} \times R^n \times Y^s \times R^s, R[Y, R])$ be defined by the formula

$$V(t, x, y, z) = z^{\mathrm{T}} \Pi(t, x, y(t)) z. \tag{4.2.15}$$

In view of Definitions 2.2.1–2.2.2 we present some definitions for stochastic matrix-valued Liapunov function.

DEFINITION 4.2.4. The stochastic matrix-valued function $\Pi\colon R_+ \times B(\rho) \times Y \to R[Y, R^{s\times s}]$ is referred to as

(i) *positive (negative) definite,* if and only if there exists a time-invariant connected neighborhood $\mathcal{N}$ of point $x = 0$ $(\mathcal{N} \subseteq R^n)$ and positive definite in the sense of Liapunov function $w(x)$ such that
   (a) Π is continuous, i.e. $\Pi \in C(R_+ \times \mathcal{N} \times Y, R[Y, R^{s\times s}])$
   (b) $\Pi(t, 0, y) = 0 \ \forall\, t \in R_+$ and $y \in Y$;
   (c) $\inf V(t, x, y, z) = w(x) \ \forall\,(t, y, z) \in R_+ \times Y \times R^s$;
   $(\sup V(t, x, y, z) = -w(x) \ \forall\,(t, y, z) \in R_+ \times Y \times R^s)$;

(ii) *positive (negative) definite on* $\mathcal{S}$, if and only if all conditions of Definition 4.2.4 (i) are satisfied for $\mathcal{N} = \mathcal{S}$;

(iii) *positive (negative) definite in the whole,* if and only if all conditions of Definition 4.2.4 (i) are satisfied for $\mathcal{N} = R^n$.

REMARK 4.2.3. If function Π does not depend on $t \in R_+$, then in Definition 4.2.4 the requirement of function $w(x)$ existence is omitted and conditions (a)–(c) are modified, and condition (c) becomes

(c$'$) $V(x, y, z) = z^{\mathrm{T}} \Pi(x, y) z > 0 \ \forall\,(x \neq 0,\ z \neq 0, y) \in \mathcal{N} \times R^s \times Y$,
$(V(x, y, z) < 0 \ \forall\,(x \neq 0,\ z \neq 0, y) \in \mathcal{N} \times R^s \times Y)$.

DEFINITION 4.2.5. The stochastic matrix-valued function $\Pi\colon R_+ \times B(\rho) \times Y \to R[Y, R^{s\times s}]$ is referred to as

(i) *positive semi-definite,* if and only if there exist a time-invariant connected neighborhood $\mathcal{N}$ of point $x = 0$ $(\mathcal{N} \subseteq R^n)$ such that
   (a) Π is continuous in $(t, x) \in R_+ \times \mathcal{N}$;
   (b) Π is non-negative on $\mathcal{N}$: $z^{\mathrm{T}} \Pi(t, x, y) z \geq 0 \ \forall\,(t, x, y) \in R_+ \times \mathcal{N} \times Y$.
   (c) Π vanishes at the origin $z^{\mathrm{T}} \Pi(t, 0, y) z = 0 \ \forall\,(z \neq 0,\ y \in Y)$;

(ii) *positive semi-definite on* $R_+ \times \mathcal{S} \times Y$ if and only if (i) holds for $\mathcal{N} = \mathcal{S}$;

(iii) *positive semi-definite in the whole* if and only if (i) holds for $\mathcal{N} = R^n$;

(iv) *negative semi-definite (in the whole)* if and only if $(-\Pi)$ is positive semi-definite (in the whole) respectively.

The following assertion is proved in the same manner as Proposition 2.6.1 from Chapter 2.

PROPOSITION 4.2.1. *The stochastic matrix-valued function* $\Pi\colon R_+ \times B(\rho) \times Y \to R[Y, R^{s\times s}]$ *is positive definite, if and only if there exists a vector* $z \in R^s$ *and a positive definite in the sense of Liapunov function* $a \in K$ *such that*

$$z^{\mathrm{T}}\Pi(t,x,y)z = z^{\mathrm{T}}\Pi_+(t,x,y)z + a(x), \tag{4.2.16}$$

where $\Pi_+(t,x,y)$ *is a stochastic positive semi-definite matrix-valued function.*

DEFINITION 4.2.6. The stochastic matrix-valued function $\Pi\colon R_+ \times B(\rho) \times Y \to R[Y, R^{s\times s}]$ is referred to as

(i) *decreasing,* if and only if there exists a time-invariant connected neighborhood $\mathcal{N}$ of point $x = 0$ and a positive definite on $\mathcal{N}$ function $b \in K$ such that

$$V(t,x,y,z) = z^{\mathrm{T}}\Pi(t,x,y)z \le b(x)$$

for all $(t,x,y) \in R_+ \times \mathcal{N} \times Y \times R^s$;

(ii) *decreasing on* $\mathcal{S}$ if and only if (i) holds for $\mathcal{N} = \mathcal{S}$;

(iii) *decreasing in the whole* if and only if (i) holds for $\mathcal{N} = R^n$.

PROPOSITION 4.2.2. *The stochastic matrix-valued function* $\Pi\colon R_+ \times B(\rho) \times Y \to R[Y, R^{s\times s}]$ *is decreasing, if and only if there exists a vector* $z \in R^s$ *and a positive definite in the sense of Liapunov function* $c \in K$ *such that*

$$z^{\mathrm{T}}\Pi(t,x,y)z = z^{\mathrm{T}}Q_-(t,x,y)z + c(x), \tag{4.2.17}$$

where $Q_-(t,x,y)$ *is a stochastic negative semi-definite matrix-valued function.*

DEFINITION 4.2.7. The stochastic matrix-valued function $\Pi\colon R_+ \times R^n \times Y \to R[Y, R^{s\times s}]$ is referred to as *radially unbounded* if and only if $z^{\mathrm{T}}\Pi(t,x,y)z \to \infty$ as $\|x\| \to +\infty$ and $y \in Y$, $t \in R_+$.

PROPOSITION 4.2.3. *The stochastic matrix-valued function* $\Pi\colon R_+ \times R^n \times Y \to R[Y, R^{s\times s}]$ *is radially unbounded, if and only if there exist a vector* $z \in R^s$ *and a function* $\gamma \in KR$ *such that*

$$z^{\mathrm{T}}\Pi(t,x,y)z = z^{\mathrm{T}}Q_+(t,x,y)z + \gamma(\|x\|) \tag{4.2.18}$$

for all $(t,x,y) \in R_+ \times R^n \times Y$, *where* $Q_+(t,x,y)$ *is a positive semi-definite in the whole matrix-valued function.*

We indicate a class of auxiliary stochastic function $v_{kl}(t,x,y(t))$, $k,\, l = 1,2,\dots,s$ using which it is possible to construct the function (4.2.15) satisfying all conditions of Definitions 4.2.4–4.2.7.

State vector $x \in R^n$ of the system (4.2.1) is represented in the form $x = (p^{\mathrm{T}}, q^{\mathrm{T}}, r^{\mathrm{T}})^{\mathrm{T}}$, where $p \in R^{n_1}$, $q \in R^{n_2}$, $r \in R^{n_3}$ and $n_1 + n_2 + n_3 = n$.

ASSUMPTION 4.2.1. There exists time-invariant connected neighborhoods $\mathcal{N}_p \subseteq R^{n_1}$, $\mathcal{N}_q \subseteq R^{n_2}$ and $\mathcal{N}_r \subseteq R^{n_3}$ of the equilibrium states $p = 0$, $q = 0$ and $r = 0$ respectively, functions $\varphi_i(\|p\|)$, $\psi_i(\|q\|)$, $\chi_i(\|r\|)$, $i = 1,2$ of class $K\,(KR)$ and constants $\underline{\alpha}_{jk}$, $\overline{\alpha}_{jk}$, $\forall\,(j,k) \in [1,3]$ and $\overline{\alpha}_{jj}$ and $\underline{\alpha}_{jj} > 0$, $j \in [1,3]$ are such that

(a) $\underline{\alpha}_{11}\varphi_1^2(\|p\|) \le v_{11}(t,x,y) \le \overline{\alpha}_{11}\varphi_2^2(\|p\|)\ \forall\,(t,x,y) \in R_+ \times \mathcal{N}_0 \times Y$,

(b) $\underline{\alpha}_{22}\psi_1^2(\|q\|) \le v_{22}(t,x,y) \le \overline{\alpha}_{22}\psi_2^2(\|q\|)\ \forall\,(t,x,y) \in R_+ \times \mathcal{N}_0 \times Y$;

(c) $\underline{\alpha}_{33}\chi_1^2(\|r\|) \le v_{33}(t,x,y) \le \overline{\alpha}_{33}\chi_2^2(\|r\|)\ \forall\,(t,x,y) \in R_+ \times \mathcal{N}_0 \times Y$;

(d) $\underline{\alpha}_{12}\varphi_1(\|p\|)\psi_1(\|q\|) \le v_{12}(t,x,y) \le \overline{\alpha}_{12}\varphi_2(\|p\|)\psi_2(\|q\|)\ \forall\,(t,x,y) \in R_+ \times \mathcal{N}_0 \times Y$;

(e) $\underline{\alpha}_{13}\varphi_1(\|p\|)\chi_1(\|r\|) \le v_{13}(t,x,y) \le \overline{\alpha}_{13}\varphi_2(\|p\|)\chi_2(\|r\|)\ \forall\,(t,x,y) \in R_+ \times \mathcal{N}_0 \times Y$;

(f) $\underline{\alpha}_{23}\psi_1(\|q\|)\chi_1(\|r\|) \le v_{23}(t,x,y) \le \overline{\alpha}_{23}\psi_2(\|q\|)\chi_2(\|r\|)\ \forall\,(t,x,y) \in R_+ \times \mathcal{N}_0 \times Y$;

(g) $\underline{\alpha}_{21}\psi_1(\|q\|)\varphi_1(\|p\|) \le v_{21}(t,x,y) \le \overline{\alpha}_{21}\psi_2(\|q\|)\varphi_2(\|p\|)\ \forall\,(t,x,y) \in R_+ \times \mathcal{N}_0 \times Y$;

(h) $\underline{\alpha}_{31}\chi_1(\|r\|)\varphi_1(\|p\|) \le v_{31}(t,x,y) \le \overline{\alpha}_{31}\chi_2(\|r\|)\varphi_2(\|p\|)\ \forall\,(t,x,y) \in R_+ \times \mathcal{N}_0 \times Y$;

(i) $\underline{\alpha}_{32}\chi_1(\|r\|)\psi_1(\|q\|) \le v_{32}(t,x,y) \le \overline{\alpha}_{32}\chi_2(\|r\|)\psi_2(\|q\|)\ \forall\,(t,x,y) \in R_+ \times \mathcal{N}_0 \times Y$,

where $\mathcal{N}_0 = \mathcal{N}_{p0} \times \mathcal{N}_{q0} \times \mathcal{N}_{r0}$; $\mathcal{N}_{p0} = \{p \in \mathcal{N}_p,\, p \ne 0\}$, $\mathcal{N}_{q0} = \{q \in \mathcal{N}_q,\, q \ne 0\}$, $\mathcal{N}_{r0} = \{r \in \mathcal{N}_r,\, r \ne 0\}$.

PROPOSITION 4.2.4. *If all conditions of Assumption 4.2.1 are satisfied, then for the function*

$$V(t,x,y,\eta) = \eta^{\mathrm{T}}\Pi(t,x,y)\eta, \tag{4.2.19}$$

with a constant positive vector $\eta \in R^s_+$ *the bilateral estimate*

$$u^{\mathrm{T}}H^{\mathrm{T}}A_1Hu \le V(t,x,y,\eta) \le w^{\mathrm{T}}H^{\mathrm{T}}A_2Hw \tag{4.2.20}$$

takes place for all $(t,x,y) \in R_+ \times \mathcal{N}_0 \times Y$, *where*

$$u^{\mathrm{T}} = (\varphi_1(\|p\|), \psi_1(\|q\|), \chi_1(\|r\|)),$$
$$w^{\mathrm{T}} = (\varphi_2(\|p\|), \psi_2(\|q\|), \chi_2(\|r\|))$$

and $A_1 = [\underline{\alpha}_{kl}]$, $A_2 = [\overline{\alpha}_{kl}]$, $H = \mathrm{diag}\,(\eta_1, \eta_2, \eta_2)$.

Estimates (4.2.20) are proved by direct substitution by estimates (a)–(i) from Assumption 4.2.1 into the form

$$V(t,x,y,\eta) = \sum_{l,k=1}^{s} \eta_l \eta_k v_{lk}(t,x,y).$$

Estimates (4.2.20) imply

PROPOSITION 4.2.5. *If in the bilateral estimate (4.2.20)*

(1) *the matrix* $H^{\mathrm{T}}A_1H$ *is positive definite (semi-definite);*
(2) *the matrix* $H^{\mathrm{T}}A_2H$ *is positive definite;*
(3) *the condition (1) is satisfied and functions* $\varphi_1, \psi_1, \chi_1$ *are of class* KR,

then stochastic function (4.2.19) is

(1) *positive definite (semi-definite);*
(2) *decreasing;*
(3) *radially unbounded*

respectively.

PROOF. Assertion (1) of Proposition 4.2.5 follows from the fact that

$$\lambda_m(\tilde{A}_1)u^{\mathrm{T}}u \le u^{\mathrm{T}}H^{\mathrm{T}}A_1Hu, \qquad \lambda_m(\tilde{A}_1) > 0,$$

where $\tilde{A}_1 = H^{\mathrm{T}}A_1H$. In fact, since $(\varphi_1, \psi_1, \chi_1) \in K$, then a function $\Phi \in K$, $\Phi = \Phi(\|x\|)$ is found such that

$$\Phi(\|x\|) \le \varphi_1^2(\|p\|) + \psi_1^2(\|q\|) + \chi_1^2(\|r\|).$$

Therefore,

$$\lambda_m(\tilde{A}_1)\Phi(\|x\|) \le u^{\mathrm{T}} H^{\mathrm{T}} A_1 H u \le V(t,x,y,\eta)$$

for all $(t,x,y) \in R_+ \times \mathcal{N}_0 \times Y$.

Assertions (2) and (3) of Proposition 4.2.5 are proved similarly.

4.2.5 Structure of the Stochastic Matrix-Valued Function Averaged Derivative

The averaged derivative, that is computed as in determined case without integrating system (2.2.1), is analogous to the total derivative of matrix-valued function for the stochastic system (4.2.1).

Let (τ, x, y) be a point in domain $B(\mathcal{T}, \rho, Y)$.

DEFINITION 4.2.8. Any of the limits

$$\begin{aligned} D^+E[\Pi] &= \limsup \big\{\{E[\Pi(t,x,y) \mid x(\tau)=x,\ y(\tau)=y] \\ &\qquad - \Pi(\tau,x,y)\}(t-\tau)^{-1}\colon t \to \tau+0\big\}; \\ D_+E[\Pi] &= \liminf \big\{\{E[\Pi(t,x,y) \mid x(\tau)=x,\ y(\tau)=y] \\ &\qquad - \Pi(\tau,x,y)\}(t-\tau)^{-1}\colon t \to \tau+0\big\}; \end{aligned} \tag{4.2.21}$$

where $E[\,\cdot \mid \cdot\,]$ is a conditional mathematical expectation, is called an *averaged derivative of stochastic matrix-valued function* $\Pi(t,x,y(t))$ *along the solution of system* (4.2.1) at point (τ,x,y). $D^*E[\Pi]$ denotes the case, when $D^+E[\Pi]$ and $D_+E[\Pi]$ are applicable.

The value $D^*E[\Pi]$ is an averaged value of the stochastic matrix-valued function $\Pi(t,x,y)$ derivative along all realizations of process $\{x(t),y(t)\}$ initiating from point (x,y) at time τ. If

$$\begin{aligned} T^+\Pi &= \int P\{\tau,x,y;t,du,dz\}\Pi(t,u,z) \\ &= E[\Pi(t,x(t),y(t)) \mid x(\tau)=x,\ y(\tau)=y], \end{aligned}$$

where $P\{\cdots\}$ is a transition function of solution to system (4.2.1) with the initial conditions $x(\tau)=x$, $y(\tau)=y$, then

$$D^+E[\Pi] = \limsup \big\{[T_\tau^t\Pi - \Pi(\tau,x,y)](t-\tau)^{-1}\colon t \to \tau+0\big\}; \tag{4.2.22}$$

$$D_+E[\Pi] = \liminf \{[T_\tau^t \Pi - \Pi(\tau, x, y)](t-\tau)^{-1} : t \to \tau + 0\} \tag{4.2.23}$$

at the point (τ, x, y).

The right-side part of (4.2.22) and (4.2.23) is a weak infinitesimal operator of process $\{x(t), y(t)\}$.

We shall present the formulas for $D^+E[\Pi]$ computation for various realizations of the random process $y(t)$.

1. Let in the system (4.2.1) the process $y(t)$ be pure discontinuous and be described by the relations (4.2.5) and (4.2.6). Then $\frac{dE[\Pi]}{dt}$ along solutions of system (4.2.1) at point (τ, x, y) is computed as

$$\begin{aligned}\frac{dE[\Pi]}{dt} &= \nabla_\tau v_{kl}(\tau, x, y) + [\nabla_x v_{kl}(\tau, x, y)]^{\mathrm{T}} f(\tau, x, y(t)) \\ &+ \sum_{\mu=1}^{r} \int [v_{kl}(\tau, x, y + \beta_\mu) - v_{kl}(\tau, x, y)] d_\beta q(\tau, y, \beta)\end{aligned} \tag{4.2.24}$$

for all $(k, l) \in [1, s]$, where β_μ is a vector, every μ-th component of which equals to β, and the others are zero.

2. Let in the system (4.2.1) $y(t)$ be a simple scalar Markov chain with a finite or countable number of states and transition probabilities satisfying the correlation

$$P\{y(t) = y_j \mid y(\tau) = y_i\} = q_{ij}(t-s) + o(t-s)$$

for all $i \neq j$. We compute $\frac{dE[\Pi]}{dt}$ by the formula

$$\begin{aligned}\frac{dE[\Pi]}{dt} &= \nabla_\tau v_{kl}(\tau, x, y) + [\nabla_x v_{kl}(\tau, x, y)]^{\mathrm{T}} f(\tau, x, y(t)) \\ &+ \sum_{j \neq i} [v_{kl}(\tau, x, y_j) - v_{kl}(\tau, x, y_i)] q_{ij}.\end{aligned} \tag{4.2.25}$$

3. Let in the system (4.2.1) $y(t)$ be a Markov process generated by the generalized differential Ito equation (4.2.8). In this case we compute $\frac{dE[\Pi]}{dt}$ at point (τ, x, y) by the formula

$$\begin{aligned}\frac{dE[\Pi]}{dt} &= \nabla_\tau v_{kl}(\tau, x, y) + [\nabla_x v_{kl}(\tau, x, y)]^{\mathrm{T}} f(\tau, x, y(t)) \\ &+ [\nabla_y v_{kl}(\tau, x, y)]^{\mathrm{T}} (a(\tau, y) - g(\tau, y)) \\ &+ \int (v_{kl}(\tau, x, y + c(\tau, y, u)) - v_{kl}(\tau, x, y)) \lambda(du) \\ &+ \frac{1}{2} \operatorname{tr} [\nabla_{xx} v_{kl}(\tau, x, y) b(\tau, y) b^{\mathrm{T}}(\tau, y)], \quad \forall\, k, l \in [1, s].\end{aligned} \tag{4.2.26}$$

where $g(\tau, y) = \int c(\tau, y, u)\lambda(du)$.

COROLLARY 4.2.1. If in the formula (4.2.26) $c(t, y, u) \equiv 0$, then $\dfrac{dE[\Pi]}{dt}$ corresponds to the case when $y(t)$ is a diffusion process.

REMARK 4.2.4. Operator $\dfrac{dE[\Pi]}{dt}$ for $c \neq 0$ is local in variable x, but non-local in y.

4. Let in the system (4.2.10) $y(t)$ be a normalized Wienner process with independent components. We compute $\dfrac{dE[\Pi]}{dt}$ at point (τ, x) by the formula

$$\begin{aligned}\frac{dE[\Pi]}{dt} &= \nabla_\tau v_{kl}(\tau, x) + [\nabla_x v_{kl}(\tau, x)]^{\mathrm{T}} f(\tau, x) \\ &+ \frac{1}{2}\operatorname{tr}[\sigma(t, x)^{\mathrm{T}} \nabla_{xx} v_{kl}(\tau, x)]\sigma(t, x)],\end{aligned} \tag{4.2.27}$$

where $k, l \in [1, s]$.

5. Let in the system (4.2.10) $y(t)$ be a normalized jump Poisson process with independent components q_i. Then $\dfrac{dE[\Pi]}{dt}$ at point (τ, x) is computed by the formula

$$\begin{aligned}\frac{dE[\Pi]}{dt} &= \nabla_\tau v_{kl}(\tau, x) + [\nabla_x v_{kl}(\tau, x)]^{\mathrm{T}} f(\tau, x) \\ &+ \sum_{i=1}^{m} \int_{q_i} [v_{kl}(\tau, x + \sigma_i(t, x)q_i) - v_{kl}(\tau, x)] p_i dP_i(dq_i),\end{aligned} \tag{4.2.28}$$

where $k, l \in [1, s]$.

Here it is assumed that during the interval Δt the jumps take place with the probability $P_i\Delta t + o(\Delta t)$ and the zero average of the jumps obeys the probability $P_i(\cdot)$.

We establish Liapunov correlation for stochastic matrix-valued function $\Pi(t, x, y(t))$. With this end we construct function (4.2.19) by means of vector $\eta \in R^s_+$. Let $V(t, x, y, \eta)$ be such that for it there exists

$$E\big[V(t, x(t), y(t), \eta) \mid x(\tau) = x,\, y(\tau) = y\big]$$

and

$$\frac{dE[V]}{dt} = H(\tau, x, y) \tag{4.2.29}$$

on the trajectories of the Markov process $\{x(t), y(t)\}$ at point (τ, x, y). Moreover, we assume that

$$\lim_{t\to\tau+0} E\big[H(t, x(t), y(t)) \mid x(\tau) = x,\ y(\tau) = y\big] = H(\tau, x, y).$$

Then we have

$$\begin{aligned} E\big[V(t, x(t), y(t), \eta) \mid x(\tau) = x,\ y(\tau) = y\big] &= V(\tau, x, y, \eta) \\ &+ \int_{\tau}^{t} E\big[H(u, x(u), y(u)) \mid x(\tau) = x, y(\tau) = y\big]\, du. \end{aligned} \tag{4.2.30}$$

Formula (4.2.30) is valid for the homogeneous Markov processes and functions V independent of time (see Dynkin [34]) and for the processes being considered here (see Kushner [90]).

Let $Q \subset R^n$ be a bounded open set and $U = Q \times Y$ be a set from which the process $\{x(t), y(t)\}$ comes out for the first time at time τ_*. It is easy to notice that $\tau_m(t) = \min\{t, \tau_*\}$ is a Markov momentum, such that $E\tau_m(t) < +\infty$. Therefore, if $\{x(s), y(s)\} \in U$, then

$$\begin{aligned} &E\big[V(\tau_m, x(\tau_m), y(\tau_m), \eta) \mid x(\tau) = x,\ \tau) = y\big] \\ &\qquad = V(\tau, x, y, \eta) + E\left[\int_{\tau}^{\tau_m} H(u, x(u), y(u))du \mid x(\tau) = x,\ \tau) = y\right] \end{aligned}$$

is valid.

It is also clear that the process $\{x(\tau_m(t)), y(\tau_m(t))\}$ is strictly Markov.

Between $\dfrac{d}{dt}E[\Pi]$ and $\dfrac{d}{dt}E[V]$ it is true that

$$\frac{d}{dt}E[V(t, x, y, \eta)] = \eta^{\mathrm{T}} \frac{d}{dt}E[\Pi(t, x, y)]\eta. \tag{4.2.31}$$

We return back to the system (4.2.1) and assume that $y(t)$ is a simple scalar Markov chain with a finite number of states. System (4.2.1) is decomposed into three subsystems

$$\begin{aligned} \frac{dp}{dt} &= X(t, p, 0, 0, y(t)) + F(t, p, q, r, y(t)); \\ \frac{dq}{dt} &= Y(t, 0, q, 0, y(t)) + G(t, p, q, r, y(t)); \\ \frac{dr}{dt} &= Z(t, 0, 0, r, y(t)) + H(t, p, q, r, y(t)); \end{aligned} \tag{4.2.32}$$

where $p \in R^{n_1}$, $q \in R^{n_2}$, $r \in R^{n_3}$, $n_1 + n_2 + n_3 = n$,

$$
\begin{array}{ll}
X \in C(R_+ \times B_1(\rho),\, R[Y, R^{n_1}]), & Y \in C(R_+ \times B_2(\rho),\, R[Y, R^{n_2}]), \\
Z \in C(R_+ \times B_3(\rho),\, R[Y, R^{n_3}]), & F \in C(R_+ \times B,\, R[Y, R^{n_1}]), \\
G \in C(R_+ \times B,\, R[Y, R^{n_2}]), & H \in C(R_+ \times B,\, R[Y, R^{n_3}]),
\end{array}
$$

and $B = B_1(\rho) \times B_2(\rho) \times B_3(\rho)$.

Vector-functions X, Y and Z and F, G and H vanish, if and only if $p = q = r = 0$ respectively.

We introduce designation

$$
\Delta(v_{kl}) = \sum_{i \neq j} \alpha_{ij}[v_{kl}(t, \cdot, i) - v_{kl}(t, \cdot, j)], \qquad k,\, l = 1, 2, 3.
$$

ASSUMPTION 4.2.2. There exist the real numbers ρ_{kr}, $k = 1, 2, 3$; $r = 1, 2, \ldots, 12$ and comparison functions $\varphi(\|p\|)$, $\psi(\|q\|)$, $\chi(\|r\|)$ of class $K(KR)$ such that

(a) $\nabla_t v_{11} + (\nabla_p v_{11})^{\mathrm{T}} X + \frac{1}{2}\Delta(v_{11}) \le \rho_{11}\varphi^2(\|p\|)\ \forall\, (t, p, y) \in R_+ \times \mathcal{N}_p \times Y$;

(b) $\nabla_t v_{12} + (\nabla_p v_{12})^{\mathrm{T}} X + \frac{1}{4}\Delta(v_{12}) \le \rho_{12}\varphi(\|p\|)\psi(\|q\|)\ \forall\, (t, p, q, y) \in R_+ \times \mathcal{N}_p \times \mathcal{N}_q \times Y$;

(c) $\nabla_t v_{13} + (\nabla_p v_{13})^{\mathrm{T}} X + \frac{1}{4}\Delta(v_{13}) \le \rho_{13}\varphi(\|p\|)\chi(\|r\|)\ \forall\, (t, p, r, y) \in R_+ \times \mathcal{N}_p \times \mathcal{N}_r \times Y$;

(d) $\nabla_t v_{22} + (\nabla_q v_{22})^{\mathrm{T}} Y + \frac{1}{2}\Delta(v_{22}) \le \rho_{21}\psi^2(\|q\|)\ \forall\, (t, q, y) \in R_+ \times \mathcal{N}_q \times Y$;

(e) $\nabla_t v_{21} + (\nabla_q v_{21})^{\mathrm{T}} Y + \frac{1}{4}\Delta(v_{21}) \le \rho_{22}\varphi(\|p\|)\psi(\|q\|)\ \forall\, (t, p, q, y) \in R_+ \times \mathcal{N}_p \times \mathcal{N}_q \times Y$;

(f) $\nabla_t v_{23} + (\nabla_q v_{23})^{\mathrm{T}} Y + \frac{1}{4}\Delta(v_{23}) \le \rho_{23}\psi(\|q\|)\chi(\|r\|)\ \forall\, (t, q, r, y) \in R_+ \times \mathcal{N}_q \times \mathcal{N}_r \times Y$;

(g) $\nabla_t v_{33} + (\nabla_r v_{33})^{\mathrm{T}} Z + \frac{1}{2}\Delta(v_{33}) \le \rho_{31}\chi^2(\|r\|)\ \forall\, (t, r, y) \in R_+ \times \mathcal{N}_r \times Y$;

(h) $\nabla_t v_{31} + (\nabla_r v_{31})^{\mathrm{T}} Z + \frac{1}{4}\Delta(v_{31}) \le \rho_{32}\varphi(\|p\|)\chi(\|r\|)\ \forall\, (t, p, r, y) \in R_+ \times \mathcal{N}_p \times \mathcal{N}_r \times Y$;

(i) $\nabla_t v_{32} + (\nabla_r v_{32})^{\mathrm{T}} Z + \frac{1}{4}\Delta(v_{32}) \le \rho_{33}\psi(\|q\|)\chi(\|r\|)\ \forall\, (t, q, r, y) \in R_+ \times \mathcal{N}_q \times \mathcal{N}_r \times Y$.

and for all $(t,p,q,r,y) \in R_+ \times \mathcal{N}_p \times \mathcal{N}_q \times \mathcal{N}_r \times Y$:

(a′) $(\nabla_p v_{11})^{\mathrm{T}} F + \frac{1}{2}\Delta(v_{11}) \le \rho_{14}\varphi^2(\|p\|) + \rho_{15}\varphi(\|p\|)\psi(\|q\|) + \rho_{16}\varphi(\|p\|)\chi(\|r\|)$;

(b′) $(\nabla_p v_{12})^{\mathrm{T}} F + \frac{1}{4}\Delta(v_{12}) \le \rho_{17}\psi^2(\|q\|) + \rho_{18}\varphi(\|p\|)\psi(\|q\|) + \rho_{19}\psi(\|q\|)\chi(\|r\|)$;

(c′) $(\nabla_p v_{13})^{\mathrm{T}} F + \frac{1}{4}\Delta(v_{13}) \le \rho_{1.10}\chi^2(\|r\|) + \rho_{1.11}\varphi(\|p\|)\chi(\|r\|) + \rho_{1.12}\psi(\|q\|)\chi(\|r\|)$;

(d′) $(\nabla_q v_{22})^{\mathrm{T}} G + \frac{1}{2}\Delta(v_{22}) \le \rho_{24}\psi^2(\|q\|) + \rho_{25}\varphi(\|p\|)\chi(\|r\|) + \rho_{26}\psi(\|q\|)\chi(\|r\|)$;

(e′) $(\nabla_q v_{21})^{\mathrm{T}} G + \frac{1}{4}\Delta(v_{21}) \le \rho_{27}\varphi^2(\|p\|) + \rho_{28}\varphi(\|p\|)\psi(\|q\|) + \rho_{29}\varphi(\|p\|)\chi(\|r\|)$;

(f′) $(\nabla_q v_{23})^{\mathrm{T}} G + \frac{1}{4}\Delta(v_{23}) \le \rho_{2.10}\chi^2(\|r\|) + \rho_{2.11}\varphi(\|p\|)\chi(\|r\|) + \rho_{2.12}\psi(\|q\|)\chi(\|r\|)$;

(g′) $(\nabla_r v_{33})^{\mathrm{T}} H + \frac{1}{2}\Delta(v_{33}) \le \rho_{34}\chi^2(\|r\|) + \rho_{35}\varphi(\|p\|)\chi(\|r\|) + \rho_{36}\psi(\|q\|)\chi(\|r\|)$;

(h′) $(\nabla_r v_{13})^{\mathrm{T}} H + \frac{1}{4}\Delta(v_{13}) \le \rho_{37}\varphi^2(\|p\|) + \rho_{38}\varphi(\|p\|)\psi(\|q\|) + \rho_{39}\varphi(\|p\|)\chi(\|r\|)$;

(i′) $(\nabla_r v_{23})^{\mathrm{T}} H + \frac{1}{4}\Delta(v_{23}) \le \rho_{3.10}\psi^2(\|q\|) + \rho_{3.11}\varphi(\|p\|)\psi(\|q\|) + \rho_{3.12}\psi(\|q\|)\chi(\|r\|)$.

PROPOSITION 4.2.6. *If for the system (4.2.1), decomposed to the form of (4.2.32), there exists a stochastic matrix-valued function $\Pi(t,x,y)$ the elements of which satisfy the conditions of Assumption 4.2.1 and all conditions of Assumption 4.2.2 are satisfied, then the structure of stochastic matrix-valued function averaged derivative $\dfrac{dE[V]}{dt}$ is defined by the inequality*

$$\frac{dE[V]}{dt} = \eta^{\mathrm{T}} \frac{dE[\Pi]}{dt}\eta \le u^{\mathrm{T}} S u \qquad \forall\,(t,x,y) \in R_+ \times \mathcal{N}_0 \times Y \tag{4.2.33}$$

where $s \times s$-matrix S has the elements expressed by formulas

$$c_{kl} = c_{lk}, \qquad (k,l) \in [1,3]:$$

$$c_{11} = \eta_1^2(\rho_{11} + \rho_{14}) + 2\eta_1(\eta_2\rho_{27} + \eta_3\rho_{37}),$$

$$c_{22} = \eta_2^2(\rho_{21} + \rho_{24}) + 2\eta_2(\eta_1\rho_{17} + \eta_3\rho_{3.10}),$$

$$c_{33} = \eta_3^2(\rho_{31} + \rho_{34}) + 2\eta_3(\eta_1\rho_{1.10} + \eta_2\rho_{2.10}),$$

$$\begin{aligned} c_{12} =& \frac{1}{2}\eta_1^2\rho_{15} + \frac{1}{2}\eta_2^2\rho_{25} + \eta_1\eta_2(\rho_{12} + \rho_{22} + \rho_{18} + \rho_{28}) \\ &+ \eta_3(\eta_1\rho_{38} + \eta_2\rho_{3.11}), \end{aligned}$$

$$\begin{aligned} c_{13} =& \frac{1}{2}\eta_1^2\rho_{16} + \frac{1}{2}\eta_3^2\rho_{35} + \eta_1\eta_3(\rho_{13} + \rho_{32} + \rho_{1.11} + \rho_{39}) \\ &+ \eta_2(\eta_1\rho_{29} + \eta_3\rho_{2.11}), \end{aligned}$$

$$\begin{aligned} c_{23} =& \frac{1}{2}\eta_2^2\rho_{26} + \frac{1}{2}\eta_3^2\rho_{36} + \eta_2\eta_3(\rho_{23} + \rho_{33} + \rho_{2.12} + \rho_{3.12}) \\ &+ \eta_1(\eta_2\rho_{19} + \eta_3\rho_{1.12}). \end{aligned}$$

THE PROOF of this proposition is similar to the proof of Proposition 2.7.3.

REMARK 4.2.5. Actually, the structure of the stochastic matrix-valued function $\Pi(t,x,y)$ averaged derivative is established by formula (4.2.33) and is based on the stochastic SL-function (see Martynyuk [120]). The structure of the stochastic matrix-valued function $\Pi(t,x,y)$ averaged derivative is somewhat different provided the stochastic VL-function is applied, i.e.

$$L(t,x,y) = A\Pi(t,x,y)b, \tag{4.2.34}$$

where A is a constant $s \times s$-matrix and b is an s-vector.

4.3 Stability to Systems in Kats–Krasovskii Form

In terms of the stochastic matrix-valued function $\Pi(t,x,y)$ constructed for system (4.2.1), the criteria of stability with respect to probability are in form similar to Theorems 2.3.1–2.3.3.

THEOREM 4.3.1. *Let the equations of perturbed motion (4.2.1) are such that:*

(1) *there exists a matrix-valued function $\Pi\colon R_+ \times B(p) \times Y \to R[Y, R^{s\times s}]$ in the time-invariant neighborhood $\mathcal{N} \subseteq R^n$ of equilibrium state $x = 0$;*

(2) *there exists a vector* $\eta \in R^s$ $(\eta \in R^s_+)$;
(3) *stochastic scalar function (4.2.19) is positive definite;*
(4) *the averaged derivative (4.2.25) is negative definite or negative semi-definite.*

Then the equilibrium state $x = 0$ *of system (4.2.1) is stable with respect to probability.*

PROOF. Let arbitrary numbers $\varepsilon \in (0, \rho)$, $\rho \in (0, 1)$ and $t_0 \in R_+$ be given. Under the conditions (1)–(2) of Theorem 4.3.1 we have the function

$$V(t, x, y, \eta) = \eta^{\mathrm{T}} \Pi(t, x, y) \eta, \quad \eta \in R^s \quad (\eta \in R^s_+),$$

that is positive definite by condition (3) of Theorem 4.3.1. Therefore, a number $\varepsilon_1 > 0$ is found, such that

$$\inf V(t, x, y, \eta) = \varepsilon_1 \quad \text{for} \quad t \in R_+, \quad \|x\| \geq \varepsilon, \quad y \in Y, \quad \eta \in R^s \quad (\eta \in R^s_+).$$

We designate $B(\varepsilon) = \{(x, y) \in R^n \times Y \colon \|x\| < \varepsilon, y \in Y\}$. Let τ_ε be the time of trajectory $(x(t), y(t))$ first leaving the domain $B(\varepsilon)$ and let $\tau_\varepsilon(\tau) = \min(\tau, \tau_\varepsilon)$. We have by condition (4)

$$\text{(4.3.1)} \qquad \begin{aligned} E[V(\tau_\varepsilon(\tau), x(\tau_\varepsilon(\tau)), y(\tau_\varepsilon(\tau)), \eta) \mid x(t_0) = x_0,\, y(t_0) = y_0] \\ \leq V(t_0, x_0, y_0, \eta). \end{aligned}$$

Now we take $\delta > 0$ so that

$$\text{(4.3.2)} \qquad \sup V(t_0, x, y_0) < p\varepsilon_1$$

whenever $\|x\| \leq \delta$.

The estimates (4.3.1) and (4.3.2) imply

$$\begin{aligned} p\varepsilon_1 > V(t_0, x_0, y_0, \eta) &\geq E[V(\tau_\varepsilon(\tau), x(\tau_\varepsilon(t)), y(\tau_\varepsilon(\tau)), \eta) \mid x_0, y_0] \\ &\geq \varepsilon_1 P\left\{ \sup_{t_0 \leq t \leq \tau} \|x(t)\| \geq \varepsilon \mid x_0, y_0 \right\}. \end{aligned}$$

Hence we get for $\tau \to +\infty$

$$P\left\{ \sup_{t \geq t_0} \|x(t)\| \geq \varepsilon \mid x_0, y_0 \right\} < p.$$

This proves the theorem.

THEOREM 4.3.2. *Let the equations of perturbed motion (4.2.1) are such that:*

(1) *hypotheses (1) and (2) of Theorem 4.3.1 are satisfied;*
(2) *the stochastic matrix-valued function $\Pi(t,x,y)$ is positive definite and decreasing;*
(3) *the averaged derivative $\dfrac{dE[V]}{dt}$ is negative definite.*

Then the equilibrium state $x = 0$ of the system (4.2.1) is asymptotically stable with probability $p(H)$, i.e. if $\|x_0\| \le H_0$ and $y_0 \in Y$, $t_0 \ge 0$ then

$$P\Big\{\sup_{t\ge t_0}\|x(t)\| < H \mid x_0, y_0\Big\} \ge 1 - p(H), \qquad H_0 < H.$$

PROOF. Let a number $p(H) < 1$ be given. Theorem 4.3.1 implies that under the conditions of Theorem 4.3.2 the equilibrium state $x = 0$ of system (4.2.1) is stable with respect to probability. Therefore, for any $\varepsilon \in (0,\rho)$ and $t_0 \ge 0$ a $\delta = \delta(t_0,\varepsilon) > 0$ can be found such that

$$P\Big\{\sup_{t\ge t_0}\|x(t)\| < \varepsilon \mid x_0, y_0\Big\} > 1 - p(H), \tag{4.3.3}$$

whenever

$$\|x_0\| < \delta \quad \text{and} \quad y_0 \in Y.$$

Let us show that the number H_0 mentioned in conditions of Theorem 4.3.2 can be taken as $H_0 = \delta$. To this end we define for arbitrary numbers $\gamma \in (0,\varepsilon)$ and $0 < q < +\infty$ the number $\gamma_1 > 0$ from the inequality

$$\begin{aligned} &\sup\big[V(t,x,y,\eta) \text{ for } t \in R_+,\ \|x\| < \gamma_1,\ y \in Y,\ \eta \in R^s_+\big] \\ &< \frac{q}{2}\inf\big[V(t,x,y,\eta) \text{ for } t \in R_+,\ \gamma_1 \le \|x\| \le \varepsilon,\ y \in Y \text{ and } \eta \in R^s_+\big]. \end{aligned} \tag{4.3.4}$$

The arguments similar to those used in the proof of Theorem 4.3.1 yield

$$P\Big\{\sup_{\tau>t}\|x(\tau)\| < \gamma \mid x(t), y(t)\Big\} > 1 - \frac{1}{2}q, \tag{4.3.5}$$

whenever

$$\|x(t)\| \le \gamma_1 \quad \text{and} \quad y(t) \in Y.$$

We claim that there exists a $\tau > t_0$ such that

$$P\{\|x(t_0+\tau)\| < \gamma_1 \mid x_0, y_0\} > 1 - \frac{1}{2}q - p(H). \tag{4.3.6}$$

If this is not true, then for trajectory $\{x(t), y(t)\}$ the inequality

$$P\{\gamma_1 \le \|x(t)\| < \varepsilon,\ t \ge t_0 \mid x_0, y_0\} > \frac{1}{2}q.$$

holds, that yields by condition (3) of Theorem 4.3.2

$$\lim_{t\to\infty} E[V(\tau_\alpha(t), x(\tau_\alpha(t)), y(\tau_\alpha(t)), \eta) \mid x_0, y_0] = -\infty. \tag{4.3.7}$$

Here $\tau_\alpha(t) = \min(\tau^*, t)$, where τ^* is a time of trajectory $(x(t), y(t))$ first leaving the set $B_1 = \{(x,y)\colon \gamma_1 < \|x\| < \varepsilon,\ y \in Y\}$.

Since the function $\Pi(t,x,y)$ is positive definite, the correlation (4.3.7) can not be satisfied. This proves inequality (4.3.6). The estimates (4.3.3), (4.3.5) and (4.3.6) imply that for arbitrary $q > 0$ a $\tau > 0$ is found so that

$$P\left\{\sup_{t\ge t_0+\tau} \|x(t)\| < \gamma \mid x_0, y_0\right\} > 1 - q - p(H),$$

whenever $\|x_0\| < H_0$ and $y_0 \in Y$.

This proves Theorem 4.3.2.

THEOREM 4.3.3. *Let the equations of perturbed motion (4.2.1) are such that:*

(1) *hypotheses (1), (2) and (3) of the Theorem 4.3.1 are satisfied for* $\mathcal{N} = R^n$;

(2) *the function* $\Pi(t,x,y)$ *is positive definite in the whole and radially unbounded;*

(3) *the averaged derivative* $\dfrac{dE[V]}{dt}$ *is negative definite in* $B(\mathcal{T}, \infty, Y)$.

Then the equilibrium state $x = 0$ *of the system (4.2.1) is stable with respect to probability in the whole.*

A theorem allowing us to find asymptotic stability with respect to probability and stability with respect to probability in the whole on the basis of negative semi-definite averaged derivative is considered.

Let an open domain G containing the origin be definite in space R^n. Function $\psi(t,x,y)\colon T_0 \times G \times Y \to R$ is referred to as positive definite on $G \times Y$ if for any numbers $r > \varepsilon > 0$ there exists a number $\delta > 0$ such that $\psi(t,x,y) \geq \delta$ holds for all $t \geq t_0$, $(x,y) \in (\mathcal{N} \cap \{\varepsilon \leq \|x\| \leq r\} \times Y)$.

Matrix-valued function $\Phi(t,x,y)\colon T_0 \times G \times Y \to R^{m\times m}$ satisfies hypotheses A if:

(a) the function Φ is bounded for all $t \geq t_0$ in any finite domain $\|x\| \leq \rho$, $y \in Y$;

(b) averaged derivative $\eta^{\mathrm T}\dfrac{dM[\Phi]}{dt}\eta$ is bounded in any finite domain due to system (4.2.1), i.e. there exists a constant K such that

$$\left|\eta^{\mathrm T}\frac{dM[\Phi]}{dt}\eta\right| \leq K;$$

(c) the function $\eta^{\mathrm T}\dfrac{dM[\Phi]}{dt}\eta$ is positive definite in domain $G \times Y$.

Then the following statement is valid.

THEOREM 4.3.4. *Let the equations of perturbed motion (4.2.1) as definite in domain $B(T_0,\infty,Y)$ and such that:*

(1) *hypotheses (1) and (2) of Theorem 4.3.3 are satisfied;*

(2) *averaged derivative (4.2.13) satisfies hypothesis*

$$\eta^{\mathrm T}\frac{dM[\Pi]}{dt}\eta \leq H(x) \leq 0,$$

where $H(x)$ is continuous in domain G;

(3) *the set $D = \{x\colon x \neq 0,\ H(x) = 0\}$ is non-empty and does not possess mutual points with bound $\partial\mathcal{N}$ in domain $\mathcal{N}$ in the sense that $\inf \|x_1 - x_2\| > K^2 > 0$ $x_1 \in \partial G$, $x_2 \in D \cap \{\varepsilon \leq \|x\| \leq r\}$;*

(4) *there exists a matrix-valued function $\Phi(t,x,y)$ satisfying hypotheses A.*

Then the equilibrium state $x = 0$ of the system (4.2.1) is stable with respect to probability in the whole.

4.4 Stability to Systems in Ito's Form

4.4.1 Decomposition of perturbed motion equations

We consider a system of the equations with random parameters in the form

$$d\omega(t) = f(t,\omega)dt + \sigma(t,\omega)d\xi(t), \tag{4.4.1}$$

where $t \in \mathcal{T}$, $\omega \in R^n$, $f\colon \mathcal{T} \times R^n \to R^n$, $\sigma\colon \mathcal{T} \times R^n \to R^{n\times m}$, and $\{\xi(t),\, t \in \mathcal{T}\}$ is an independent measurable random Markov process.

Assume that the system (4.4.1) allows decomposition into l interconnected subsystems that can be described by equations in the form

$$\begin{aligned} d\omega_i =& f_i(t,\omega_i)dt + \sigma_{ii}(t,\omega_i)d\xi_i \\ &+ g_i(t,\omega)dt + \sum_{j=1}^{l} \sigma_{ij}(t,\omega_j)d\xi_j, \qquad i \in [1,l]. \end{aligned} \tag{4.4.2}$$

Each interconnected subsystem (4.4.2) consists of the independent subsystem

$$d\omega_i = f(t,\omega_i)dt + \sigma_{ii}(t,\omega_i)d\xi_i, \qquad i \in [1,l], \tag{4.4.3}$$

and link functions

$$g_i(t,\omega)dt + \sum_{j=1}^{l} \sigma_{ij}(t,\omega_j)d\xi_j, \qquad i \in [1,l]. \tag{4.4.4}$$

Here $\omega_i \in R^{n_i}$, $\omega \in R^n$, $\omega = (\omega_1^{\mathrm{T}}, \omega_2 T, \dots, \omega_l^{\mathrm{T}})^{\mathrm{T}}$, $\xi_i \in R^{m_i}$, $f_i\colon \mathcal{T}_0 \times R^{n_i} \to R^{n_i}$, $\sigma_{ij}\colon \mathcal{T} \times R^{n_j} \to R^{n_i\times m_j}$, $g_i\colon \mathcal{T} \times R^{n_1} \times \dots \times R^{n_l} \to R^{n_i}$, and $\{\xi_i(t),\, t \in \mathcal{T}\}$ are independent measurable Markov processes.

We assume on function f_i and σ_{ii} that they satisfy the existence condition for solutions to subsystems (4.4.3), and link functions (4.4.4) vanish, if and only if $\omega_j = 0$ and $\omega = 0$. Thus, the points $\omega = 0$ and $\omega_j = 0$, $j \in [1,l]$ are the only equilibrium states of systems (4.4.1), (4.4.2) and (4.4.3) respectively.

The transformation of systems (4.4.1) to (4.4.2) is referred to as the decomposition of stochastic Ito system of the first level. Suppose that from system (4.4.1) couples (i, j) of interconnected subsystems are taken in the form

$$(4.4.5)\quad \begin{aligned} d\omega_i =& f_{ij}(t, \omega_i, \omega_j)dt + \sigma_{ii}(t, \omega_i)d\xi_i + \sigma_{ij}(t, \omega_j)d\xi_j \\ &+ g_{ij}(t, \omega)dt + \sum_{\substack{k=1 \\ (k \neq i,j)}}^{l} \sigma_{ik}(t, \omega_k)d\xi_k, \qquad i \in [1, l]. \\ d\omega_j =& f_{ji}(t, \omega_j, \omega_i)dt + \sigma_{jj}(t, \omega_j)d\xi_j + \sigma_{ji}(t, \omega_i)d\xi_i \\ &+ g_{ji}(t, \omega)dt + \sum_{\substack{k=1 \\ (k \neq i,j)}}^{l} \sigma_{jk}(t, \omega_k)d\xi_k, \qquad (i \neq j) \in [1, l]. \end{aligned}$$

Here $f_{ij}: \mathcal{T} \times R^{n_i} \times R^{n_j} \to R^{n_i} \times R^{n_j}$, $g_{ij}: \mathcal{T} \times R^n \to R^{n_i} \times R^{n_j}$. We introduce following designations $\omega_{ij} = (\omega_i^{\mathrm{T}}, \omega_j^{\mathrm{T}})^{\mathrm{T}}$, $\overline{f}_{ij}(t, \omega_{ij}) = (f_{ij}^{\mathrm{T}}, f_{ji}^{\mathrm{T}})^{\mathrm{T}}$; $\overline{g}_{ij}(t, \omega) = (g_{ij}^{\mathrm{T}}, g_{ji}^{\mathrm{T}})^{\mathrm{T}}$, $\sigma_{ij}^k = [\sigma_{ik}^{\mathrm{T}}, \sigma_{jk}^{\mathrm{T}}]^{\mathrm{T}}$, $d\xi_{ij} = (d\xi_i^{\mathrm{T}}, d\xi_j^{\mathrm{T}})^{\mathrm{T}}$, and

$$\overline{\sigma}_{ij} = \begin{pmatrix} \sigma_{ii} & \sigma_{ij} \\ \sigma_{ji} & \sigma_{jj} \end{pmatrix}.$$

Then the (i, j) couple (4.4.5) can be represented as

$$(4.4.6)\quad \begin{aligned} d\omega_{ij} =& \overline{f}_{ij}(t, \omega_{ij})dt + \overline{\sigma}_{ij}d\xi_{ij} + \overline{g}_{ij}(t, \omega)dt \\ &+ \sum_{\substack{k=1 \\ (k \neq i,j)}}^{l} \sigma_{ij}^k d\xi_k, \qquad (i \neq j) \in [1, l]. \end{aligned}$$

Besides, the free (i, j) couple has the form

$$(4.4.7)\quad d\omega_{ij} = \overline{f}_{ij}(t, \omega_{ij})dt + \overline{\sigma}_{ij}d\xi_{ij} \qquad (i \neq j) \in [1, l].$$

and the link functions are represented by the formulas

$$(4.4.8)\quad \overline{g}_{ij}(t, \omega)dt + \sum_{\substack{k=1 \\ (k \neq i,j)}}^{l} \sigma_{ij}^k d\xi_k, \qquad (i \neq j) \in [1, l].$$

Further we need the following assumptions.

ASSUMPTION 4.4.1. There exists a time invariant open connected neighborhood $\mathcal{N}_i \subseteq R^{n_i}$, a function $v_{ii}(t, \omega_i)\colon \mathcal{T} \times \mathcal{N}_i \to R_+$, the comparison functions ψ_{i1}, ψ_{i2} and ψ_{i3} and the positive real numbers ρ_i such that for all $i \in [1, l]$ estimates

(a) $\psi_{i1}(\|\omega_i\|) \le v_{ii}(t, \omega_i) \le \psi_{i2}(\|\omega_i\|)$;

(b) $\dfrac{dE_i[v_{ii}(t, \omega_i)]}{dt} \le p_i \psi_{i3}(\|\omega_i\|)$

are satisfied for any $\omega_i \in \mathcal{N}_i$ and $t \in \mathcal{T}$.

DEFINITION 4.4.1. The isolated subsystems (4.4.3) possesses *property* $A(\mathcal{N}_i)$, provided all conditions of Assumption 4.4.1 are satisfied for each of the subsystems.

DEFINITION 4.4.2. If in Assumption 4.4.1 $\psi_{i1}(\|\omega_i\|) = c_{i1}\|\omega_i\|^2$, $\psi_{i2}(\|\omega_i\|) = c_{i2}\|\omega_i\|^2$ and $\psi_{i3}(\|\omega_i\|) = \dfrac{c_{ii}}{\rho_i}\|\omega_i\|^2$, where c_{i1} and c_{i2} are positive constants, and c_{ii} constants $i \in [1, l]$, then isolated subsystem (4.4.3) is said to possess *property* $B(\mathcal{N}_i)$.

DEFINITION 4.4.3. If in Assumption 4.4.1 $\mathcal{N}_i = R^{n_i}$ for all $i \in [1, l]$ and functions ψ_{11}, $\psi_{12} \in KR$, then isolated subsystems (4.4.3) are said to possesses *property* $B_i(\infty)$.

ASSUMPTION 4.4.2. There exist a time-invariant open connected products of neighborhood $\mathcal{N}_i \times \mathcal{N}_j \subseteq R^{n_i} \times R^{n_j}$ of point $\omega_{ij} = 0$, functions $v_{ij}(t, \omega_{ij})\colon \mathcal{T} \times \mathcal{N}_i \times \mathcal{N}_j \to R_+$, a functions ψ^1_{ij}, ψ^2_{ij} and ψ_{i3} of class K and positive real numbers β^1_{ij}, β^2_{ij} and β^3_{ij} such that for all $(i < j) \in [1, l]$ the estimates

(a) $\psi^1_{ij}(\|\omega_{ij}\|) \le v_{ij}(t, \omega_{ij}) \le \psi^2_{ij}(\|\omega_{ij}\|)$;

(b) $\dfrac{dE_{ij}[v_{ij}]}{dt} \le \beta^1_{ij}\psi_{i3}(\|\omega_i\|) + 2\beta^2_{ij}\psi^{1/2}_{i3}(\|\omega_i\|)\psi^{1/2}_{j3}(\|\omega_j\|) + \beta^3_{ij}\psi_{i3}(\|\omega_j\|)$

are satisfied for any $\omega_{ij} \in \mathcal{N}_i \times \mathcal{N}_j$ and $t \in \mathcal{T}$.

DEFIINITION 4.4.4. Isolated couples (i, j) of subsystems (4.4.7) possesses *property* $A(\mathcal{N}_i \times \mathcal{N}_j)$, if for every of them all conditions of Assumption 4.4.2 are satisfied.

DEFINITION 4.4.5. If in Assumption 4.4.2 $\psi^1_{ij} = c^1_{ij}\|\omega_{ij}\|^2$, $\psi^2_{ij} = c^2_{ij} \times \|\omega_{ij}\|^2$ and $\beta^1_{ij}\psi_{i3}(\|\omega_i\|) + 2\beta^2_{ij}\psi^{1/2}_{i3}(\|\omega_i\|)\psi^{1/2}_{j3}(\|\omega_j\|) + \beta^3_{ij}\psi_{j3}(\|\omega_j\|) =$

$c_{ij}^3\|\omega_{ij}\|^2$, $(i < j) \in [1, l]$, where c_{ij}^1, c_{ij}^2, c_{ij}^3 are constant, then the independent couples (i, j) of subsystems (4.4.7) are said to possess *property* $B(\mathcal{N}_i \times \mathcal{N}_j)$.

DEFINITION 4.4.6. If in Assumption 4.4.2 $\mathcal{N}_i = R^{n_i}$ and the functions Ψ_{ij}^1, $\Psi_{ij}^2 \in KR$, then the independent couples (i, j) of the subsystems (4.4.7) are said to possess *property* $A_{ij}(\infty)$.

REMARK 4.4.1. In Assumptions 4.4.1 and 4.4.2 the constants ρ_i, $i \in [1, l]$ and c_{ij}^3 $(i < j) \in [1, l]$ are negative if independent subsystems (4.4.3) and independent couples (i, j) of subsystems (4.4.7) are exponentially stable with respect to probability.

REMARK 4.4.2. The matrix B_{ij}, defined by the expression

$$B_{ij} = \begin{pmatrix} \beta_{ij}^1 & \beta_{ij}^2 \\ \beta_{ij}^2 & \beta_{i3}^3 \end{pmatrix} \qquad (i < j) \in [1, l]$$

is negative semi-definite (negative definite), if the independent couples (i, j) of subsystems (4.4.7) are stable (asymptotically stable) with respect to probability.

4.4.2 Structure of the Hierarchical Matrix-Valued Function Averaged Derivative

We construct for subsystems (4.4.3) the functions $v_{ii}(t, \omega_i)$, $i \in [1, l]$ and for couples (i, j) of subsystems (4.4.7) the functions $v_{ij}(t, \omega_{ij})$ $(i < j) \in [1, l]$. Let us construct from the above mentioned elements the matrix-valued function.

$$\Pi(t, \omega) = [v_{ij}(t, \cdot)], \tag{4.4.9}$$

where $\Pi\colon \mathcal{T} \times R^{n_i} \times R^{n_j} \times Y \to R[Y, R^{l \times l}]$.

The function (4.4.9) reflects the hierarchy of stochastic subsystems (4.4.3) and (4.4.7) in the large-scale system (4.4.1).

The application of formula (4.2.27) to systems (4.4.2) and (4.4.6) yields the following expressions for $\dfrac{dE[\cdot]}{dt}$

$$
\begin{aligned}
\frac{dE[v_{ii}(t,\omega)]}{dt} &= \sum_{j=1}^{l}[f_j(t,\omega_j)+g_j(t,\omega)]^{\mathrm{T}}\nabla_{\omega_j}v_{ii}(t,\omega_i) \\
&\quad + \frac{1}{2}\mathrm{tr}[\sigma_{ii}^{\mathrm{T}}(t,\omega_i)\nabla_{\omega_i\omega_i}v_{ii}(t,\omega_i)\sigma_{ii}(t,\omega_i)] \\
&\quad + \frac{1}{2}\sum_{\substack{j,k,m=1\\(j\neq i)}}^{l}\mathrm{tr}[\sigma_{kj}^{\mathrm{T}}(t,\omega_j)\nabla_{\omega_k\omega_m}v_{ii}(t,\omega_i)\sigma_{mj}(t,\omega_j)] \\
&\quad + \nabla_t v_{ii}(t,\omega_i) \\
&= \sum_{j=1}^{l}[f_j(t,\omega_j)+g_j(t,\omega)]^{\mathrm{T}}\delta_{ij}\nabla_{\omega_j}v_{ii}(t,\omega_i) \\
&\quad + \frac{1}{2}\mathrm{tr}[\sigma_{ii}^{\mathrm{T}}(t,\omega_i)\nabla_{\omega_i\omega_i}v_{ii}(t,\omega_i)\sigma_{ii}(t,\omega_i)] \\
&\quad + \frac{1}{2}\sum_{\substack{j,k,m=1\\(j\neq i)}}^{l}\mathrm{tr}[\sigma_{kj}^{\mathrm{T}}(t,\omega_j)\delta_{ki}\delta_{mi}\nabla_{\omega_i\omega_i}v_{ii}(t,\omega_i)\sigma_{mj}(t,\omega_j)] \\
&\quad + \nabla_t v_{ii}(t,\omega_i) \\
&= [f_i(t,\omega_i)+g_i(t,\omega)]^{\mathrm{T}}\nabla_{\omega_i}v_{ii}(t,\omega_i) \\
&\quad + \frac{1}{2}\mathrm{tr}[\sigma_{ii}^{\mathrm{T}}(t,\omega_i)\nabla_{\omega_i\omega_i}v_{ii}(t,\omega_i)\sigma_{ii}(t,\omega_i)] \\
&\quad + \frac{1}{2}\sum_{\substack{j=1\\(j\neq i)}}^{l}\mathrm{tr}[\sigma_{ij}^{\mathrm{T}}(t,\omega_j)\nabla_{\omega_i\omega_i}v_{ii}(t,\omega_i)\sigma_{ij}(t,\omega_j)] + \nabla_t v_{ii}(t,\omega_i) \\
&= \frac{dE_i[v_{ii}(t,\omega_i)]}{dt} + g_i^{\mathrm{T}}(t,\omega)\nabla_{\omega_i}v_{ii}(t,\omega_i) \\
&\quad + \frac{1}{2}\sum_{\substack{j=1\\(j\neq i)}}^{l}\mathrm{tr}[\sigma_{ij}^{\mathrm{T}}(t,\omega_j)\nabla_{\omega_i\omega_i}v_{ii}(t,\omega_i)\sigma_{ij}(t,\omega_j)], \\
&\quad (i\neq j)\in[1,l].
\end{aligned}
\tag{4.4.10}
$$

Similarly we have

$$
\begin{aligned}
\frac{dE[v_{ij}(t,\omega)]}{dt} &= [\overline{f}_{ij}(t,\omega_{ij}) + \overline{g}_{ij}(t,\omega)]^{\mathrm{T}}\nabla_{\omega_{ij}} v_{ij}(t,\omega_{ij}) \\
&+ \frac{1}{2}\mathrm{tr}[\overline{\sigma}_{ij}^{\mathrm{T}}(t,\omega_{ij})\nabla_{\omega_{ij}\omega_{ij}} v_{ij}(t,\omega_{ij})\overline{\sigma}_{ij}(t,\omega_{ij})] \\
&+ \frac{1}{2}\sum_{\substack{k=1 \\ (k\neq i,j)}}^{l} \mathrm{tr}[\sigma_{ij}^{k}(t,\omega_k)^{\mathrm{T}}\nabla_{\omega_{ij}\omega_{ij}} v_{ij}(t,\omega_{ij})\sigma_{ij}^{k}(t,\omega_k)] \\
&+ \nabla_t v_{ij}(t,\omega_{ij}) \\
&= \frac{dE_{ij}[v_{ij}(t,\omega_{ij})]}{dt} + \overline{g}_{ij}^{\mathrm{T}}(t,\omega)\nabla_{\omega_{ij}} v_{ij}(t,\omega_{ij}) \\
&+ \frac{1}{2}\sum_{\substack{k=1 \\ (k\neq i,j)}}^{l} \mathrm{tr}[\sigma_{ij}^{kT}(t,\omega_k)\nabla_{\omega_{ij}\omega_{ij}} v_{ij}(t,\omega_{ij})\sigma_{ij}^{k}(t,\omega_k)], \\
&(i<j) \in [1,l],
\end{aligned}
\tag{4.4.11}
$$

where $\nabla_u = \dfrac{\partial}{\partial u}$, and δ_{ij} is the Kronecker symbol.

REMARK 4.4.3. If, in particular, $\sigma_{ij}(t,\omega) = 0$ for all $i \neq j$, then (4.4.10) and (4.4.11) become

$$
\frac{dE[v_{ii}(t,\omega_i)]}{dt} = \frac{dE_i[v_{ii}(t,\omega_i)]}{dt} + (g_i(t,\omega))^{\mathrm{T}}\nabla_{\omega_i} v_{ii}(t,\omega_i), \quad i \in [1,l];
\tag{4.4.12}
$$

$$
\frac{dE[v_{ij}(t,\omega)]}{dt} = \frac{dE_{ij}[v_{ij}(t,\omega_{ij})]}{dt} + (\overline{g}_{ij}(t,\omega))^{\mathrm{T}}\nabla_{\omega_{ij}} v_{ij}(t,\omega_{ij}), \quad (i<j) \in [1,l]
\tag{4.4.13}
$$

Thus, the structure of averaged derivative (4.4.10), (4.4.11) represents adequately the hierarchical dependence of subsystems in large-scale system (4.4.1).

4.4.3 Sufficient Conditions for Stability to Probability of Stochastic Ito System

To formulate sufficient conditions for stability with respect to the probability of system (4.4.1) we make some assumptions on the system.

ASSUMPTION 4.4.3. The system (4.4.1) allows first and second level decompositions and

(1) independent subsystems (4.4.3) possess property $A(\mathcal{N}_i)\ \forall\, i \in [1,l]$;
(2) independent couples (i,j) of subsystems (4.4.7) possess property $A(\mathcal{N}_i \times \mathcal{N}_j)\ \forall\,(i \neq j) \in [1,l]$.

REMARK 4.4.4. If for the system (4.4.1) there exist p and q $(p < q) \in [1,l]$ for which no free couple (p,q) of (4.4.7) can be found, then we take $v_{pq}(t,x_{pq}) \equiv 0$.

ASSUMPTION 4.4.4. There exist time-invariant neighborhoods $\mathcal{N}_i \subseteq R^{n_i}$ and $\mathcal{N}_i \times \mathcal{N}_i \subseteq R^{n_i} \times R^{n_j}$ of states $\omega_i = 0$ and ω_{ij} respectively, constants b_{ij}, d_i, $\gamma_{qp}^{ij} = \gamma_{qp}^{ji}$, α_{ij}, ν_{ij}^k, μ_{ij}^k and functions $\varphi_{i3} \in K$ such that estimates

(1) $g_i^{\mathrm{T}} \nabla_{\omega_i} v_{ii}(t,\omega_i) \leq \psi_{i3}^{1/2}(\|\omega_i\|) \sum_{k=1}^{l} b_{ik} \psi_{k3}^{1/2}(\|\omega_k\|)$,

(2) $\overline{g}_{ij}^{\mathrm{T}} \nabla_{\omega_{ij}} v_{ij}(t,\omega_{ij}) \leq \sum_{\substack{k=1\\p=k}}^{l} \gamma_{kp}^{ij} \psi_{k3}^{1/2}(\|\omega_k\|) \psi_{p3}^{1/2}(\|\omega_p\|)$;

(3) $(u^i)^{\mathrm{T}} \nabla_{\omega_i\omega_i} v_{ii}(t,\omega_i) u^i \leq d_i \|u^i\|^2$;

(4) $(u_k^i)^{\mathrm{T}} \nabla_{\omega_{ij}\omega_{ij}} v_{ij}(t,\omega_{ij}) u_k^i \leq \nu_{ij}^k \|u_k^i\|^2$;

(5) $\|\sigma_{ij}(t,\omega_j)\|^2 \leq \alpha_{ij} \psi_{j3}(\|\omega_j\|)$;

(6) $\|\sigma_{ij}^k(t,\omega_k)\|^2 \leq \mu_{ij}^k \psi_{k3}(\|\omega_k\|)$,

are satisfied for all u_k^i, $\omega_i \in R^{n_i}$, $\omega_{ij} \in R^{n_i} \times R^{n_j}$, $t \in \mathcal{T}$, $(i \neq j) \in [1,l]$, p, $k = 1,2,\dots,l$.

An important part in the structure of averaged derivative of the function (4.2.15) is played by a symmetric $l \times l$ matrix

$$S = \frac{1}{2}(\overline{S} + \overline{S}^{\mathrm{T}}),$$

where $\overline{S}$ is an upper triangle matrix with elements $\overline{s}_{pq}$ defined as

$$\overline{s}_{pp} = \eta_p^2\Big(\rho_p + b_{pp} + \frac{1}{2}\sum_{\substack{i=1\\i\neq p}}^{l} d_p \alpha_{pi}\Big) + 2\eta_p \sum_{i=p+1}^{l} \eta_i \beta_{pi}^1$$
$$+ 2\eta_p \sum_{i=p+1}^{l} \gamma_{pp}^{pi} \eta_i + 2\eta_p \sum_{i=1}^{l} \beta_{pi}^3 \eta_i + \frac{1}{2}\eta_p \sum_{k,j=1}^{l} \mu_{pj}^k \nu_{pj}^k \eta_k;$$

$$\bar{s}_{pq} = \eta_p^2 b_{pq} + 4\beta_{pq}^2 \eta_p \eta_q + \sum_{k=1}^{l} \sum_{j=k+1}^{l} \gamma_{pq}^{kj} \eta_k \eta_j,$$

$$\bar{s}_{qp} = 0, \quad (p < q) \in [1, l], \quad \eta \in R_+^l, \quad \eta > 0.$$

Sufficient conditions for stability with probability of the system (4.4.1) are obtained in terms of the function (4.4.9) being applied in construction of the function

$$V(t, \omega) = \eta^{\mathrm{T}} \Pi(t, \omega) \eta, \quad \eta \in R_+^l, \quad \eta > 0 \tag{4.4.14}$$

Namely, we shall prove the following result.

THEOREM 4.4.1. *Let the perturbed motion of the equation (4.4.1) are such that:*

(1) $\{\xi(t), t \in \mathcal{T}\}$ *is a normalized Wienner process and* $\sigma_{ij}(t, \omega) \neq 0$, $\forall\, (i, j) \in [1, l]$;
(2) *all conditions of Assumptions 4.4.1–4.4.4 are satisfied*
(3) *the matrix S is*
    (a) *negative semi-definite;*
    (b) *negative definite.*

Then the equilibrium state $\omega = 0$ *of system (4.4.1) is*

(a) *stable in probability;*
(b) *asymptotically stable in probability.*

PROOF. We take the functions $v_{ij}(t, \cdot)$ according to Assumption 4.4.1 and a vector $\eta \in R_+^l$, $\eta > 0$. The function (4.4.14) in coordinate form is

$$\begin{aligned} V(t, \omega) &= \sum_{i=1}^{l} \eta_i^2 v_{ii}(t, \omega_i) + \sum_{\substack{i,j=1 \\ (i \neq l)}}^{l} \eta_i \eta_j v_{ij}(t, \omega_{ij}) \\ &= \sum_{i=1}^{l} \eta_i^2 v_{ii}(t, \omega_i) + 2 \sum_{i=1}^{l} \sum_{j=i+1}^{l} \eta_i \eta_j v_{ij}(t, \omega_{ij}) \end{aligned} \tag{4.4.15}$$

Assumption 4.4.1 implies that at the presence of properties $\mathcal{A}(\mathcal{N}_i)$ and $\mathcal{A}(\mathcal{N}_i \times \mathcal{N}_j)$ the bilateral estimate

$$\sum_{i=1}^{l} \eta_i^2 \psi_{i1}(\|\omega_i\|) + \sum_{\substack{i,j=1 \\ (i \neq j)}}^{l} \eta_i \eta_j \psi_{ij}^1(\|\omega_{ij}\|) \leq V(t,\omega)$$
$$\leq \sum_{i=1}^{l} \eta_i^2 \psi_{i2}(\|\omega_i\|) + \sum_{\substack{i,j=1 \\ (i \neq j)}}^{l} \eta_i \eta_j \psi_{ij}^2(\|\omega_{ij}\|)$$

is valid for function (4.4.15) when all $\omega_i \in \mathcal{N}_i$, $\omega_{ij} \in \mathcal{N}_i \times \mathcal{N}_j$ and $t \in \mathcal{T}$.

Since $\psi_{i1}, \psi_{i2} \in K$ and $\psi_{ij}^1, \psi_{ij}^2 \in K$, then the function $V(t,\omega)$ is positive definite and decreasing. Moreover, functions $\Psi_1(\|\omega\|)$ and $\Psi_2(\|\omega\|) \in K$ can be found such that

$$\Psi_1(\|\omega\|) \leq V(t,\omega) \leq \Psi_2(\|\omega\|) \tag{4.4.16}$$

for all $\omega \in \mathcal{N} = \mathcal{N}_1 \times \cdots \times \mathcal{N}_l$, $t \in \mathcal{T}$.

For the function (4.4.14) the averaged derivative $\dfrac{dE[V]}{dt}$ along the solutions of (4.4.1) is

$$\frac{dE[V(t,\omega)]}{dt} = \eta^{\mathrm{T}} \frac{dE[\Pi(t,\omega)]}{dt} \eta = \sum_{i=1}^{l} \eta_i^2 \left(\frac{dE_i[v_{ii}(t,\omega)]}{dt} + g_i^{\mathrm{T}} \nabla_{\omega_i} v_{ii}(t,\omega_i) \right)$$
$$+ \frac{1}{2} \sum_{\substack{j=1 \\ (j \neq l)}}^{l} \operatorname{tr} \left[\sigma_{ij}^{\mathrm{T}}(t,\omega_j) \nabla_{\omega_i \omega_i} v_{ii}(t,\omega_i) \sigma_{ij}(t,\omega_j) \right]$$
$$+ 2 \sum_{i=1}^{l} \sum_{j=i+1}^{l} \eta_i \eta_j \left(\frac{dE_{ij}[v_{ij}(t,\omega_{ij})]}{dt} \right.$$
$$\left. + \frac{1}{2} \sum_{\substack{k=1 \\ (k \neq i,j)}}^{l} \operatorname{tr} \left[(\sigma_{ij}^k)^{\mathrm{T}} \nabla_{\omega_{ij} \omega_{ij}} v_{ij}(t,\omega_{ij}) \sigma_{ij}^k \right] \right).$$

for all $t \in \mathcal{T}$ and $\omega_i \in \mathcal{N}_i$, $\omega_{ij} \in \mathcal{N}_i \times \mathcal{N}_j$.

In view of conditions (c) of Assumptions 4.4.3 and 4.4.4 we get the estimate

$$\frac{dE[V(t,\omega)]}{dt} \le \sum_{i=1}^{l} \eta_i^2 [P_i \Psi_{i3}(\|\omega_i\|)$$

$$+ \Psi_{i3}^{1/2}(\|\omega_i\|) \sum_{k=1}^{l} b_{ik} \Psi_{k3}^{1/2}(\|\omega_k\|) + \frac{1}{2} \sum_{\substack{k=1 \\ (k \ne i)}}^{l} d_k \alpha_{ik} \Psi_{k3}(\|\omega_k\|)]$$

$$+ 2 \sum_{i=1}^{l} \sum_{j=i+1}^{l} \eta_i \eta_j [\beta_{ij}^1 \Psi_{i3}(\|\omega_i\|) + 2\beta_{ij}^2 \Psi_{i3}^{1/2} \Psi_{j3}^{1/2} \tag{4.4.17}$$

$$+ \beta_{ij}^3 \Psi_{j3}(\|\omega_j\|) + \sum_{k=1, p=k}^{l} \gamma_{kp}^{ij} \Psi_{k3}^{1/2}(\|\omega_k\|) \Psi_{p3}^{1/2}(\|\omega_p\|)$$

$$+ \frac{1}{2} \sum_{\substack{k=1 \\ (k \ne j, i)}}^{l} \nu_{ij}^k \mu_{ij}^k \Psi_{k3}(\|\omega_k\|)]$$

for all $t \in \mathcal{T}$ and $\omega_i \in \mathcal{N}_i$.

With regard to the structure of the matrix S we get from estimate (4.4.12)

$$\frac{dE[V(t,\omega)]}{dt} \le \Psi^{\mathrm{T}}(\|\omega\|) S \Psi(\|\omega\|) \tag{4.4.18}$$

where $\Psi(\|\omega\|) = \left(\psi_{13}^{1/2}(\|\omega_1\|), \dots, \psi_{l3}^{1/2}(\|\omega_l\|)\right)^{\mathrm{T}}$.

Since by condition (3)(a) of Theorem 4.4.1 the matrix S is negative semi-definite, then $\lambda_M(S) \le 0$ and

$$\frac{dE[V(t,\omega)]}{dt} \le \lambda_M(s) \sum_{i=1}^{l} \psi_{i3}(\|\omega_i\|)$$

for all $t \in \mathcal{T}$ and $\omega_i \in \mathcal{N}_i$.

Since $\varphi_{i3} \in K$, there exists a comparison function $\Psi_3(\|\omega\|) \in K$ such that

$$\sum_{i=1}^{l} \psi_{i3}(\|\omega_i\|) \le \Psi_3(\|\omega\|)$$

for all $\omega_i \in \mathcal{N}_i$ and $\omega \in \mathcal{N} = \mathcal{N}_1 \times \dots \mathcal{N}_l$.

Hence

$$\frac{dE[V(t,\omega)]}{dt} \le \lambda_M(S)\Psi_3(\|\omega\|) \tag{4.4.19}$$

is negative semi-definite for all $t \in \mathcal{T}$ and $\omega \in \mathcal{N}$.

Thus all conditions of Theorem 4.3.1 from Section 4.3 are satisfied, and the equilibrium state $\omega = 0$ of system (4.4.1) is stable in probability.

To verify assertion (b) of Theorem 4.4.1 it is sufficient to note that under condition (3)(b) in the estimate (4.4.18) $\lambda_M < 0$. Then according to inequality (4.4.19) all hypotheses of Theorem 4.3.2 are satisfied and the equilibrium state $\omega = 0$ of system (4.4.1) is asymptotically stable in probability.

The Theorem 4.4.1 is proved.

ASSUMPTION 4.4.5. The system (4.4.1) allows the first and the second level decompositions and

(1) independent subsystems (4.4.3) possess the property $B_j(\infty)$, $j \in [1,l]$;
(2) independent couples (i,j) of the subsystems (4.4.7) possess the property $A_{ij}(\infty)$, $\forall\,(i \ne j) \in [1,l]$.

THEOREM 4.4.2. *Let the perturbed motion of the equations (4.4.1) are such that*

(1) $\{\xi(t), t \in \mathcal{T}\}$ *is a normalized process and* $\sigma_{ij}(t,\omega_j) \ne 0$ $\forall\,(i,j) \in [1,l]$;
(2) *all conditions of Assumption 4.4.5 are satisfied;*
(3) *the conditions of Assumption 4.4.4 are satisfied for* $\mathcal{N}_i = R^{n_i}$, $\mathcal{N}_i \times \mathcal{N}_j = R^{n_i} \times R^{n_j}$ *with functions* $\varphi_{i3} \in KR$, $i \in [1,l]$;
(4) *the matrix* S *is negative definite.*

Then, the equilibrium state $\omega = 0$ *of system (4.4.1) is asymptotically stable in probability in the whole.*

PROOF. Under the conditions of Assumption 4.4.5 the function (4.4.14) satisfies estimates (4.2.20) and its averaged derivative (4.2.27) satisfies inequality (4.4.18) where functions $\psi_{i3} \in KR$. In consequence of condition (4) of Theorem 4.4.2 and estimate (4.4.19), $\dfrac{dE[V(t,\omega)]}{dt}$ is negative definite for all $t \in \mathcal{T}$ and $\omega \in R^n$. Thus, all conditions of Theorem 4.3.3 are satisfied and the equilibrium state $\omega = 0$ of system (4.4.1) is asymptotically stable in probability in the whole.

REMARK 4.4.5. If for the perturbed motion the equations (4.4.1) are such that $\sigma_{ij}(t,\omega_j) = 0$ for all $(i,j) \in [1,l]$, i.e. random interconnections between the subsystems are absent, then the structure of matrix S is simplified and its elements are:

$$\overline{s}_{pp} = \eta_p^2(\rho_p + b_{pp}) + 2\eta_p \sum_{i=p+1}^{l} \eta_i\beta_{pi}^1 + 2\eta_p \sum_{i=p+1}^{l} \gamma_{pp}^{pi}\eta_i + 2\eta_p \sum_{i=1}^{l} \beta_{pi}^3\eta_i,$$

$$\overline{s}_{pq} = \eta_p^2 b_{pq} + 4\beta_{pq}^2\eta_p\eta_q + \sum_{k=1}^{l}\sum_{j=k+1}^{l} \gamma_{pq}^{kj}\eta_k\eta_j, \qquad \forall\,(p<q) \in [1,l];$$

$$\overline{s}_{qp} = 0, \qquad p < q.$$

Here η_p, $p \in [1,l]$ are components of vector $\eta \in R_+^l$.

4.5 Applications

In this section general results on stochastic stability are applied in the investigation of some real processes models.

4.5.1 Stochastic Version of the Lefschetz Problem

The following problem is a development of the Lefschetz [100] problem we dealt with in Chapter 2.

Let us decompose system (4.2.1) into two subsystems

$$\begin{aligned}\frac{dp}{dt} &= X(t,p,0,y(t)) + F(t,p,q,y(t)),\\ \frac{dq}{dt} &= X(t,0,q,y(t)) + G(t,p,q,y(t)),\end{aligned} \tag{4.5.1}$$

where $p \in R^{n_1}$, $q \in R^{n_2}$, $X \in C[T_0 \times B_p, R[\Omega, R^{n_1}]]$, $Y \in C[T_0 \times B_q, R[\Omega, R^{n_2}]]$, $F \in C[T_0 \times B, R[\Omega, R^{n_1}]]$, $G \in C[T_0 \times B, R[\Omega, R^{n_2}]]$, X, F, Y, G vanish if and only if $p = 0$ and $q = 0$ respectively.

ASSUMPTION 4.5.1. There exist time-invariant neighborhoods $\mathcal{N}_p \subseteq R^{n_1}$, $\mathcal{N}_q \subseteq R^{n_2}$ of the equilibrium states $p = 0$, $q = 0$ respectively, and a matrix-valued function $\Pi(t,x,y)$ with elements v_{kl} $k,\, l=1,2$ such that

$$
(4.5.2)\qquad
\begin{aligned}
&\underline{\alpha}_{11}\zeta_1^2(\|p\|)\le v_{11}(t,p,y)\le \overline{\alpha}_{11}\zeta_2^2(\|p\|) \quad \forall p\in\mathcal{N}_{p0}, \forall y\in Y;\\
&\underline{\alpha}_{22}\zeta_1^2(\|q\|)\le v_{22}(t,q,y)\le \overline{\alpha}_{22}\zeta_2^2(\|q\|) \quad \forall q\in\mathcal{N}_{q0}, \forall y\in Y;\\
&\underline{\alpha}_{12}\zeta_1(\|p\|)\psi_1(\|q\|)\le v_{12}(t,p,q,y)\le \overline{\alpha}_{12}\zeta_2(\|p\|)\psi_2(\|q\|)\\
&\qquad\qquad \forall (p,q,y)\in\mathcal{N}_{p0}\times\mathcal{N}_{q0}\times Y
\end{aligned}
$$

where $\mathcal{N}_{p0}=\{p\in\mathcal{N}_p,\ p\ne 0\}$, $\mathcal{N}_{q0}=\{q\in\mathcal{N}_q,\ q\ne 0\}$, $\overline{\alpha}_{kk},\underline{\alpha}_{kk}=\text{const}>0$, $\underline{\alpha}_{12},\overline{\alpha}_{12}=\text{const}$, $k=1,2$; ζ_k,ψ_k are functions of class K.

If conditions of the Assumption 4.5.1 are satisfied, properties of the function (4.4.14) (property of having a fixed sign, the existence of an infinitely small upper bound; an infinitely large lower bound) are defined by properties of matrices $A=H^{\mathrm{T}}A_1H$; $B=H^{\mathrm{T}}A_2H$ where

$$
(4.5.3)\qquad A_1=[\underline{\alpha}_{kl}],\quad A_2=[\overline{\alpha}_{kl}],\quad H=\operatorname{diag}(\eta_1,\eta_2),\qquad k,\,l=1,2.
$$

We introduce the designation

$$
\Delta(v_{kl})=\sum_{i\ne j}\alpha_{ij}[v_{kl}(t,\cdot,i)-v_{kl}(t,\cdot,j)],\qquad k,\,l=1,2
$$

ASSUMPTION 4.5.2. There exist constants ρ_{kr}, $k=1,2;\, r=1,2,\dots,10$ and functions $\zeta(\|p\|)$, $\psi(\|q\|)$ of class $K(KR)$ such that

$$
\begin{aligned}
\nabla_t v_{11}+(\nabla_p v_{11}^{\mathrm{T}})X+\frac{1}{2}\Delta(v_{11})&\le \rho_{11}\zeta^2+h_{11}(\zeta,\psi),\\
\nabla_t v_{22}+(\nabla_q v_{22}^{\mathrm{T}})Y+\frac{1}{2}\Delta(v_{22})&\le \rho_{12}\psi^2+h_{21}(\zeta,\psi),\\
(\nabla_p v_{11}^{\mathrm{T}})F+\frac{1}{2}\Delta(v_{11})&\le \rho_{12}\zeta^2+\rho_{13}\zeta\psi+\rho_{14}\psi^2+h_{12}(\zeta,\psi),\\
(\nabla_q v_{22}^{\mathrm{T}})G+\frac{1}{2}\Delta(v_{22})&\le \rho_{22}\zeta^2+\rho_{23}\zeta\psi+\rho_{24}\psi^2+h_{22}(\zeta,\psi),\\
\nabla_t v_{12}+(\nabla_p v_{12}^{\mathrm{T}})X+\frac{1}{4}\Delta(v_{12})&\le \rho_{15}\zeta^2+\rho_{16}\zeta\psi+\rho_{17}\psi^2+h_{13}(\zeta,\psi),\\
(\nabla_q v_{12}^{\mathrm{T}})Y+\frac{1}{4}\Delta(v_{12})&\le \rho_{25}\zeta^2+\rho_{25}\zeta\psi+\rho_{27}\psi^2+h_{23}(\zeta,\psi),\\
(\nabla_p v_{12}^{\mathrm{T}})F+\frac{1}{4}\Delta(v_{12})&\le \rho_{18}\zeta^2+\rho_{19}\zeta\psi+\rho_{110}\psi^2+h_{14}(\zeta,\psi),\\
(\nabla_q v_{12}^{\mathrm{T}})G+\frac{1}{4}\Delta(v_{12})&\le \rho_{28}\zeta^2+\rho_{29}\zeta\psi+\rho_{210}\psi^2+h_{24}(\zeta,\psi),
\end{aligned}
$$

where $h_{sk}(\zeta,\psi)$, $k=1,2$; $s=1,2,3,4$ are polynomials with respect to ζ, ψ containing additives of power higher than two.

PROPOSITION 4.5.1. *If all hypotheses of Assumption 4.5.2 are satisfied and*

(a) *the matrix* $C = [c_{ij}]$, $c_{ij} = c_{ji}$, $i \neq j$; $i, j = 1, 2$ *with elements*

$$c_{11} = \eta_1^2(\rho_{11} + \rho_{12}) + \eta_2^2\rho_{22} + 2\eta_1\eta_2(\rho_{15} + \rho_{18} + \rho_{25} + \rho_{28}),$$
$$c_{22} = \eta_1^2\rho_{14} + \eta_2^2(\rho_{21} + \rho_{24}) + 2\eta_1\eta_2(\rho_{17} + \rho_{110} + \rho_{27} + \rho_{210}),$$
$$c_{12} = \frac{1}{2}(\eta_1^2\rho_{13} + \eta_2^2\rho_{23}) + \eta_1\eta_2(\rho_{16} + \rho_{19} + \rho_{26} + \rho_{29}),$$

is negative definite, then due to system (4.5.1) averaged derivative

$$\eta^{\mathrm{T}}\frac{dE[\Pi]}{dt}\eta = \frac{dE[V]}{dt}, \qquad \eta \in R_+^2 \tag{4.5.4}$$

is a negative definite function.

If besides hypothesis (a), hypothesis (b) is satisfied, hypotheses of Assumptions 4.5.1 and 4.5.2 are satisfied

$$\begin{aligned} h(\zeta, \psi) &= \eta_1^2(h_{11} + h_{12}) + \eta_2^2(h_{21} + h_{22}) \\ &\quad + 2\eta_1\eta_2(h_{13} + h_{14} + h_{23} + h_{24}) \leq 0 \end{aligned} \tag{4.5.5}$$

for $p \in R^{n_1}$, $q \in R^{n_2}$, $n_1 + n_2 = n$ *and for functions* $\zeta(\|p\|) \in KR$, $\psi(\|q\|) \in KR$, *then the averaged derivative (4.5.4) is negative definite in the whole.*

THEOREM 4.5.1. *If the system of equations of perturbed motion (4.5.1) is such that all hypotheses of Assumptions 4.5.1 and 4.5.2 (a) are satisfied and matrices A and B are positive definite and matrix C is negative definite, the equilibrium state* $p = 0$, $q = 0$ *of the system (4.5.1) is asymptotically stable with respect to probability.*

If in Assumption 4.5.1 and 4.5.2 $\mathcal{N}_p = R^{n_1}$, $\mathcal{N}_q = R^{n_2}$ *the functions* $\zeta(\|p\|)$ *and* $\psi(\|q\|)$ *are of class KR, the equilibrium state* $p = 0$, $q = 0$ *is asymptotically stable with respect to probability in the whole.*

The assertions of Theorem 4.5.1 are implied by estimate

$$\frac{dE[V]}{dt} \leq \xi^{\mathrm{T}}C\xi + h(\xi) \tag{4.5.6}$$

where $\xi = (\zeta, \psi)^{\mathrm{T}}$ and by the fact that if the hypotheses of Theorem 4.5.1 are satisfied, the hypotheses of Theorems 4.3.2 and 4.3.3 are satisfied respectively.

4.5.2 Stability in Probability of Oscillating System

Let us consider for an oscillating system the perturbed motion equations which are of the form

$$\begin{aligned} \frac{dp}{dt} &= A_1(y)p + f_1(p,q,r,y(t)), \\ \frac{dq}{dt} &= A_2(y)q + f_2(p,q,r,y(t)), \\ \frac{dr}{dt} &= A_3(y)r + f_3(p,q,r,y(t)). \end{aligned} \tag{4.5.7}$$

Here $p,\ q,\ r \in R^2,\ \ f_i \in C(B, R[Y, R^2])$,

$$A_i(y) = \begin{pmatrix} 0 & 1 \\ -b_i(y) & -a_i(y) \end{pmatrix}, \qquad i = 1,2,3$$

The functions $a_i(y)$ and $b_i(y)$ are bounded and $y(t)$ is a homogeneous Markov chain with a finite number of states $Y = \{y_1, \dots, y_r\}$ and with transitional probabilities

$$p_{ij}(\tau) = \alpha_{ij}\tau + o(\tau), \qquad \alpha_{ij} = \text{const} \quad (i \neq j) \in [1, r].$$

We designate $b_i(y_k) = b_k^i,\ a_i(y_k) = a_k^i$ and assume that $b_k^i > 0$. Matrix-valued function $\Pi(p,q,r,y(t))$ elements $v_{ik}(\cdot)$ are taken in the form

$$\begin{aligned} v_{11}(p, y_k) &= p^{\mathrm{T}} \operatorname{diag}\left(1, \frac{1}{b_k^1}\right) p, \\ v_{22}(q, y_k) &= q^{\mathrm{T}} \operatorname{diag}\left(1, \frac{1}{b_k^2}\right) q, \\ v_{33}(r, y_k) &= r^{\mathrm{T}} \operatorname{diag}\left(1, \frac{1}{b_k^3}\right) r, \\ v_{12}(p, q, y_k) &= p^{\mathrm{T}} \operatorname{diag}\left(1, \frac{0,1}{b_k^1}\right) q, \\ v_{13}(p, r, y_k) &= p^{\mathrm{T}} \operatorname{diag}\left(1, \frac{0,1}{b_k^3}\right) r, \\ v_{23}(q, r, y_k) &= q^{\mathrm{T}} \operatorname{diag}\left(1, \frac{0,1}{b_k^2}\right) r, \\ v_{ij}(\cdot) &= v_{ij}(\cdot) \quad \forall\, (i \neq j) \in [1, 3]. \end{aligned} \tag{4.5.8}$$

It is easy to notice that for functions (4.5.8) the following estimates are valid

(a) if $0 < b_k^i \le 1,\ i = 1,2,3$, then

$$v_{11}(p, y_k) \ge ||p||^2; \qquad v_{22}(q, y_k) \ge ||q||^2;$$
$$v_{33}(r, y_k) \ge ||r||^2; \qquad v_{12}(p, q, y_k) \ge -0,1||p||||q||;$$
$$v_{13}(p, r, y_k) \ge -0,1||p||||r||; \qquad v_{23}(q, r, y_k) \ge -0,1||q||||r||;$$

(b) if $b_k^i > 1,\ i = 1,2,3$, then

$$v_{11}(p, y_k) \ge \frac{1}{b_k^1}||p||^2; \qquad v_{22}(q, y_k) \ge \frac{1}{b_k^2}||q||^2;$$
$$v_{33}(r, y_k) \ge \frac{1}{b_k^3}||r||^2; \qquad v_{12}(p, q, y_k) \ge -\frac{0,1}{b_k^1}||p||||q||;$$
$$v_{13}(p, r, y_k) \ge -\frac{0,1}{b_k^3}||p||||r||; \qquad v_{23}(q, r, y_k) \ge -\frac{0,1}{b_k^2}||q||||r||.$$

For the function

$$V(p, q, r, y(t)) = \eta^{\mathrm{T}}\Pi(p, q, r, y(t))\eta, \tag{4.5.9}$$

where $\eta \in R_+^3$ the matrix A_1 in the estimate of (4.2.20) has the form

$$\overline{A}_1(y_k) = \begin{cases} \begin{pmatrix} 1 & -0,1 & -0,1 \\ -0,1 & 1 & -0,1 \\ -0,1 & -0,1 & 1 \end{pmatrix}, & \text{if } 0 < b_k^i \le 1, \quad i = 1,2,3; \\[2ex] \begin{pmatrix} \frac{1}{b_k^1} & -\frac{0,1}{b_k^1} & -\frac{0,1}{b_k^3} \\ -\frac{0,1}{b_k^1} & \frac{1}{b_k^2} & -\frac{0,1}{b_k^2} \\ -\frac{0,1}{b_k^1} & -\frac{0,1}{b_k^2} & \frac{1}{b_k^3} \end{pmatrix}, & \text{if } b_k^i > 1, \quad i = 1,2,3. \end{cases}$$

The matrix $\overline{A}_1$ is positive definite, if

$$\frac{b_k^1}{b_k^3} + \frac{b_k^3}{b_k^2} + \frac{b_k^2}{b_k^1} < 99,8, \qquad k = 1,2,\dots,r. \tag{4.5.10}$$

For the averaged derivative $\dfrac{dE[v_{ij}(\cdot)]}{dt}$ of the function $\Pi(p, q, r, y(t))$ with elements (4.5.8) it is easy to establish estimate in the form

$$\frac{dE[V(p, q, r, y(t))]}{dt} \le u^{\mathrm{T}}Su \tag{4.5.11}$$

where $u = (\|p\|, \|q\|, \|r\|)^{\mathrm{T}}$, $\eta = (1,1,1)^{\mathrm{T}}$ and matrix S elements are

(4.5.12)

$$c_{ii}(y_k) = -\Big(\frac{2a_k^i}{b_k^i} - \Delta b^i - \sum_{l=1}^{3} d_{ii}^l\Big), \quad i = 1,2,3; \quad k = 1,2,\dots,r;$$

$$c_{12}(y_k) = \frac{1}{2}\Big(\sum_{l=1}^{6} d_{12}^l + \frac{0,1}{b_k^1}|a_k^1 + a_k^2| + 0,1|\Delta b^1|\Big), \quad k = 1,2,\dots,r;$$

$$c_{13}(y_k) = \frac{1}{2}\Big(\sum_{l=1}^{6} d_{13}^l + \frac{0,1}{b_k^3}|a_k^1 + a_k^3| + 0,1|\Delta b^3|\Big), \quad k = 1,2,\dots,r;$$

$$c_{23}(y_k) = \frac{1}{2}\Big(\sum_{l=1}^{6} d_{23}^l + \frac{0,1}{b_k^2}|a_k^2 + a_k^3| + 0,1|\Delta b^2|\Big), \quad k = 1,2,\dots,r,$$

where

$$\Delta b^i = \sum_{j\neq k}^{r}\Big(\frac{1}{b_j^i} - \frac{1}{b_k^i}\Big)\alpha_{kj}, \quad i = 1,2,3; \quad k = 1,2,\dots,r.$$

Here d_{ij}^k, $i, j \in [1,3]$, $k \in [1,6]$ are constants that are found when estimating $\dfrac{dE[v_{ij}(\cdot)]}{dt}$.

The matrix S with elements (4.5.12) is negative definite if

(a) $$\frac{2a_k^i}{b_k^i} - \Delta b^i > \sum_{l=1}^{3} d_{ii}^l, \quad i = 1,2,3; \quad k = 1,2,\dots,r;$$

(b) $$\prod_{i=1}^{2}\Big(\frac{2a_k^i}{b_k^i} - \Delta b^i - \sum_{l=1}^{3} d_{ii}^l\Big) > \frac{1}{4}\Big(\sum_{l=1}^{6} d_{12}^l + W_k^1(a_k^1, a_k^2, b_k^1, b^1)\Big);$$

(c) $$\prod_{i=1}^{3}\Big(\frac{2a_k^i}{b_k^i} - \Delta b^i - \sum_{l=1}^{3} d_{ii}^l\Big) + \frac{1}{4}\Big(\sum_{l=1}^{6} d_{12}^l + W_k^1(a_k^1, a_k^2, b_k^1, b^1)\Big)\Big(\sum_{l=1}^{6} d_{13}^l + W_k^3(a_k^1, a_k^3, b_k^3, b^3)\Big)$$
$$< \frac{1}{4}\Big(\sum_{l=1}^{6} d_{13}^l + W_k^3(a_k^1, a_k^3, b_k^3, b^3)\Big)\Big(\sum_{l=1}^{3} d_{22}^l + \Delta b^2 - \frac{2a_k^2}{b_k^2}\Big)$$
$$+ \frac{1}{4}\Big(\sum_{l=1}^{6} d_{23}^l + W_k^2(a_k^2, a_k^3, b_k^2, b^2)\Big)\Big(\sum_{l=1}^{3} d_{22}^l + \Delta b^1 - \frac{2a_k^1}{b_k^1}\Big)$$

$$+\frac{1}{4}\Big(\sum_{l=1}^{6} d_{12}^{l} + W_k^1(a_k^1, a_k^2, b_k^1, b^1)\Big)^2 \Big(\sum_{l=1}^{3} d_{33}^{l} + \Delta b^3 - \frac{2a_k^3}{b_k^3}\Big),$$
$$k = 1, 2, \dots, r,$$

where

$$W_k^1(a_k^1, a_k^2, b_k^1, b^1) = \frac{0,1}{b_k^1}|a_k^1 + a_k^2| + 0,1|\Delta b^1|;$$

$$(4.5.13)\quad W_k^2(a_k^2, a_k^3, b_k^2, b^2) = \frac{0,1}{b_k^2}|a_k^2 + a_k^3| + 0,1|\Delta b^2|;$$

$$W_k^3(a_k^1, a_k^3, b_k^3, b^3) = \frac{0,1}{b_k^3}|a_k^1 + a_k^3| + 0,1|\Delta b^3|, \qquad k = 1, 2, \dots, r.$$

Thus, under conditions (4.5.10) the function (4.5.9) is positive definite, and when inequalities (a)–(c) are satisfied its averaged derivative (4.5.11) is negative definite.

Applying Theorem 4.3.3 we conclude that conditions (4.5.10) and (a)–(c) are sufficient for stability in probability in the whole of the equilibrium state $p = q = r = 0$ of oscillating system (4.5.7).

4.5.3 Stability in Probability of a Regulation System

We consider an autonomous stochastic regulation system

$$(4.5.14)\qquad d\omega_i = \sum_{j=1}^{l} A_{ij}\omega_j dt + \sigma_i(\omega_i)dz_i + b_i f_i(\theta_i)dt, \qquad i \in [1, l],$$

where $\theta_i = \sum\limits_{k=1}^{l} c_{ik}^{\mathrm{T}}\omega_k$, $b_i, \omega_i \in R^{n_i}$, $c_{ik} \in R^{n_k}$, A_{ij} are constant matrices of the corresponding to vector ω_j dimensions, $\{z_i(t),\ t \in \mathcal{T}\}$, is a m_i-dimensional Wienner process. Besides, $f_i(\theta_i) = 0$, if and only if $\theta_i = 0$, $0 \le f_i(\theta_i) < k_i\theta_i^2$ provided $\theta_i \ne 0$.

First level decomposition results in the system

$$(4.5.15)\ d\omega_i = A_{ii}\omega_i dt + \sigma_i(\omega_i)dz_i + \sum_{\substack{j=1 \\ (j\ne i)}}^{l} A_{ij}\omega_j dt + b_i f_i(\theta_i)dt, \quad i \in [1, l],$$

with the independent subsystems

$$d\omega_i = A_{ii}\omega_i dt + \sigma_i(\omega_i)dz_i \quad i \in [1,l], \tag{4.5.16}$$

and link functions

$$g_i(\omega) = \sum_{\substack{j=1\\(j\neq i)}}^{l} A_{ij}\omega_j dt + b_i f_i(\theta_i)dt, \qquad i \in [1,l]. \tag{4.5.17}$$

The second level decomposition yields

$$\begin{aligned} d\omega_i =& A_{ii}\omega_i dt + A_{ij}\omega_j dt + \sigma_i(\omega_i)dz_i \\ &+ \sum_{\substack{k=1\\(k\neq i,j)}}^{l} A_{ik}\omega_k dt + b_i f_i(\theta_i)dt; \\ d\omega_j =& A_{jj}\omega_j dt + A_{ji}\omega_i dt + \sigma_j(\omega_j)dz_j \\ &+ \sum_{\substack{k=1\\(k\neq i,j)}}^{l} A_{jk}\omega_k dt + b_j f_j(\theta_j)dt. \end{aligned} \tag{4.5.18}$$

Equations (4.5.18) are represented as

$$\begin{aligned} d\omega_{ij} = \overline{A}_{ij}\omega_{ij}dt + \sigma_{ij}dz_{ij} + \sum_{\substack{k=1\\(k\neq i,j)}}^{l} \overline{A}_{ij}^{k}\omega_k dt + B_{ij}dt, \\ (i<j) \in [1,l]. \end{aligned} \tag{4.5.19}$$

Here $\omega_{ij} = (\omega_i^{\mathrm{T}}, \omega_j^{\mathrm{T}})^{\mathrm{T}}$, $\omega_{ij} \in R^{n_i} \times R^{n_j}$ and matrices $\overline{A}_{ij}$, $\overline{A}_{ij}^{k}$, σ_{ij}, B_{ij} with dimensions $(n_i+n_j)\times(n_i+n_j)$, $(n_i+n_j)\times n_k$, $(n_i+n_j)\times(m_i+m_j)$, $(n_i+n_j)\times 1$ $\forall\,(i,j,k) \in [1,l]$ respectively, are defined by formulas

$$\overline{A}_{ij}^{k} = (A_{ik}^{\mathrm{T}}, A_{jk}^{\mathrm{T}})^{\mathrm{T}}; \qquad B_{ij} = (b_i^{\mathrm{T}} f_i(\theta_i), b_j^{\mathrm{T}} f_j(\theta_i));$$

$$\sigma_{ij} = \operatorname{diag}(\sigma_i(\omega_i), \sigma_j(\omega_j)); \qquad \overline{A}_{ij} = \begin{pmatrix} A_{ii} & A_{ij} \\ A_{ji} & A_{jj} \end{pmatrix}.$$

Alongside the systems (4.5.15) and (4.5.19) we shall consider the matrix-valued function

(4.5.20)

$$\Pi(\omega) = [\operatorname{diag}(v_{ii}(\omega_i)) + (v_{ij}(\omega_{ij}))], \quad i<j \in [1,l], \quad i = 1,2,\dots,l$$

with the elements

$$v_{ii}(\omega_i) = \omega_i^{\mathrm{T}} P_{ii} \omega_i \qquad i \in [1, l];$$
$$v_{ij}(\omega_{ij}) = \omega_{ij}^{\mathrm{T}} P_{ij} \omega_{ij} \qquad (i < j) \in [1, l].$$

Matrices P_{ii} are found by Liapunov equations

$$A_{ii}^{\mathrm{T}} P_{ii} + P_{ii} A_{ii} = -G_{ii}, \qquad i \in [1, l] \tag{4.5.21}$$

where G_{ii} are symmetric positive definite matrices of dimensions $n_i \times n_i$.

Matrices P_{ij} are also found by Liapunov equations

$$\overline{A}_{ij}^{\mathrm{T}} P_{ij} + P_{ij} \overline{A}_{ij} = -G_{ij}, \qquad (i < j) \in [1, l] \tag{4.5.22}$$

where G_{ij} are symmetric positive definite matrices of dimensions $(n_i + n_j) \times (n_i + n_j)$.

The functions $v_{ii}(\omega_i)$ and $v_{ij}(\omega_{ij})$ are positive definite if matrices A_{ii} and $\overline{A}_{ij}$ are stable. We shall suppose that this condition is satisfied for the systems (4.5.15) and (4.5.19).

Now we introduce symmetric matrices of dimensions $n_p \times n_p$:

$$\begin{aligned}\Xi_{pp} &= \eta_p^2 G_{pp} + 2\eta_p^2 c_{pp} b_p^{\mathrm{T}} k_p^* P_{pp} + \eta_p \sum_{\substack{j=1 \\ (j \neq p)}}^{l} \eta_j (G_{pj}^p + G_{jp}^j) \\ &+ 4\eta_p \sum_{j=p+1}^{l} \eta_j \big[c_{pp} b_p^{\mathrm{T}} k_p^* P_{pj}^{\mathrm{T}} + c_{jp} b_j^{\mathrm{T}} k_j^* \overline{P}_{pj}^{\mathrm{T}} \\ &+ c_{jp} b_p^{\mathrm{T}} k_p^* \overline{P}_{pj} + c_{jj} b_j^{\mathrm{T}} k_j^* P_{jp}^j \big], \qquad p \in [1, l],\end{aligned}$$

and matrices of dimensions $n_p \times n_q$, $(p < q) \in [1, l]$:

$$\begin{aligned}\Xi_{pq} &= \eta_q^2 A_{qp}^{\mathrm{T}} P_{qq} + \eta_p^2 P_{pp} b_p^{\mathrm{T}} k_p^* c_{pq}^{\mathrm{T}} + 2\eta_p \eta_q \overline{G}_{pq} \\ &+ \eta_p \sum_{\substack{j=q \\ (j \neq p)}}^{l} \eta_j (A_{qp}^{\mathrm{T}} P_{qj}^q + A_{jp}^{\mathrm{T}} \overline{P}_{qj}^{\mathrm{T}}) + \eta_q \sum_{j=q}^{l} \eta_j (A_{jp}^{\mathrm{T}} \overline{P}_{jq} + A_{qp}^{\mathrm{T}} P_{jq}^j) \\ &+ \eta_p \sum_{\substack{j=p \\ (j \neq q)}}^{l} (P_{pj}^{p\,\mathrm{T}} A_{pq} + \overline{P}_{pj}^{\mathrm{T}} A_{jq}) \eta_j + \eta_p \sum_{\substack{j=p \\ (j \neq q)}}^{l} \eta_j (\overline{P}_{jp}^{\mathrm{T}} A_{jq} + P_{jq}^{j\,\mathrm{T}} A_{pq}) \\ &+ 2\eta_p \sum_{\substack{k=1 \\ (k \neq p,q)}}^{l} (c_{pk} b_p^{\mathrm{T}} k_p^* P_{pq}^p + c_{qk} b_q^{\mathrm{T}} k_q^* \overline{P}_{pq}^{\mathrm{T}} \\ &+ c_{pk} b_p^{\mathrm{T}} k_p^* \overline{P}_{pq} + c_{qk} b_q^{\mathrm{T}} k_q^* P_{pq}^q) \eta_q, \qquad \eta_p \in R_+, \quad \eta_p > 0.\end{aligned}$$

Here

$$k_i^* = \begin{cases} k_i, & \text{for } \sum_{k=1}^{l} \omega_k^{\mathrm{T}} c_{ik} b_i^{\mathrm{T}} P_{pq}^i \omega_i > 0 \\ & (\text{or } \theta_i b_i^{\mathrm{T}} \overline{P}_{pq} \omega_i > 0, \text{ or } \theta_i b_i^{\mathrm{T}} P_{ii} \omega_i > 0); \\ 0, & \text{in the other cases.} \end{cases}$$

We designate by $\lambda_M(\Xi_{pp})$ and $\lambda_M^{1/2}(\Xi_{pq}\Xi_{pq})$ the maximal eigenvalue of matrix Ξ_{pp} and the norm of matrix $\Xi_{pq}^{\mathrm{T}}\Xi_{pq}$ respectively. For system (4.5.14) the following result is valid.

THEOREM 4.5.2. *If system of the equations (4.5.14) is such that*

(1) *the first and second level decompositions are described by equations (4.5.15) and (4.5.19) respectively;*
(2) *the matrices A_{ii} and $\overline{A}_{ij}$ in systems (4.5.15) and (4.5.19) are stable;*
(3) *the matrix S with elements*

$$s_{pq} = \begin{cases} \lambda_M(\Xi_{pp}) + \sigma_{pp}, & p = q; \\ \lambda_M^{1/2}(\Xi_{pq}\Xi_{pq}), & p < q; \\ s_{qp}, & p > q, \quad (p, q) \in [1, l], \end{cases}$$

   (a) *negative semi-definite;*
   (b) *negative definite.*

Then, the equilibrium state $\omega = 0$ of system (4.5.14) is

(a) *uniformly stable in probability;*
(b) *uniformly asymptotically stable in probability.*

PROOF. We construct by means of the function (4.5.20) the function

$$V(\omega) = \eta^{\mathrm{T}} \Pi(\omega) \eta, \quad \eta \in R_+^l, \quad \eta > 0. \tag{4.5.23}$$

By condition (2) of Theorem 4.5.2 the function $V(\omega)$ is positive definite and radially unbounded. For the averaged derivative

$$\frac{dE[V(\omega)]}{dt} = \eta^{\mathrm{T}} \frac{dE[\Pi(\omega)]}{dt} \eta, \qquad \eta \in R_+^l$$

it is easy to obtain the estimate

$$\begin{aligned} \frac{dE[V(\omega)]}{dt} &\le \sum_{i=1}^{l} (\lambda_M(\Xi_{ii}) + \sigma_{ii}) \|\omega_i\|^2 \\ &\quad + 2 \sum_{i=1}^{l} \sum_{j=i+1}^{l} \lambda_M^{1/2}(\Xi_{ij}^{\mathrm{T}} \Xi_{ij}) \|\omega_i\| \|\omega_j\| = u^{\mathrm{T}} S u, \end{aligned} \tag{4.5.24}$$

where $u = (\|\omega_1\|, \dots, \|\omega_l\|)^{\mathrm{T}}$. Under the condition (3) of Theorem 4.5.2 the averaged derivative (4.5.24) is negative semi-definite or negative definite. According to Theorem 4.3.4 the Theorem 4.5.2 is proved.

4.6 Notes

4.1. General outlines on probability theory and theory of stochastic processes can be found in the books by Doob [31], Gikhman and Skorokhod [42], Dynkin [34], etc. The problems of stochastic stability are presented in a number of monographs (see e.g. Kushner [90], Arnold [5], Khasminskii [83], Michel and Miller [143], Ladde and Lakshmikantham [91], etc.). In these investigations the second Liapunov method is further developed with interesting applications.

4.2. Stochastic system in the form of (4.2.1) is called here the Kats–Krasovskii form with reference to Kats and Krasovskii [82] where it was introduced.

Basic definitions of stochastic stability are formulated as the generalization of Definitions 2.2.1—2.2.4 from Chapter 2 for stochastic systems. Stochastic matrix-valued function is introduced according to Martynyuk [115] and averaged derivative is due to Kats and Krasovskii [82] and Martynyuk [115].

4.3. Theorems 4.3.1–4.3.4 are due to Martynyuk [115].

4.4. Theorems 4.4.1, 4.4.2 are taken from Azimov and Martynyuk [8] and Azimov [6].

4.5. Stochastic version of the Lefschetz [100] problem is presented according to Martynyuk [115]. Oscillating system (4.5.7) was investigated by Azimov and Martynyuk [8], and system of automatic control (4.5.14) was considered by Azimov [7].

5

SOME MODELS OF REAL WORLD PHENOMENA

5.1 Introduction

This chapter contains several examples of real world phenomena that illustrate the versatility and applicability of the matrix-valued Liapunov functions in stability investigation of its equilibrium state.

Section 5.2 deals with mathematical models in population. The neighborhood of the non-trivial equilibrium state is investigated in the general case for a predator-prey system and estimates of stability, asymptotic stability and instability domains are found in this section.

In Section 5.3 the model of an orbital astronomical observatory is considered. Conditions are established under which the whole system is stable even though its separate subsystem are unstable.

In Section 5.4 we discuss a power system model consisting of N generators. General conditions are specified for asymptotic stability of the equilibrium state of such a system to be applicable in the case of $3, 5$ and 7 generators to obtain the system parameters such that the system is asymptotically stable, while the method of scalar or vector Liapunov functions have failed to work herein.

Finally, in Section 5.5 the motion in space of winged aircraft is treated.

5.2 Population Models

We shall discuss in this section mathematical models in population dynamics. In particular, we consider mathematical models of population growth of competing as well as predator-prey species as prototype models of our analysis. The models are based on certain simplifying assumptions as stated below.

(1) The density of a species, that is, the number of individuals per unit

area, can be represented by a single variable, when differences of age, sex and genotype are ignored.

(2) Crowding affects all population members equally. This is unlikely to be true if the members of the species occur in clumps rather than being evently distributed throughout the available space.

(3) The affects of interactions within and between species are instantaneous. This means that there is no delayed action on the dynamics of the population.

(4) Abiotic environmental factors are sufficiently constant.

(5) Population growth rate is density-dependent even at the lowest densities. It may be more reasonable to suppose that there is some threshold density below which individuals do not interfere with one another.

(6) The females in a sexually reproducing population always find mates, even though the density may be low.

The assumptions relative to density dependency and crowding affects the fact that the growth of any species in a restricted environment must eventually be limited by a shortage of resources.

5.2.1 Competition

For simplicity, let us first consider a two-species community model living together and competing with each other for the same limiting resources. Under assumptions (1)–(6), a mathematical model of population growth of two competing species is described by

$$\begin{aligned} \frac{dx_1}{dt} &= x_1(a_1 - b_{11}x_1 - b_{12}x_2), \\ \frac{dx_2}{dt} &= x_2(a_2 - b_{21}x_1 - b_{22}x_2), \end{aligned} \tag{5.2.1}$$

where x_i is the population density of species i for $i = 1, 2$ and for $i, j = 1, 2$, a_i, b_{ij} are positive constants. These equations are derived from the Verhulst-Pearl logistic equation

$$\frac{dx_i}{dt} = x_i(a_i - b_{ii}x_i), \quad i = 1, 2, \tag{5.2.2}$$

by including the additional terms $-b_{ij}x_j$ for $i, j = 1, 2$ and $i \neq j$ to describe the inhibiting effects of each species on its competior. The logistic equation is best regarded as a purely descriptive equation.

The important features of (5.2.2) are:

(a) The species increase exponentially whenever they are isolated.

(b) They approach their equilibrium without oscillations in the absence of its competitor.

In (5.2.1), for $i = 1, 2$ $a_i x_i$ can be interpreted as the potential rate of increase that the ith species would grow if the resources were unlimited and intra/inter-specific effects are neglected. Here a_i is the intrinsic rate of natural increase of the ith species, $a_i/b_{ii} = k_i$ is referred as the carrying capacity if the ith species. From this (5.2.2) can be written as

$$\frac{dx_i}{dt} = a_i x_i \left(1 - \frac{x_i}{k_i}\right). \tag{5.2.3}$$

We observe that the per capita growth rate $\left(\frac{dx_i}{dt}\right) \Big/ x_i$ will be negative or positive depending on the population density $x_i > k_i$ or $x_i < k_i$. Thus the constants k_i determine the saturation level of population densities.

5.2.2 Predator-Prey

In the community of competing species, each species inhibits the multiplication of the other species. In a community of two species in which one species is a parasite or predator and the other its host or prey, a different form of interaction between these two species takes place. The mathematical models for host-parasite and predator-prey systems are equivalent. Obviously, the more abundant the prey, the more opportunities there are for the predator to breed. However, as the predator population grows, the number of prey eaten by the predator increases. To formulate the mathematical model describing the predator-prey interaction between two species, we assume the following: (a) in the absence of a predator, the prey species satisfies assumptions (1)–(6) and (b) the predator cannot survive without the presence of prey and the rate at which prey are eaten is proportional to the product of the densities of predator and prey. Under these assumptions, a mathematical model describing the predator-prey interaction between a prey and a predator in a given community is given by

$$\begin{aligned} \frac{dx_1}{dt} &= x_1(a_1 - b_{11}x_1 - b_{12}x_2), \\ \frac{dx_2}{dt} &= x_2(-a_2 + b_{21}x_1), \end{aligned} \tag{5.2.4}$$

where x_1 is prey density and x_2 is predator density and a_1, a_2, b_{11}, b_{21} are positive constants.

From the foregoing discussion with regard to the two-species competition model and the predator-prey model, we can readily generalize to n interacting species so that the general model is described by

$$\frac{dx_i}{dt} = x_i\left(a_i + \sum_{j=1}^{n} b_{ij}x_j\right), \qquad x_i(0) = x_{i0} \geq 0, \tag{5.2.5}$$

where x_i is density of the ith species in the community, a_i, $-b_{ii}$ are positive constants and b_{ij}, $i \neq j$, are constants with any sign. Any arbitrary sign for b_{ij}, $i \neq j$, allows us a greater flexibility for the interactions between the ith and j^{th} species in the community. For example, in a competitive model, b_{ij}, b_{ji}, $i \neq j$, are both negative, while for a predator-prey model, b_{ij}, b_{ji}, $i \neq j$, are of opposite signs. In a model for commensalism (symbiosis), b_{ij}, b_{ji}, $i \neq j$, are both positive.

The system (5.2.5) is represented in the vector form

$$\frac{dx}{dt} = X(a + Bx), \qquad x(0) = x_0 \leq 0 \tag{5.2.6}$$

and decomposed into two subsystems

$$\begin{gathered}\frac{dx_s}{dt} = X_s(a_s + A_{s1}x_1 + A_{s2}x_2),\\ x_s(0) = x_s \leq 0, \quad s = 1, 2.\end{gathered} \tag{5.2.7}$$

Here $x = (x_1^{\mathrm{T}}, x_2^{\mathrm{T}})^{\mathrm{T}} \in R_+^n$, $x_s \in R_+^{n_s}$, $(a_1^{\mathrm{T}}, a_2^{\mathrm{T}})^{\mathrm{T}} \in R^n$, $B = [A_{sj}]$, $s, j = 1, 2$; $a_s = (a_{s1}, a_{s2}, \ldots, a_{sn_s})^{\mathrm{T}} \in R^{n_s}$, A_{sj} are constant matrices $n_s \times n_j$, $X = \operatorname{diag}(X_1, X_2)$, $X_s = \operatorname{diag}(x_{s1}, \ldots, x_{sn_s})$, $s = 1, 2$.

Equilibrium population are determined by

$$X(a + Bx) = 0. \tag{5.2.8}$$

From (5.2.8) it is easy to conclude that $x = 0$ is an equilibrium which is not interesting and so, we must assume that $X \neq 0$. In this case (5.2.8) reduces to

$$a + Bx = 0, \tag{5.2.9}$$

where B is an n by n matrix and a is an n-vector.

We assume that there exists an equilibrium population $x^* > 0$ as a positive solution

$$x^* = -B^{-1}a \tag{5.2.10}$$

of (5.2.9). This assumption is consistent with consideration of community stability. In the case when b has all off-diagonal elements non-negative, that is B is a Metzler matrix, then it is known that stability of B implies $x^* > 0$. It is possible to show that for a Metzler matrix B, the quasi-dominant diagonal condition

$$d_j|b_{jj}| > \sum_{\substack{i=1\\ i\neq j}}^{n} d_i|b_{ij}| \tag{5.2.11}$$

with $d_i > 0$, is equivalent to saying that $-B^{-1}$ is non-negative and since B^{-1} cannot have a row of zeros, positivity of the vector a implies positivity of x^*.

If B is a Metzler matrix, then an elegant solution of the problem on stability of state x^* is obtained by means of the function

$$V(x) = \sum_{i=1}^{n} d_i \left[x_i - x_i^* - x_i^* \ln\left(\frac{x_i}{x_i^*}\right)\right], \qquad d_i > 0.$$

Our aim is to establish stability conditions for system (5.2.6) without assuming matrix B being Metzler. This may be achived by decomposition of system (5.2.6) with further application of the matrix-valued function.

By means of the Liapunov transformation

$$y = x - x^* \tag{5.2.12}$$

we reduce the system (5.2.7) to the form

$$\frac{dy_s}{dt} = X_s^*(A_{s1}y_1 + A_{s2}y_2) + Y_s(A_{s1}Y_1 + A_{s2}y_2) \tag{5.2.13}$$

where

$$X_s^* = \operatorname{diag}\{x_{s1}^*, x_{s2}^*, \dots, x_{sn_s}^*\}, \qquad s = 1, 2,$$
$$Y_s = \operatorname{diag}\{y_{s1}, y_{s2}, \dots, y_{sn_s}\}, \qquad s = 1, 2.$$

For the system (5.2.13) the matrix-valued function

$$U(y) = [v_{sj}(y_s, y_j)], \qquad s, j = 1, 2, \tag{5.2.14}$$

is constructed with the elements

$$\begin{aligned} v_{ss}(y_s) &= y_s^{\mathrm{T}} P_s y_s, \qquad s = 1, 2, \\ v_{sj}(y_s, y_j) &= v_{js}(y_j . y_s) = y_1^{\mathrm{T}} P_3 y_2. \end{aligned} \tag{5.2.15}$$

Here P_s are positive definite symmetric matrices of the dimensions $n_s \times n_s$, $s = 1, 2$, and P_3 is a constant matrix n_1 by n_2.

For the function

$$V(y, \eta) = \eta^{\mathrm{T}} U(y) \eta, \qquad \eta \in R^2, \tag{5.2.16}$$

the following estimates are valid

$$u^{\mathrm{T}} H^{\mathrm{T}} D_1 H u \leq V(y, \eta) \leq u^{\mathrm{T}} H^{\mathrm{T}} D_2 H u, \tag{5.2.17}$$

where

$$u^{\mathrm{T}} = (\|y_1\|, \|y_2\|), \qquad H = \operatorname{diag}\{\eta_1, \eta_2\},$$

$$D_1 = \begin{pmatrix} \lambda_m(P_1) & -\operatorname{sign}(\eta_1 \eta_2) \lambda_M^{1/2}(P_3 P_3^{\mathrm{T}}) \\ -\operatorname{sign}(\eta_1 \eta_2) \lambda_M^{1/2}(P_3 P_3^{\mathrm{T}}) & \lambda_m((P_2) \end{pmatrix},$$

$$D_2 = \begin{pmatrix} \lambda_M(P_1) & -\operatorname{sign}(\eta_1 \eta_2) \lambda_M^{1/2}(P_3 P_3^{\mathrm{T}}) \\ -\operatorname{sign}(\eta_1 \eta_2) \lambda_M^{1/2}(P_3 P_3^{\mathrm{T}}) & \lambda_M(P_2) \end{pmatrix}.$$

We have for the function $D^+ V(y, \eta) = \eta^{\mathrm{T}} D^+ U(y) \eta$:

$$\begin{aligned} D^+ V(y, \eta) &= \eta^{\mathrm{T}} D^+ U(y) \eta = \eta_1^2 D^+ v_{11}(y_1) + 2\eta_1 \eta_2 D^+ v_{12}(y_1, y_2) \\ &+ \eta_2^2 D^+ v_{22}(y_2) = y_1^{\mathrm{T}} [F_{11} + G_{11}] y_1 + 2 y_1^{\mathrm{T}} F_{12} y_2 \\ &+ y_2^{\mathrm{T}} [F_{22} + G_{22}] y_2 + 2 y_1^{\mathrm{T}} G_{12} y_2. \end{aligned} \tag{5.2.18}$$

Here

$$F_{11} = \eta_1^2 \left[P_1 X_1^* A_{11} + (X_1^* A_{11})^{\mathrm{T}} P_1 \right] + \eta_1 \eta_2 \left[P_3 X_2^* A_{21} + (X_2^* A_{21})^{\mathrm{T}} P_3^{\mathrm{T}} \right];$$

$$F_{12} = \eta_1^2 P_1 X_1^* A_{12} + \eta_2^2 (X_2^* A_{21})^{\mathrm{T}} P_2 + \eta_1 \eta_2 \left[(X_1^* A_{11})^{\mathrm{T}} P_3 + P_3^{\mathrm{T}} X_2^* A_{22} \right];$$

$$F_{22} = \eta_2^2 \left[P_2 X_2^* A_{22} + (X_2^* A_{22})^{\mathrm{T}} P_2 \right] + \eta_1 \eta_2 \left[(X_1^* A_{12})^{\mathrm{T}} P_3 + P_3^{\mathrm{T}} X_1^* A_{12} \right];$$

$$G_{11} = \eta_1^2 \left[P_1 Y_1 A_{11} + (Y_1 A_{11})^{\mathrm{T}} P_1 \right] + \eta_1 \eta_2 \left[P_3 Y_2 A_{21} + (Y_2 A_{21})^{\mathrm{T}} P_3^{\mathrm{T}} \right];$$

$$G_{12} = \eta_1^2 P_1 Y_1 A_{12} + \eta_2^2 (Y_2 A_{21})^{\mathrm{T}} P_2 + \eta_1 \eta_2 \left[P_3 Y_2 A_{22} + (Y_1 A_{22})^{\mathrm{T}} P_3 \right];$$

$$G_{22} = \eta_2^2 \left[P_2 Y_2 A_{22} + (Y_2 A_{22})^{\mathrm{T}} P_2 \right] + \eta_1 \eta_2 \left[P_3^{\mathrm{T}} Y_1 A_{12} + (Y_1 A_{12})^{\mathrm{T}} P_3 \right].$$

We have for (5.2.18) the estimate

$$D^+ V(y, \eta) \le u^{\mathrm{T}} \left[C + G(y) \right] u, \tag{5.2.19}$$

where

$$\begin{aligned} u^{\mathrm{T}} &= (\|y_1\|, \|y_2\|), \\ C &= [c_{sj}], \quad s, j = 1, 2, \quad c_{12} = c_{21}, \\ G(y) &= [\sigma_{sj}(y)], \quad \sigma_{12}(y) = \sigma_{21}(y). \end{aligned}$$

Here c_{11}, c_{22} are maximal eigenvalues of the matrices F_{11}, F_{22}; c_{12} is the norm of matrix F_{12} and $\sigma_{sj}(y)$ is the norm of matrix G_{sj}, $s, j = 1, 2$.

It follows from (5.2.18) that

$$D^+ V(y, \eta) \ge u^{\mathrm{T}} [C^* - G(y)] u, \tag{5.2.20}$$

where

$$C^* = \begin{pmatrix} c_{11}^* & -c_{12} \\ -c_{21} & c_{22}^* \end{pmatrix}$$

and c_{11}^*, c_{22}^* are minimal eigenvalues of the matrices F_{11}, F_{22} respectively.

Let us introduce the following notations

$$\begin{aligned} \Pi_1 = \{ & y \in R_+^n : \sigma_{11}(y) + c_{11} \le 0, \quad \sigma_{22}(y) + c_{22} \le 0, \\ & (\sigma_{11}(y) + c_{11})(\sigma_{22}(y) + c_{22}) - (\sigma_{12}(y) + c_{12})^2 \ge 0 \}; \\ \Pi_2 = \{ & y \in R_+^n : \sigma_{11}(y) + c_{11} < 0, \quad \sigma_{22}(y) + c_{22} < 0, \\ & (\sigma_{11}(y) + c_{11})(\sigma_{22}(y) + c_{22}) - (\sigma_{12}(y) + c_{12})^2 > 0 \}; \\ \Pi_3 = \{ & y \in R_+^n : c_{11}^* - \sigma_{11}(y) > 0, \quad c_{22}^* - \sigma_{22}(y) > 0, \\ & (c_{11}^* - \sigma_{11}(y))(c_{22}^* - \sigma_{22}(y)) - (\sigma_{12}(y) + c_{12})^2 > 0 \}. \end{aligned}$$

Estimates (5.2.17), (5.2.19) and (5.2.20) yield the following assertion.

PROPOSITION 5.2.1. *The equilibrium state x^* of the system (5.2.6) is:*

(1) *Stable (asymptotically) in the domain Π_1 (Π_2) if the matrix D_1 is positive definite and the matrix C is negative definite.*
(2) *Unstable in the domain Π_3 if the matrices D_1 and C^* are positive definite.*

PROOF. The fact that the matrix D_1 is positive definite yields that the function $V(y)$ if positive definite for all $y \in R^n_+$. Since the matrix C is negative definite, then by estimate (5.2.19) the function $D^+V(y)$ is non-positive in the domain Π_1. Hence all conditions of Theorem 2.3.3 are satisfied, and the equilibrium state x^* is stable.

The other assertion of Proposition 5.2.1 follows from Theorem 2.3.7.

5.3 Model of Orbital Astronomic Observatory

According to Geiss, Cohen et al. [40] the orbital astronomic observatory consists of following blocks:

(1) observatory vehicle
(2) observatory body
(3) compensation system
(4) engine
(5) system of data (error) processing.

The subsystems (1)–(4) are physycal and its states are characterized by the variables y_1, y_2, y_3 and y_4 respectively. Under some assumptions the mathematical model of the motion control system for the observatory is described by the equations

$$
\begin{aligned}
\frac{dy_1}{dt} &= F_1(y_1)y_2 + F_1(y_1)d_1 + c_1y_2,\\
\frac{dy_2}{dt} &= Y(y_2)d_2 - \beta_1 f_2(\sigma, y_3) + Y(y_2)f_2(\sigma, y_3) + (\beta_2 I + c_2)y_4,\\
\frac{dy_3}{dt} &= -\beta_3 f_1(\sigma) - \beta_4 y_3,\\
\frac{dy_4}{dt} &= -\beta_1 f_2(\sigma, y_3) - \beta_2 y_4,\\
\sigma &= F_2(y_1)y_1 + c_2y_1 + F_2(y_1)d_3.
\end{aligned}
\tag{5.3.1}
$$

Here $y_1 = (y_{11}, y_{12}, y_{13}, y_{14})^{\mathrm{T}}$, $y_i = (y_{i1}, y_{i2}, y_{i3})^{\mathrm{T}}$, $i = 2, 3, 4$

$$\sigma = (\sigma_1, \sigma_2, \sigma_3)^{\mathrm{T}},$$

(5.3.2)
$$F_1(y_1) = \begin{pmatrix} 0 & f_{12} & f_{13} \\ 0 & f_{22} & f_{23} \\ 0 & f_{32} & f_{33} \\ 0 & f_{42} & f_{43} \end{pmatrix}$$

and

(5.3.3)
$$\begin{aligned}
f_{12} &= \sin(y_{13} + \alpha_3) - \sin\alpha_3, \\
f_{13} &= \cos(y_{13} + \alpha_3) - \cos\alpha_3, \\
f_{22} &= -\sin(y_{14} + \alpha_4) + \sin\alpha_4, \\
f_{23} &= -\cos(y_{14} + \alpha_4) + \cos\alpha_4, \\
f_{32} &= -\mathrm{tg}\,(y_{11} + \alpha_1)\cos(y_{13} + \alpha_3) + \mathrm{tg}\,\alpha_1 \cos\alpha_3, \\
f_{33} &= \mathrm{tg}\,(y_{11} + \alpha_1)\sin(y_{13} + \alpha_3) - \mathrm{tg}\,\alpha_1 \sin\alpha_3, \\
f_{42} &= \mathrm{tg}\,(y_{12} + \alpha_2)\cos(y_{14} + \alpha_4) - \mathrm{tg}\,\alpha_2 \cos\alpha_4, \\
f_{43} &= -\mathrm{tg}\,(y_{12} + \alpha_2)\sin(y_{14} + \alpha_4) + \mathrm{tg}\,\alpha_2 \sin\alpha_4,
\end{aligned}$$

$$F_2(y_1) = \begin{pmatrix} 0 & 0 & 0 & 0 \\ g_{21} & g_{22} & 0 & 0 \\ g_{31} & g_{32} & 0 & 0 \end{pmatrix},$$

where

$$\begin{aligned}
g_{32} &= -\delta[\sin(y_{13} + \alpha_3) - \sin\alpha_3], \\
g_{21} &= \delta[\cos(y_{14} + \alpha_4) - \cos\alpha_4], \\
g_{22} &= \delta[\cos(y_{13} + \alpha_3) - \cos\alpha_3], \\
g_{31} &= -\delta[\sin(y_{14} + \alpha_4) - \sin\alpha_4].
\end{aligned}$$

Furthermore

(5.3.4)
$$Y(y_2) = J^{-1} \begin{pmatrix} 0 & y_{23} & -y_{22} \\ -y_{23} & 0 & y_{21} \\ y_{22} & -y_{21} & 0 \end{pmatrix},$$

(5.3.5)
$$Z_1(\zeta) = \begin{pmatrix} 100\,\mathrm{sign}\,\zeta, & |\zeta| \geq 100 \\ \zeta, & |\zeta| \leq 100 \end{pmatrix},$$

(5.3.6)
$$Z_2(\zeta) = \begin{pmatrix} 26\,\mathrm{sign}\,\zeta, & |\zeta| \geq 26 \\ \zeta, & |\zeta| \leq 26 \end{pmatrix},$$

$$f_1(\sigma) = \begin{pmatrix} z_1(\sigma_1 + \alpha_8) - z_1(\alpha_8) \\ z_1(\sigma_2 + \alpha_9) - z_1(\alpha_9) \\ z_1(\sigma_3 + \alpha_{10}) - z_1(\alpha_{10}) \end{pmatrix}, \tag{5.3.7}$$

$$f_2(\sigma, y_3) = \begin{pmatrix} z_2[\beta_1 z_1(\sigma_1 + \alpha_8) + y_{31} + \alpha_{11}] - z_2[\beta_1 z_1(\alpha_8) + \alpha_{11}] \\ z_2[\beta_1 z_1(\sigma_2 + \alpha_9) + y_{32} + \alpha_{12}] - z_2[\beta_1 z_1(\alpha_3) + \alpha_{12}] \\ z_2[\beta_1 z_1(\sigma_3 + \alpha_{10}) + y_{33} + \alpha_{13}] - z_2[\beta_1 z_1(\alpha_{10} + \alpha_{13}] \end{pmatrix}, \tag{5.3.8}$$

$$C_1 = \begin{pmatrix} 0 & \sin\alpha_3 & \cos\alpha_3 \\ 0 & -\sin\alpha_4 & -\cos\alpha_4 \\ 1 & -\operatorname{tg}\alpha_1 \cos\alpha_3 & \operatorname{tg}\alpha_1 \sin\alpha_3 \\ 1 & \operatorname{tg}\alpha_2 \cos\alpha_4 & -\operatorname{tg}\alpha_2 \sin\alpha_4 \end{pmatrix}, \tag{5.3.9}$$

$$C_2 = J^{-1} \begin{pmatrix} \alpha_{14} & \alpha_{15} & 1 & 0 \\ \delta\cos\alpha_4 & \delta\cos\alpha_3 & 0 & 0 \\ -\delta\sin\alpha_4 & -\delta\sin\alpha_3 & 0 & 0 \end{pmatrix}, \tag{5.3.10}$$

$$d_i = \begin{pmatrix} d_{i1} \\ d_{i2} \\ d_{i3} \end{pmatrix}, \qquad i = 1, 2, 3. \tag{5.3.11}$$

In the neighborhood of the equilibrium state

$$y_i = 0, \quad i = 1, 2, 3, 4, \quad \sigma = 0 \tag{5.3.12}$$

under some additional assumptions the system (5.3.1) is reduced to the form

$$\begin{gathered} \frac{dx_i}{dt} = A_{i1}x_1 + A_{i2}x_2 + A_{i3}x_3 + \nu B_i f(\Sigma), \\ \Sigma = Cx, \qquad \forall i = 1, 2, 3, \end{gathered} \tag{5.3.13}$$

besides, x_i, $i = 1, 2, 3$, is determined as

$$x_1 = \begin{pmatrix} \Delta\varphi \\ \Delta V_\varphi \\ \Delta\omega_\varphi \end{pmatrix}, \qquad x_2 = \begin{pmatrix} \Delta\theta \\ \Delta V_\theta \\ \Delta\omega_\theta \end{pmatrix}, \qquad x_3 = \begin{pmatrix} \Delta\psi \\ \Delta V_\psi \\ \Delta\omega_\psi \end{pmatrix},$$

and (φ, θ, ψ) are Euler anglers specifying the rotating motion of the observatory, $(\omega_\varphi, \omega_\theta, \omega_\psi)$ are the velocities of its changing, $V_\varphi, V_\omega, V_\psi$ are the components of vector V that determines the velocity of plane-parallel motion, x_1, x_2, x_3 specify the observatory deviation from the directed position

$$\begin{array}{lll} \Delta\varphi = \varphi^* - \varphi, & \Delta\theta = \theta^* - \theta, & \Delta\psi = \psi^* - \psi; \\ \Delta V_\varphi = V_\varphi^* - V_\varphi, & \Delta V_\theta = V_\theta^* - V_\theta, & \Delta V_\psi^* - V_\psi; \\ \Delta\omega_\varphi = \omega_\varphi^* - \omega_\varphi, & \Delta\omega_\theta = \omega_\theta^* - \omega_\theta, & \Delta\omega_\psi = \omega_\psi^* - \omega_\psi. \end{array}$$

Here φ^*, θ^*, ψ^*; ω^*_φ, ω^*_θ; ω^*_ψ; $V^*{}_\varphi$, V^*_θ, V^*_ψ are the parameters of the observatory directed position. The matrices A_{ij} and B_i are

$$A_{11} = \begin{pmatrix} 0 & a_1 & 0 \\ a_2 & -a_3 & a_4 \\ -a_5 & 0 & -a_5 \end{pmatrix}, \qquad A_{12} = -A_{13} = \begin{pmatrix} 0 & 0 & 0 \\ -a_6 & 0 & 0 \\ -a_7 & 0 & 0 \end{pmatrix},$$

$$B_i = \begin{pmatrix} 0 & 0 & 0 \\ \delta_{i1} & \delta_{i2} & \delta_{i3} \\ 0 & 0 & 0 \end{pmatrix}, \qquad A_{22} = A_{33} = \begin{pmatrix} 0 & -a_1 & 0 \\ 2a_2 & -a_3 & a_4 \\ -2a_5 & 0 & -a_5 \end{pmatrix}$$

δ_{ij} is a Kronecker delta, $A_{ij} = 0,\ i = 2,3;\ j = 1,2,3\ \forall\,(i \neq j)$,

$$C = \begin{pmatrix} r^{\mathrm{T}}_{11} & r^{\mathrm{T}}_{12} & r^{\mathrm{T}}_{13} \\ 0 & r^{\mathrm{T}}_{22} & 0 \\ 0 & 0 & r^{\mathrm{T}}_{33} \end{pmatrix},$$

$$r^{\mathrm{T}}_{1i} = \left(\rho^1_{1i}, \rho^2_{1i}, \rho^3_{1i}\right), \quad i = 1,2,3; \quad r^{\mathrm{T}}_{ji} = \left(\rho_{j1}, \rho_{j2}, \rho_{j3}\right), \quad j = 2,3;$$

$$f(\Sigma) = (\varphi_1(\sigma_1), \varphi_2(\sigma_2), \varphi_3(\sigma_3))^{\mathrm{T}}, \quad \Sigma = (\sigma_1, \sigma_2, \sigma_3)^{\mathrm{T}},$$

$$\frac{\varphi_i(\sigma_i)}{\sigma_i} \in [0,1] \quad \forall\,\sigma_i \in R, \quad \varphi_i(\sigma_i) \in C(R,R).$$

The elements a_s, $s = 1,2,\dots,7$, of the matrices A_{ij} as well as the values r^k_{1i} $(i.k) \in [1,3]$, r_{ik}, $i = 2,3$, $k \in [1,3]$ are known real constants.

System (5.3.13) has a unique equilibrium state $(x = 0) \in R^3$.

The problem is to establish conditions for asymptotic stability in the whole of system (5.3.13).

Let us use the algorithm of constructing the hierarchical Liapunov function (see Martynyuk and Krapivny [124]). The first level decomposition of system (5.3.13) results in the independent subsystems

$$\frac{dx_i}{dt} = A_{ii}x_i + (1 - \delta_{1i})\nu B_i f(\Sigma), \qquad i = 1,2,3 \tag{5.3.14}$$

and the relation functions

$$\begin{aligned} g_1(x) &= A_{12}x_2 + A_{13}x_3 + \nu B_1 f(\Sigma), \\ g_i(x) &= 0, \qquad i = 2,3. \end{aligned} \tag{5.3.15}$$

The second level decomposition yields three couples of the independent subsystems

$$\frac{dx_{ij}}{dt} = \bar{A}_{ij}x_{ij} + \nu B_{ij} f(\Sigma), \qquad (i < j) = 1,2,3, \tag{5.3.16}$$

where $x_{ij} = \left(x_i^{\mathrm{T}}, x_j^{\mathrm{T}}\right)^{\mathrm{T}}$ and the matrices $\bar{A}_{ij}$ and B_{ij} are

$$\bar{A}_{ij} = \begin{pmatrix} A_{11} & A_{1j} \\ 0 & A_{jj} \end{pmatrix}, \quad \bar{A}_{23} = \begin{pmatrix} A_{22} & 0 \\ 0 & A_{33} \end{pmatrix}, \quad B_{ij} = \begin{pmatrix} \delta_{2i} B_i \\ B_j \end{pmatrix}.$$

The relation functions between them are

$$\begin{aligned} &\bar{g}_{1j}(x) = A_{1j}^k + \nu \overline{B}_1 f(\Sigma), \qquad (i \neq k) = 2, 3, \\ &\bar{g}_{23}(x) = 0, \end{aligned} \tag{5.3.17}$$

where

$$A_{1j}^k = \begin{pmatrix} A_{1k} \\ A_{k1} \end{pmatrix} = \begin{pmatrix} A_{1k} \\ 0 \end{pmatrix}, \qquad \overline{B}_1 = \begin{pmatrix} B_1 \\ 0 \end{pmatrix}.$$

We construct for the subsystem (5.3.14) the function

$$v_{ii}(x_i) = x_i^{\mathrm{T}} H_{ii} x_i, \qquad i = 1, 2, 3, \tag{5.3.18}$$

where $H_{ii} > 0$ satisfy the algebraic Liapunov equations

$$A_{ii}^{\mathrm{T}} H_{ii} + H_{ii} A_{ii} = G_{ii}, \qquad i = 1, 2, 3, \tag{5.3.19}$$

where $G_{ii} < 0$ if and only if the subsystems

$$\frac{dx_i}{dt} = A_{ii} x_i$$

are asymptotically stable. For functions (5.3.18) the estimates

$$\begin{aligned} &\lambda_m(H_{ii}) \|x_i\|^2 \leq v_{ii}(x_i) \leq \lambda_M(H_{ii}) \|x_i\|^2 \\ &\forall x_i \in R^{n_i}, \quad i = 1, 2, 3, \quad n_1 = n_2 = n_3 = 3, \end{aligned} \tag{5.3.20}$$

are known.

Assume that for all $x_i \in R^3$ for the functions $v_{ii}(x_i)$ time-derivative along the solutions of subsystems (5.3.14) the estimates

$$\left.\frac{dv_{ii}(x_i)}{dt}\right|_{(5.3.14)} \leq \rho_{ii}^0 \|x_i\|^2, \qquad i = 1, 2, 3 \tag{5.3.21}$$

are satisfied and for (5.3.15)

$$\left(\frac{\partial v_{ii}(x_i)}{\partial x_i}\right)^{\mathrm{T}} g_1(x) \leq \|x_i\|^{1/2} \sum_{k=1}^{3} \mu_{ik} \|x_k\|^{1/2}, \tag{5.3.22}$$

where

$$
\begin{aligned}
\rho_{ii}^{0} &= \lambda_M(G_{ii}) + 2(1-\delta_{1i})\nu\|H_{ii}\|\|r_i\|, \quad i = 1,2,3;\\
\mu_{11} &= 2\nu\|H_{11}\|\,\|r_{11}\|;\\
\mu_{12} &= 2\|H_{11}\|\,[\|A_{12}\| + \nu\|r_{12}\|]\,;\\
\mu_{13} &= 2\|H_{11}\|\,[\|A_{13}\| + \nu\|r_{13}\|]\,;\\
\mu_{ik} &= 0, \quad i = 2,3; \quad k = 1,2,3.
\end{aligned}
\tag{5.3.23}
$$

We construct for (i,j)-couples of subsystem (5.3.16) the functions

$$
v_{ij}(x_{ij}) = x_{ij}^{\mathrm{T}} H_{ij} x_{ij}, \quad (i<j) = 1,2,3, \tag{5.3.24}
$$

where the matrices $H_{ij} > 0$ satisfy the algebraic Liapunov equations

$$
\bar{A}_{ij}^{\mathrm{T}} H_{ij} + H_{ij}\bar{A}_{ij} = G_{ij}, \quad (i<j) = 1,2,3, \tag{5.3.25}
$$

for $G_{ij} < 0$ if and only if (i,j)-couples

$$
\frac{dx_{ij}}{dt} = \bar{A}_{ij} x_{ij}, \quad (i<j) = 1,2,3,
$$

are asymptotically stable.

For functions $v_{ij}(x_{ij})$ the estimates

$$
\begin{gathered}
\lambda_m(H_{ij})\|x_{ij}\|^2 \le v_{ij}(x_{ij}) \le \lambda_M(H_{ij})\|x_{ij}\|^2\\
\forall x_{ij} \in R^{n_i \times n_j}, \quad (i<j) = 1,2,3,
\end{gathered}
\tag{5.3.26}
$$

take place.

We assume now that for the functions $v_{ij}(x_{ij})$ time-derivative along the solutions of subsystems (5.3.16) the estimates

$$
\left.\frac{dv_{ij}(x_{ij})}{dt}\right|_{(5.3.16)} \le \rho_{ij}^{1}\|x_i\| + 2\rho_{ij}^{2}\|x_i\|^{1/2}\|x_j\|^{1/2} + \rho_{ij}^{3}\|x_j\| \tag{5.3.27}
$$

are satisfied for all $x_i \in R^3$ and for (5.3.17)

$$
\left(\frac{\partial v_{ij}(x_{ij})}{\partial x_{ij}}\right)^{\mathrm{T}} g_{1j}(x) \le \sum_{\substack{k=1\\ p=k}}^{3} \nu_{kp}^{ij}\|x_k\|^{1/2}\|x_p\|^{1/2}. \tag{5.3.28}
$$

The contstants ρ_{ij}^1, ρ_{ij}^2, ρ_{ij}^3 can be determined as follows

$$\rho_{ij}^1 = \lambda_M(G_{ij}) + 2\nu\delta_{2i}\|H_{22}^j\|\,\|r_{22}\|,$$
$$\rho_{ij}^2 = \nu\|\overline{H}_{ij}\|\,\|r_{ij}\| + \nu\delta_{2i}\|\overline{H}_{23}\|\,\|r_{22}\|, \tag{5.3.29}$$
$$\rho_{ij}^3 = \lambda_M(G_{ij}) + 2\nu\|H_{ij}^i\|\,\|r_{jj}\|, \quad (i<j) = 1,2,3$$

and the constants ν_{kp}^{ij} as follows

$$\begin{aligned}
\nu_{11}^{1j} &= 2\nu\|H_{11}^j\|\|r_{11}\|,\\
\nu_{jj}^{1j} &= 2\nu\|\overline{H}_{ij}\|(\delta_{2j}\|r_{12}\| + \delta_{31}\|r_{13}\|),\\
\nu_{23}^{1j} &= 2\|\overline{H}_{ij}\|[\|A_{1k} + \nu(\delta_{3j}\|r_{12}\| + \delta_{2j}\|r_{13}\|)];\\
\nu_{1k}^{1j} &= 2\|H_{11}^j\|\,[(1-\delta_{jk})\|A_{1k}\| + \nu\|r_{1k}\|]\\
&\quad + 2\delta_{kj}\nu\|\overline{H}_{ij}\|\|r_{11}\|, \qquad k,\, j = 2,3,\\
\nu_{33}^{12} &= \nu_{22}^{13} = \nu_{kp}^{23} = 0 \qquad \forall\,(k \le p) = 1,2,3.
\end{aligned} \tag{5.3.30}$$

Here the matrices H_{ij}^j and $\overline{H}_{ij}$, $(i<j) = 1,2,3$, of the dimensions 3×3 are the blocks of the matrix H_{ij} so that

$$H_{ij} = \begin{pmatrix} H_{ii}^j & \overline{H}_{ij} \\ \overline{H}_{ij}^{\mathrm{T}} & H_{jj}^i \end{pmatrix}.$$

Using the matrix-valued function $U(x)$ with elements (5.3.18) and (5.3.24), and by virtue of (5.3.21), (5.3.22), (5.3.27) and (5.3.28) we see that

$$\frac{dV(x,\eta)}{dt} \le \varphi^{\mathrm{T}}(\|x\|)S\varphi(\|x\|), \tag{5.3.31}$$

where

$$V(x,\eta) = \eta^{\mathrm{T}}U(x)\eta, \quad \eta \in R_+^3, \quad \eta > 0,$$
$$\varphi(\|x\|) = \left(\|x_1\|^{1/2}, \dots, \|x_3\|^{1/2}\right).$$

The matrix S in (5.3.31) has the form

$$S = \frac{1}{2}(\Pi + \Pi^{\mathrm{T}}),$$

where Π is the upper triangle matrix with the elements

$$
\begin{aligned}
\pi_{kk} &= \eta_k^2(\rho_{kk}^0 + \mu_{kk}) + 2\eta_k \sum_{i=1}^{k-1} \eta_i \rho_{ik}^3 \\
&\quad + 2\eta_k \sum_{i=k+1}^{3} \eta_i \rho_{ki}^1 + \sum_{\substack{i,j=1 \\ i\neq j}}^{3} \eta_i \eta_j \nu_{kk}^{ij}, \\
\pi_{kp} &= \eta_k^2 \mu_{kp} + 4\eta_k \eta_p \rho_{kp}^2 + 2\sum_{i=1}^{s} \sum_{j=i+1}^{s} \eta_i \eta_j \nu_{kp}^{ij}, \quad k < p, \\
\pi_{pk} &= 0, \qquad k < p.
\end{aligned}
\tag{5.3.32}
$$

The matrix S in the estimate (5.3.31) is negative definite, if

$$s_{11} < 0, \quad s_{22} < 0, \quad s_{33} < 0 \tag{5.3.33}$$

and

$$s_{11}s_{22} - s_{12}^2 > 0, \quad \det S < 0 \tag{5.3.34}$$

since $s_{ij} > 0 \ \forall (i \neq j) \in [1,3]$.

Stability conditions (5.3.33), (5.3.34) are analyzed for two cases, first, for the case when only the first level decomposition is made. This corresponds to the approach based on the vector Liapunov function, applied by Grujić, Martynyuk and Ribbens-Pavella [57].

In this case the elements of matrix Π for system (5.3.13) are in view of (5.3.23)–(5.3.30) and (5.3.32)

$$
\begin{aligned}
\pi_{ii} &= \eta_i^2 \left[\lambda_M(G_{ii}) + 2\nu \|H_{ii}\| \, \|r_{ii}\|\right], & i &= 1,2,3; \\
\pi_{1j} &= 2\eta_1^2 \|H_1\| \left[\|A_{1j}\| + \nu \|r_{1j}\|\right], & j &= 2,3; \\
\pi_{23} &= \pi_{ji} = 0 & \forall\,(j &> i) = 1,2,3.
\end{aligned}
$$

We introduce the designations

$$Q = \operatorname{diag}(\eta_1^2, \eta_2^2, \eta_3^2)$$

and the matrix $D = [d_{ij}]$ the elements of which are expressed via the elements of matrix Π as follows

$$d_{ij} = \frac{\pi_{ij}}{\eta_i^2}, \qquad (i,j) = 1,2,3.$$

Therefore we have

$$S = \frac{1}{2}(\overline{\Pi} + \overline{\Pi}^{\mathrm{T}}) = \frac{1}{2}(QD + D^{\mathrm{T}}Q). \tag{5.3.35}$$

The matrix S is negative definite if and only if the matrix D is an M-matrix. The matrix D is an upper triangular and $d_{ij} \geq 0$ $(i,j) = 1,2,3$, hence, if $d_{ii} < 0$, then D is the M-matrix. Therefore, the conditions for matrix S being negative definite are

$$\lambda_M(G_{ii}) + 2\nu\|H_{ii}\|\,\|r_{ii}\| < 0 \qquad \forall\, i = 1,2,3. \tag{5.3.36}$$

These are the well-known conditions for the asymptotic stability in the whole of system (5.3.13).

Let us show conditions (5.3.33), (5.3.34) for the asymptotic stability in the whole of the system (5.3.13) to be more general than the conditions (5.3.36).

The conditions (5.3.36) are satisfied if $\lambda_M(G_{ii}) < 0$. This means that the subsystems

$$\frac{dx_i}{dt} = A_{ii}x_i, \qquad i = 1,2,3, \tag{5.3.37}$$

obtained from (5.3.14) must be asymptotically stable.

Therefore, if one of the subsystems (5.3.37) is unstable, the conditions (5.3.36) are not satisfied and the approach based on the vector function does not work.

Assume the 3rd subsystem from (3.5.37) is unstable, i.e. $\lambda_M(G_{33}) > 0$. In view of the second level decomposition one of conditions (3.5.33), namely $s_{33} < 0$ becomes

$$\begin{aligned}\eta_3^2(\lambda_M(G_{33}) + 2\nu\|H_{33}\|\,\|r_{33}\|) + 2\eta_1\eta_3\lambda_M(G_{13})\\ + 2\eta_2\eta_3(\lambda_M(G_{33}) + 2\nu\|H_{33}^2\|\,\|r_{33}\|) < 0.\end{aligned} \tag{5.3.38}$$

It is clear that, if the 3rd subsystem forms asymptotically stable couples $(2,3)$ and $(1,3)$, then $\lambda_M(G_{13}) < 0$ and $\lambda_M(G_{23}) < 0$. This may prove to be sufficient for inequality (5.3.36) to be satisfied. However this inequality may be derived by means of the matrix-valued function only.

Thus, the application of the matrix-valued function and two-level decomposition yields less strict conditions for the asymptotic stability in the whole of the system (5.3.13) as compared with conditions (5.3.36) established by means of the vector Liapunov function.

5.4 Power System Model

The dynamical and structural complexity combined with the high order of the power system make many methods developed in theory of differential equations inapplicable in the investigation of these systems. The method of Liapunov functions (scalar, vector or matrix) is one of the methods used in the analysis of stability and the estimation of asymptotic stability domains. In this section we shall show the application of the matrix-valued Liapunov function to be advantageous as compared with the results by the vector Liapunov function.

5.4.1 Description of the Power System

Considered is the N-machine power system with uniform mechanical damping λ. The ith machine motion is modeled by the equations

$$M_i\ddot{\delta}_i + D_i\dot{\delta} = P_{mi} - P_{ei}, \qquad i = 1, 2, \dots, N, \tag{5.4.1}$$

where

$$P_{ei} = E_i^2 Y_{ii}\cos\theta_{ii} + \sum_{j\neq i}^{n} E_i E_j Y_{ij}\cos(\delta_{ij} - \theta_{ij}), \tag{5.4.2}$$

and $M_i \in R$ is the inertia coefficient of the ith machine, $D_i \in R$ is the mechanical damping of the ith machine, $P_{mi} \in R$ is the mechanical power delivered by the ith machine, $E_i \in R$ is the modulus of the internal voltage, $Y_{ij} \in R$ is the magnitude of the (i,j)-th element of the reduced admittances matrix Y, $\delta_i \in R$ is the absolute rotor angle: $\delta_{ij} = \delta_i - \delta_j = \delta_{iN} - \delta_{iN}$, $\delta_{ij}^0 = \delta_i^0\delta_j^0$, $\theta_{ij} \in R$ is the angle of the (i,j)-th element of the reduced admittances matrix.

Let us take the Nth machine as a standard one and introduce $(2N-1)$ state variables

$$\begin{aligned} \sigma_{iN} &= \delta_{iN} - \delta_{iN}^0, && i \neq N; \\ \omega_i &= \dot{\delta}_i, && i = 1, 2, \dots, N, \end{aligned} \tag{5.4.3}$$

where $\sigma_{ij} \in R$ is a subsidiary variable, $\omega_i \in R$ is the absolute angular speed of the ith machine rotor. Here δ_{iN}^0 are the solutions of the system of equations

$$E_i^2 Y_{ii} \cos\theta_{ii} + \sum_{j\neq i}^{N} E_i E_j Y_{ij} \cos(\delta_{iN}^0 - \delta_{jN}^0 - \theta_{ij}) = P_{mi}, \qquad i = 1, 2, \dots, N. \tag{5.4.4}$$

The motion of the whole N-machine system can be described by the equations

$$\begin{aligned} \dot{\sigma}_{iN} &= \omega_{iN}, \\ \dot{\omega}_i &= -\lambda\omega_i - M_i^{-1} \sum_{j\neq i}^{N} A_{ij} f_{ij}(\sigma_{ij}), \qquad i = 1, 2, \dots, N, \end{aligned} \tag{5.4.5}$$

where $A_{ij} = E_j E_i Y_{ij}$, $A_i = A_{iN}$, f_{ij} are non-linear functions

$$f_{ij}(\sigma_{ij}) = \cos(\sigma_{ij} + \sigma_{ij}^0 - \theta_{ij}) - \cos(\delta_{ij}^0 - \theta_{ij}), \tag{5.4.6}$$

satisfying the conditions

$$f_{ij}(0) = 0, \quad 0 \leq \frac{f_{ij}(\sigma_{ij})}{\sigma_{ij}} \leq \xi_{ij}, \quad \sigma_{ij} \neq 0, \tag{5.4.7}$$

as soon as σ_{ij} take the value on compact intervals J_{ij}:

$$J_{ij} = \left\{ \sigma_{ij} : -2(\pi - \theta_{ij} + \delta_{ij}^0) \leq \sigma_{ij} \leq 2(\theta_{ij} - \delta_{ij}^0) \right\}. \tag{5.4.8}$$

The constants ξ_{ij} in (5.4.7) are determined as follows

$$\xi_{ij} = \left. \frac{\partial f_{ij}(\sigma_{ij})}{\partial \sigma_{ij}} \right|_{\sigma_{ij}=0} .$$

5.4.2 Mathematical Decomposition of the Power system model

The state vector of the whole system is designated as

$$\hat{x} = (\sigma_{1N}, \omega_1, \sigma_{2N}, \omega_2, \dots, \sigma_{N-1,N}, \omega_{N-1}, \omega_N)^{\mathrm{T}},$$

and the subvectors

$$x_i = (\sigma_{iN}, \omega_{iN})^{\mathrm{T}} = (x_{i1}, x_{i2})^{\mathrm{T}}, \qquad i = 1, 2, \dots, N-1, \tag{5.4.9}$$

are introduced.

System (5.4.5) is represented as

$$\begin{aligned} \frac{dx_i}{dt} &= P_i x_i + B_i F_i(\sigma_i) + h_i(x), \\ \sigma_i &= C_i^{\mathrm{T}} x_i, \qquad i = 1, 2, \dots, s. \end{aligned} \tag{5.4.10}$$

Each subsystem of (5.4.10) consist of free subsystems

$$\begin{aligned} \frac{dx_i}{dt} &= P_i x_i + B_i F_i(\sigma_i), \\ \sigma_i &= C_i^{\mathrm{T}} x_i, \qquad i = 1, 2, \dots, s, \end{aligned} \tag{5.4.11}$$

and relation functions

$$h_i(x) = \begin{pmatrix} 0 \\ \sum\limits_{j \neq i}^{N-1} (-M_i^{-1} A_{ij} f_{ij}(\sigma_{ij}) + M_N^{-1} A_{Nj} f_{Nj}(\sigma_{Nj})) \end{pmatrix}. \tag{5.4.12}$$

The vector of nonlinearities $F_i(\sigma_i)$ is a decomposition of two nonlinearities

$$\begin{aligned} f_{i1}(\sigma_{i1}) &= \cos(\sigma_{iN} + \delta_{iN}^0 - \theta_{iN}) - cos(\delta_{iN}^0 - \theta_{iN}), \\ f_{i2}(\sigma_{i2}) &= \cos(\sigma_{Ni} + \delta_{Ni}^0 - \theta_{iN}) - cos(\delta_{Ni}^0 - \theta_{iN}). \end{aligned} \tag{5.4.13}$$

The other matrices and functions appearing in the system (5.4.14) are

$$P_i = \begin{pmatrix} 0 & 1 \\ 0 & -\lambda \end{pmatrix},$$

$\lambda = D_i M_i^{-1}$ is a uniform damping, $i - 1, 2, \dots, N$;

$$B_i = \begin{pmatrix} 0 & 0 \\ -M_i^{-1} A_i & M_N^{-1} A_i \end{pmatrix}, \qquad C_i^{\mathrm{T}} = \begin{pmatrix} 1 & 0 \\ -1 & 0 \end{pmatrix}.$$

5.4.3 Application Algorithm of the Matrix-Valued Function

The elements v_{ij} of the matrix-valued function $U(x)$ are taken as

$$v_{ii}(x_i) = x_i^{\mathrm{T}} H_i x_i + \sum_{k=1}^{2} \gamma_{ik} \int_0^{\sigma_{ik}} f_{ik}(\sigma_{ik})\, d\sigma_{ik}, \tag{5.4.14}$$
$$i = 1, 2, \dots, s,$$

$$v_{ij}(x_i, x_j) = \alpha_{ij} \int_0^{\sigma_{ij}} f_{ij}(\sigma_{ij})\, d\sigma_{ij} \tag{5.4.15}$$
$$(i \neq j), \quad i, j = 1, 2, \dots, s.$$

Here H_i are 2×2 symmetric positive definite matrices, γ_{ik} and α_{ij} are arbitrary positive numbers.

Let $\eta = (1, \dots, 1)^{\mathrm{T}} \in R_+^s$ and

$$\dot V(x, \eta) = \eta^{\mathrm{T}} \dot U(x) \eta, \qquad \dot U(x) = [\dot v_{ij}(x_i, x_j)].$$

The function v_{ij} time-derivative along the solutions of the ith interconnected subsystem is

$$\frac{dv_{ii}}{dt} = \left.\frac{dv_{ii}}{dt}\right|_{(5.4.11)} + \left.\frac{dv_{ii}}{dt}\right|_{(5.4.12)}, \tag{5.4.16}$$

where

$$\left.\frac{dv_{ii}}{dt}\right|_{(5.4.11)} = 2x_i^{\mathrm{T}} H_i [P_i x_i + B_i F_i(\sigma_i)] + \sum_{k=1}^{2} \gamma_{ik} f_{ik}(\sigma_{ik}) \dot\sigma_{ik}, \tag{5.4.17}$$

$$\left.\frac{dv_{ii}}{dt}\right|_{(5.4.12)} = 2x_i^{\mathrm{T}} H_i h_i(x). \tag{5.4.18}$$

Further we introduce the following matrices

$$r_i = \operatorname{diag}\{\gamma_{i1}, \gamma_{i2}\}, \tag{5.4.19}$$

$$\Phi_i = \operatorname{diag}\left\{ \frac{f_{i1}(\sigma_{i1})}{\sigma_{i1}}, \frac{f_{i1}(\sigma_{i1})}{\sigma_{i1}} \right\} \in [a_i, b_i], \tag{5.4.20}$$

where $a_i = \operatorname{diag}\{\varepsilon_{i1}, \varepsilon_{i2}\}$ and $b_i = \{\xi_{i1}, \xi_{i2}\}$ are prescribed values.

The expressions (5.4.18) and (5.4.18) are transformed as

$$\left.\frac{dv_{ii}}{dt}\right|_{(5.4.11)} = -x_i^{\mathrm{T}}[G_i - (aH_iB_i + P_i^{\mathrm{T}}C_ir_i)\Phi_iC_i^{\mathrm{T}}]x_i, \tag{5.4.21}$$

where

$$-G_i = H_iP_i + P_i^{\mathrm{T}}H_i$$

and

$$\left.\frac{dv_{ii}}{dt}\right|_{(5.4.12)} = 2x_i^{\mathrm{T}}H_iD_{ia}x_i + 2x_i^{\mathrm{T}}H_i\sum_{j\neq i}^{s} D_{ib}x_j, \tag{5.4.22}$$

where

$$D_{ia} = \begin{pmatrix} 0 & 0 \\ -M_i^{-1}\sum\limits_{j\neq i}^{s} A_{jj}\Phi_{ij} & 0 \end{pmatrix},$$

$$D_{ib} = \begin{pmatrix} 0 & 0 \\ M_i^{-1}A_{ij}\Phi_{ij} - M_N^{-1}A_N\Phi_N & 0 \end{pmatrix}$$

and

$$\begin{aligned} &\Phi_{ij}(0) = 0, \\ &\Phi_{ij}(\sigma_{ij}) = \frac{f_{ij}(\sigma_{ij})}{\sigma_{ij}}, \quad \sigma_{ij} \neq 0. \end{aligned} \tag{5.4.23}$$

Combining (5.4.21) and (5.4.22) yields

$$\begin{aligned} \frac{dv_{ii}}{dt} &= -x_i^{\mathrm{T}}\left\{G_i - (2H_iB_i + P_i^{\mathrm{T}}C_ir_i)\Phi_iC_i^{\mathrm{T}} - 2H_iD_{ia}\right\}x_i \\ &\quad + 2x_i^{\mathrm{T}}H_i\sum_{j\neq i}^{s} D_{ib}x_j. \end{aligned} \tag{5.4.24}$$

For functions v_{ij} defined by (5.4.15) we have

$$\begin{aligned} \frac{dv_{ii}}{dt} &= \alpha_{ij}\Phi_{ij}x_i^{\mathrm{T}}dd^{\mathrm{T}}P_ix_i - \alpha_{ij}\Phi_{ij}x_i^{\mathrm{T}}(dd^{\mathrm{T}}P_j + P_j^{\mathrm{T}}dd^{\mathrm{T}})x_j \\ &\quad + \alpha_{ij}\Phi_{ij}x_j^{\mathrm{T}}dd^{\mathrm{T}}P_jx_i, \qquad i \neq j, \end{aligned} \tag{5.4.25}$$

where $d = (1, 0)^{\mathrm{T}}$.

We have for function

$$\dot{V}(x, \eta) = -\sum_{i=1}^{s} x_i^{\mathrm{T}} D_{ii} x_i + \sum_{i=1}^{s} \sum_{j \neq i}^{s} x_i^{\mathrm{T}} D_{ij} x_j, \tag{5.4.26}$$

where

$$\begin{aligned} D_{ii} = G_i &- (2H_i B_i + P_i^{\mathrm{T}} C_i^{\mathrm{T}} r_i)\Phi_i C_i^{\mathrm{T}} - 2H_i D_{ia} \\ &- \sum_{j \neq i}^{s} (\alpha_{ij}\Phi_{ij} + \alpha_{ij}\Phi_{ji}) dd^{\mathrm{T}} P_i \end{aligned} \tag{5.4.27}$$

and

$$D_{ij} = 2H_i D_{ib} - \alpha_{ij}\Phi_{ij}(dd^{\mathrm{T}} P_j + P_i^{\mathrm{T}} dd^{\mathrm{T}}). \tag{5.4.28}$$

Further we show that the right-hand part of (5.4.26) can be estimated by the expression $w^{\mathrm{T}}(x) A w(x)$, i.e.

$$\dot{V}(x, \eta) \leq w^{\mathrm{T}}(x) A w(x), \tag{5.4.29}$$

where $w(x) = (\|x_1\|, \dots, \|x_s\|)^{\mathrm{T}}$, $A = [a_{ij}]$, $i, j = 1, 2, \dots, s$. Here a_{ij} is a computed in terms of estimate of the right-hand part of (5.4.26).

If we set $W(x) = \operatorname{diag}\{\|x_1\|, \|x_2\|, \dots, \|x_s\|\}$, then

$$\dot{V}(x, \eta) \leq \eta^{\mathrm{T}} W(x) A W(x) \eta.$$

It should be noted that $\dot{U}(x)$ is not estimated by the expression $W(x)AW(x)$ in view of (5.4.24)–(5.4.28).

Then the matrices H_i are taken in the form

$$H_i = \begin{pmatrix} \lambda h_{12}^i & h_{12}^i \\ h_{12}^i & \dfrac{1+k_i}{\lambda} h_{12}^i \end{pmatrix}, \tag{5.4.30}$$

where k_i are arbitrary positive constants and matrices G_i are computed

$$G_i = \begin{pmatrix} 0 & 0 \\ 0 & 2k_i h_{12}^i \end{pmatrix}. \tag{5.4.31}$$

We take the constants

$$
\begin{aligned}
\gamma_{i1} &= 2M_i^{-1} A_i h_{22}^i,\\
\gamma_{i2} &= 2M_N^{-1} A_i h_{22}^i,\\
\alpha_{ij} &= \alpha_{ji} = M_i^{-1} A_{ij} h_{22}
\end{aligned} \tag{5.4.32}
$$

and transform the expression $-x_i^{\mathrm{T}} D_{ii} x_i$ as

$$
\begin{aligned}
-x_i^{\mathrm{T}} D_{ii} x_i = &-2h_{i2}^i \Big\{ A_i \left(M_i^{-1} \frac{f_{i1}(\sigma_{i1})}{\sigma_{i1}} + M_N^{-1} \frac{f_{i2}(\sigma_{i2})}{\sigma_{i2}} \right)\\
&+ M_i^{-1} \sum_{j \neq i}^{s} A_{ij} \Phi_{ij} \Big\} x_{i1}^2 - 2k_i h_{i2}^i x_{i2}^2\\
&+ \sum_{j \neq i}^{s} M_i^{-1} h_{22}^i A_{ij} (\Phi_{ji} + \Phi_{ij}) x_{i1} x_{i2}.
\end{aligned} \tag{5.4.33}
$$

The right-hand part of (5.4.33) may be estimated by the value $-\lambda_{im}(Q_i)\|x_i\|^2$:

$$
-x_i^{\mathrm{T}} D_{ii} x_i \leq -\lambda_{im}(Q_i)\|x_i\|^2, \qquad i = 1, 2, \dots, s, \tag{5.4.34}
$$

where $\lambda_{im}(Q_i)$ is the minimal eigenvalue of the matrix Q_i, the elements of which are determined as

$$
\begin{aligned}
q_{11}^i = q_{22}^i &= 2h_{i2}^i \Big\{ A_i (M_i^{-1} \varepsilon_{i1} + M_N^{-1} \varepsilon_{i2}) + M_i^{-1} \sum_{j \neq i}^{s} A_{ij} \varepsilon_{ij} \Big\},\\
q_{12}^i &= -\frac{1}{2} M_i^{-1} h_{22}^i \sum_{j \neq i}^{s} \max(\xi_{ij}, \xi_{ji}).
\end{aligned} \tag{5.4.35}
$$

We note that $\varepsilon_{ij} \in (0, \xi_{ij})$ and the constants k_i are taken according to

$$
k_i = A_i (M_i^{-1} \varepsilon_{i1} + M_N^{-1} \varepsilon_{i2}) + M_i^{-1} \sum_{j \neq i}^{s} A_{ij} \varepsilon_{ij}. \tag{5.4.36}
$$

We have in view of (5.4.28)

$$
\begin{aligned}
x_i^{\mathrm{T}} D_{ij} x_j = &\, 2h_{i2}(M_i^{-1} A_{ij} \Phi_{ij} - M_N^{-1} A_{Nj} \Phi_{Nj}) x_{i1} x_{j1}\\
&- \alpha_{ij} \Phi_{ij} x_{i1} x_{i2}\\
&+ \left\{ 2h_{22}^i (M_i^{-1} A_{ij} \Phi_{ij} - M_N^{-1} A_{Nj} \Phi_{Nj} - \alpha_{ij} \Phi_{ij} \right\} x_{i2} x_{j1}.
\end{aligned} \tag{5.4.37}
$$

To estimate the right-hand part of (5.4.37) the functions $Z_1: R^2 \to R$ and $Z_2: R^3 \to R$ are introduced by the formulas

$$
\begin{aligned}
Z_1(\alpha,\beta) &= \min\left\{\sqrt{2}\max(|\alpha|,|\beta|),\ (|\alpha|+|\beta|)\right\},\\
Z_2(\alpha,\beta,\gamma) &= \min\Big\{\sqrt{2}\max(|\alpha|,|\beta|,|\gamma|),\ (|\alpha|+|\beta|+|\gamma|),\\
&\qquad Z_1(\alpha,\beta)+|\gamma|,\ Z_1(\alpha,\beta)+|\beta|,\ Z_2(\beta,\gamma)+|\alpha|\Big\}.
\end{aligned}
$$

Having noted that the expressions $x_{i1}x_{j1}$, $x_{i1}x_{j2}$, $x_{i2}x_{j1}$ can be treated as the components of the 3-dimensional subspace, where each of the expressions may take either positive, negative or zero value, the estimate of the righ-hand part of (5.4.37) can be obtained in the form

$$
\begin{aligned}
x_i^{\mathrm{T}} D_{ij} x_j \le Z_2\,\{&2h_{12}^i \max\left(M_i^{-1}A_{ij}\xi_{ij},\ M_N^{-1}A_{Nj}\xi_{nj}\right),\\
&M_i^{-1}A_{ij}h_{22}^i\xi_{ij},\ h_{22}^i \max\left(M_i^{-1}A_{ij}\xi_{ij},\right.\\
&\left.2M^{-1}A_{Nj}\xi_{Nj}\right)\}\ \|x_i\|\,\|x_j\|.
\end{aligned}
\tag{5.4.38}
$$

In view of (5.4.34) and (5.4.38) we get for the elements a_{ij} of matrix A:

$$
\hat{a}_{ij} = \begin{cases} -\lambda_{im} \quad (i=j);\\ Z_2\{2h_{12}^i \max(M_i^{-1}A_{ij}\xi_{ij},\ M_N^{-1}A_{Nj}\xi_{Nj},\\ \quad M_i^{-1}A_{ij}h_{22}^i\xi_{ij},\ h_{22}^i\max(M_i^{-1}A_{ij}\xi_{ij},\\ \quad 2M_N^{-1}A_{Nj}\xi_{Nj}\} \quad (i\ne j) \end{cases}
\tag{5.4.39}
$$

and

$$
a_{ij} = \frac{1}{2}(\hat{a}_{ij}+\hat{a}_{ji}), \qquad i,j=1,2,\dots,s.
\tag{5.4.40}
$$

We formulate now the following assertion.

PROPOSITION 5.4.1. *In order for the equilibrium state $x=0$ of system (5.4.10) to be asymptotically stable it is sufficient that the inequalities*

$$
(-1)^k \begin{vmatrix} a_{11} & \dots & a_{1s}\\ \dots & \dots & \dots\\ a_{s1} & \dots & a_{ss} \end{vmatrix} > 0, \qquad k=1,2,\dots,s,
\tag{5.4.41}
$$

be satisfied.

PROOF. Let the matrix A in estimate (5.4.29) be constructed according to (5.4.39) and (5.4.40). When inequalities (5.4.41) are satisfied, the matrix A is negative definite, and by (5.4.40) $A = A^{\mathrm{T}}$. The function $V(x,\eta) = \eta^{\mathrm{T}}U(x)\eta$ is positive definite, since $H_i = H_i^{\mathrm{T}}$ is positive definite, $\gamma_{ik} > 0$ and $\alpha_{ij} > 0$ and the integral terms in (5.4.14) and (5.4.15) are non-negative in the neighborhood of $x = 0$. Thus, function $V(x,\eta)$ for system (5.4.10) is positive definite and $\dot{V}(x,\eta)$ is negative definite in the neighborhood of $x = 0$ due to inequalities (5.4.41). By Theorem 2.3.3 the equilibrium state $x = 0$ of system (5.4.10) is asymptotically stable.

5.4.4 Numerical examples

5.4.4.1 Example. The proposed algorithm of the power system stability analysis is applicable to the 3-machine power system considered by Jocić, Ribbens-Pavella and Šiljak [79]. We admit the following parameter values for the system (5.4.10):

$$N = 3;\quad E_1 = 1.017;\quad E_2 = 1.005;\quad E_3 = 1.033;\quad \delta_{12} = 5^\circ;$$
$$\delta_{13} = 2^\circ;\quad \delta_{23} = -3^\circ;\quad Y_{12} = 0.98\times 10^{-3}\angle 86^\circ;\quad Y_{13} = 0.114\angle 88^\circ;$$
$$Y_{23} = 0.106\angle 89^\circ;\quad M_1 = M_2 = 0.01;\quad M_3 = 2.0.$$

Treating the third machine as a standard one we get two subsystems. Let us take the constants $\lambda = 0.3$, $\varepsilon_{11} = \varepsilon_{21} = 0.06$ and $\varepsilon_{12} = \varepsilon_{23} = \xi_{12} = \xi_{21} = 0.001$. The matrix $\hat{A} = [\hat{a}_{ij}]$, defined by formula (5.4.39) is of the form

$$\hat{A} = \begin{pmatrix} -1.1506 & 1.0814 \\ 1.0671 & -1.0437 \end{pmatrix}.$$

The matrix $2A = \hat{A} + \hat{A}^{\mathrm{T}}$ satisfies conditions (5.4.41) and therefore the equilibrium state $x = 0$ is asymptotically stable. It is important to note that in this case Jocić, Ribbens-Pavella and Šiljak [79] established the conditions of asymptotic stability for $\lambda = 100$, $\varepsilon = 0,99$. In a paper by Shaaban and Grujić [164] the asymptotic stability of the system in question was stated for $\lambda = 0.45$, $\varepsilon_{11} = \varepsilon_{21} = 0.10$.

The asymptotic stability conditions for the equilibrium state $x = 0$ obtained herein are the least value for the parameters λ and ε.

5.4.4.2 Example. Let in system (5.4.10) $N = 4$ and the parameter values are the following (see El-Abiad and Nagappan [35]):

$$E_1 = 1.057/5.7^\circ, \quad E_2 = 1.152/14.4^\circ, \quad E_3 = 1.095/2.3^\circ, \quad E_4 = 1.0/0.1^\circ,$$
$$Y_{11} = 0.88/-88.1^\circ, \quad Y_{22} = 0.873/-83.2^\circ, \quad Y_{33} = 1.014/-75.5^\circ,$$
$$Y_{44} = 2.447/-69,7^\circ, \quad Y_{12} = 0.124/82.1^\circ, \quad Y_{13} = 0.065/82.4^\circ,$$
$$Y_{23} = 0.064/88.2^\circ, \quad Y_{24} = 0.655/96.8^\circ,$$
$$Y_{34} = 0.754/99^\circ, \quad Y_{14} = 0.658/91.1^\circ;$$
$$M_1 = 1130, \quad M_2 = 2260, \quad M_3 = 1508, \quad M_4 = 75\,350.$$

Choosing the fourth machine as a standard one we get three subsystems. For the values $\lambda = 0.8$, $\varepsilon_{11} = \varepsilon_{21} = \varepsilon_{31} = 0.5$ the matrix $\hat{A}$ (see formula (5.4.39)) is

$$\hat{A} = \begin{pmatrix} -4.9087 & 3.7790 & 1.8484 \\ 1.8109 & -2.7037 & 0.9811 \\ 1.4073 & 1.4898 & -4.8370 \end{pmatrix}.$$

The matrix $a = \frac{1}{2}\left(\hat{A} + \hat{A}^{\mathrm{T}}\right)$ satisfies the conditions (5.4.41) and therefore, the state $x = 0$ of the system is asymptotically stable. Earlier it has been stated (see Grujić and Shaaban [61]) that the asymptotic stability of the equilibrium state $x = 0$ of the system takes place provided that $\lambda = 1.0$ and $\varepsilon_1 = \varepsilon_2 = \varepsilon_3 = 0.60$.

Therefore, this case as well the proposed algorithm allows us to establish the conditions of asymptotic stability for smaller valies of λ and ε.

5.4.4.3 Example. Let in system (5.4.10) $N = 7$ and the parameter values are taken following Shaaban and Grujić [164]. Taking the seventh machine as a standard one we get six subsystems. For the values $\lambda = 2.0$, $\varepsilon_{i1} = 0.80$, $i = 1, 2, 3$; $\varepsilon_{j1} = 0.85$, $j = 4, 5, 6$, the matrix $\hat{A}$ (see (5.4.39)) is

$$\hat{A} = \begin{pmatrix} -2.0176 & 1.0286 & 0.2408 & 0.2521 & 0.2876 & 0.2730 \\ 1.3301 & -2.3742 & 0.2660 & 0.2785 & 0.3177 & 0.2952 \\ 0.2944 & 0.3111 & -1.8805 & 0.8070 & 0.2744 & 0.2594 \\ 0.2910 & 0,2714 & 0.7547 & -1.9315 & 0.2848 & 0.2577 \\ 0.3022 & 0.2949 & 0.2357 & 0.2505 & -1.9757 & 0.7701 \\ 0.3155 & 0.2941 & 0.2461 & 0.2577 & 0.8847 & -2.1405 \end{pmatrix}$$

and $a = \frac{1}{2}\left(\hat{A} + \hat{A}^{\mathrm{T}}\right)$ satisfies the conditions (5.4.41). Then the equilibrium state $x = 0$ of the system is asymptotically stable. In the above mentioned paper by Shaaban and Grujić [164] the asymptotic stability of the equilibrium state was established for $\lambda = 3.0$ and $\varepsilon_{i1} = 0.95$, $i = 1, 2, \dots, 6$. This applies to the smaller values of λ and ε as well as to the asymptotic stability of the equilibrium state $x = 0$.

The application of the approach to three–four and seven-machine system enables us to conclude as follows (see Grujić and Shaaban [61]):

(1) We can decrease the value of the parameter λ for which asymptotic stability of $x = 0$ of the system is assured (value of λ is decreased from 100 to only 0.3 for the three-machine system, and decreased by 33% of that in Shaaban and Grujić [164] for the four and seven machine systems). Noting that the smaller value of λ means that the generator is less damped and that it is more difficult to assure stability, we can deduce that the developed approach is more powerful then those developed so far via vector Liapunov functions.

(2) Smaller value of the parameter ε can be assumed and the asymptotic stability assured by applying the developed approach (value of ε is assumed to be 85% of that in Shaaban and Grujić [164] for the four and seven machine systems, and it is decreased from 0.10 to only 0.06 for the three-machine system). This essentially means that the developed approach can lead to larger asymptotic stability domain estimates.

(3) Using the developed approach, we can decrease the conservativeness of the decomposition-aggregation method.

(4) The matrix-valued Liapunov function methodology leads to more adequate scalar Liapunov functions for power systems and simplifies their construction via the vector Liapunov function concept.

(5) The stability test computation is reduced to only the negative definiteness test of a single elementwise constant aggregation symmetric matrix. Its dimension is reduced to the number $s = N - 1$ of the subsystems of an N-machines power system.

5.5 The Motion in Space of Winged Aircraft

According to Aminov and Sirazetdinov [2] we will consider the case when the aircraft, moving with fixed absolute value of the velocity, performs a manouvre with constant load factor. Thus, to the undistrturbed motion there corresponds constant values of the angles of attack α_0 and of side-slip β_o, and angular velocities of pitch ω_{z0}, yaw ω_{y0} and rotation ω_{x0}. Their deviations from the perturbed values will be called α, β, ω_z, ω_y, ω_x respectively. The deviations of the angular velocities of side-slip, yaw and rotation must not exceed given limits.

We consider the equations of the perturbed motion in the form (see Byushgens and Studnev [18])

$$
\begin{aligned}
\frac{d\alpha}{dt} &= \mu\omega_z - \frac{1}{2}c_y^{\alpha}\alpha - \mu\beta\omega_x - \frac{1}{2}c_y^{\delta_e}\delta_e, \\
\frac{d\omega_z}{dt} &= m_z^{\alpha}\alpha + m_x^{\omega_z}\omega_z - \mu A\omega_x\omega_y + m_z^{\delta_e}\delta_e, \\
\frac{d\beta}{dt} &= \mu\omega_y + \frac{1}{2}c_z^{\beta}\beta + \mu\alpha\omega_x + \frac{1}{2}c_z^{\delta_r}\delta_r, \\
\frac{d\omega_y}{dt} &= m_y^{\beta}\beta + m_y^{\omega_y}\omega_y + \mu B\omega_x\omega_z + m_y^{\delta_r}\delta_r, \\
\frac{d\omega_x}{dt} &= m_x^{\beta}\beta + m_x^{\omega_x}\omega_x - \mu C\omega_y\omega_z + m_x^{\delta_a}\delta_a,
\end{aligned}
\tag{5.5.1}
$$

where

$$
A = \frac{J_y - J_x}{J_z} > 0, \quad B = \frac{J_z - J_x}{J_y} > 0, \quad C = \frac{J_z - J_y}{J_x} > 0,
$$

and μ is the aircraft relative density, c_u are the coefficients of the aerodynamic forces, m_u are the coefficients of the aerodynamic moments, δ_e, δ_r, δ_a are the deviations of the elevator, aileron and rudder, and J_x, J_y, J_z are the aircraft moments of inertia with respect to the connected coordinate system.

We take the law of stabilization in the form

$$
\begin{aligned}
\delta_e = k_e^{\alpha}\alpha + k_e^{z}\omega_z, \qquad \delta_r = k_r^{\beta}\beta + k_r^{y}\omega_y, \\
\delta_a = k_a^{\beta}\beta + k_a^{x}\omega_x.
\end{aligned}
\tag{5.5.2 }
$$

We substitute the values (5.5.2) into equations (5.5.1). We use the notations

$$
(5.5.3)\qquad
\begin{aligned}
&x_1 = \omega_x, \quad x_2 = \omega_y, \quad x_3 = \omega_z, \quad x_4 = \alpha, \quad x_5 = \beta,\\
&a_{11} = m_x^\beta + k_a^\beta m_x^{\delta_a}, \qquad a_{15} = m_x^{\omega_x} + k_a^x m_x^{\delta_a},\\
&a_{22} = m_y^\beta + k_r^\beta m_y^{\delta_r}, \qquad a_{25} = m_y^{\omega_y} + k_r^y m_y^{\delta_r},\\
&a_{33} = m_z^\alpha + k_e^\alpha m_z^{\delta_e}, \qquad a_{34} = m_z^{\omega_z} + k_e^z m_z^{\delta_e},\\
&a_{44} = \frac{1}{2}\left(c_y^\alpha + k_e^\alpha c_y^{\delta_e}\right), \quad a_{43} = \mu - \frac{1}{2}k_e^z c_y^{\delta_e},\\
&a_{55} = \frac{1}{2}\left(c_z^\beta + k_r^\beta c_z^{\delta_r}\right), \quad a_{52} = \mu + \frac{1}{2}k_r^y c_z^{\delta_r},\\
&b_1 = -\mu C, \quad b_2 = \mu B, \quad b_3 = -\mu A, \quad b_4 = -\mu, \quad b_5 = \mu.
\end{aligned}
$$

Using this notation we can write system (5.5.1) as

$$
(5.5.4)\qquad
\begin{aligned}
\frac{dx_1}{dt} &= a_{11}x_1 + a_{15}x_5 + b_1x_2x_3,\\
\frac{dx_2}{dt} &= a_{22}x_2 + a_{25}x_5 + b_2x_1x_3.\\
\frac{dx_3}{dt} &= a_{33}x_3 + a_{34}x_4 + b_3x_1x_2,\\
\frac{dx_4}{dt} &= a_{43}x_3 + a_{44}x_4 + b_4x_1x_5,\\
\frac{dx_5}{dt} &= a_{52}x_2 + a_{55}x_5 + b_5x_1x_4.
\end{aligned}
$$

We shall find the conditions connected to the coefficients of the system (5.5.4) under which the solution of the system $x = 0$ is multistability, i.e., asymptotically stable with respect to (x_4, x_5), and stable with respect to (x_1, x_2, x_3).

We use the Theorem 2.6.1. In our example $N = 2$, i.e., there are two groups of variables (x_1, x_2, x_3) and (x_4, x_5). We consider the matrix-valued Liapunov function

$$
U(x) = \frac{1}{2}\,\mathrm{diag}\,[-b_2b_3x_1^2,\ 2b_1b_3x_2^2,\ -b_1b_2x_3^2,\ x_4^2,\ x_5^2],
$$

and $\eta \in R_+^5$, $\eta_i = 1$, $i = 1, 2, \dots, 5$.

The function

$$
(5.5.5)\qquad \eta^{\mathrm{T}}U(x)\eta = V(x,\eta) = \frac{1}{2}\left(-b_2b_3x_1^2 + 2b_1b_3x_2^2 - b_1b_2x_3^2 + x_4^2 + x_5^2\right)
$$

is positive definite, decreasing and radially unbounded. In view of the system (5.5.4) the derivative of the function (5.5.5) is

$$\begin{aligned} DV(x,\eta) = &- b_2b_3a_{11}x_1^2 - b_2b_3a_{15}x_1x_5 + 2b_1b_3a_{22}x_2^2 \\ &+ (2b_1b_3a_{25} + a_{52})x_2x_5 - b_1b_2a_{33}x_3^2 \\ &+ (a_{43} - b_1b_2a_{34})x_3x_4 + a_{44}x_4^2 + a_{55}x_5^2. \end{aligned} \tag{5.5.6}$$

In order to solve our problem we have to find the conditions whereby function (5.5.6) is non-positive with respect to (x_1, x_2, x_3) and negative definite with respect to (x_4, x_5).

The method of finding these conditions is given by Aminov and Sirazetdinov [3] and is as follows. We equate the derivative $DV(x, \eta)$ of (5.5.6) to the function

$$\begin{aligned} W(x) = &- (c_{11}x_1 + c_{15}x_5)^2 - (c_{22}x_2 + c_{25}x_5)^2 \\ &- (c_{33}x_3 + c_{34}x_4)^2 - (c_4x_4)^2 - (c_5x_5)^2 \end{aligned} \tag{5.5.7}$$

and, comparing coefficient of like terms of (5.5.6) and (5.5.7), we find the conditions for the existence of the coefficients of function (5.5.7) which are in fact the required conditions for the function (5.5.6) to be non-positive with respect to (x_1, x_2, x_3) and negative definite with respect to (x_4, x_5). These conditions are

$$\begin{gathered} a_{11} < 0, \quad a_{22} < 0, \quad a_{33} < 0, \quad a_{44} + \frac{(a_{43} - b_1b_2a_{34})^2}{b_1b_2a_{33}} < 0, \\ a_{55} + \frac{a_{15}^2b_2b_3}{a_{11}} - \frac{(2b_1b_3a_{25} + a_{52})^2}{2b_1b_3a_{22}} < 0. \end{gathered} \tag{5.5.8}$$

On substituting the values of the coefficients (5.5.3) into inequality (5.5.8) we obtain the sufficient conditions that solve the aircraft space manouvre problem.

5.6 Notes

5.2. The basic result of this section (Proposition 5.2.1) is new. The description of model and the competition discussion is due to Lakshmikantham, Leela and Martynyuk [94]. For the large number of references on this topic see Freedman [36]. The application of the Metzler matrix theory and vector

Liapunov functions in the investigation of thise problems is due to Šiljak [167], Grujić and Burgat [56], etc.

5.3. The description of the model of an orbital astronomical observatory is taken from Geiss, Cohen et al. [40] and Grujić [55]. The results of investigation of this model are cited following Krapivny supervised by A. A. Martynyuk. The comparison of the obtained results with those by Grujić, Martynyuk and Ribbens-Pavella [57] has displayed the advantages of the matrix-valued function application. For other results on the subject see Šiljak [167], Abdullin, Anapolskii et al [1], etc.

5.4. The results of this section are due to Grujić and Shaaban [61]. The scalar Liapunov functions are applied by El-Abiad and Nagappan [35], Michel, Fouad and Vittal [142]. For the application of vector Liapunov functions see Pai and Narayana [151], Grujić, Martynyuk and Ribbens-Pavella [57], Grujić and Ribbens-Pavella [58], [59], Grujić, Ribbens-Pavella and Bouffioux [60], Jocić, Ribbens-Pavella and Šiljak [79], Michel, Nam and Vittal [144], Shaaban and Grujić [164], [165], etc. Matrix-valued Liapunov functions are applied by Miladzhanov [145] including the systems with structural perturbations.

5.5. The results of this section are due to Martynyuk [111] and Aminov and Sirazetdinov [2].

REFERENCES

1. Abdullin, P.Z., Anapolskii, L.Yu., etc, *Method of Vector Lyapunov Functions in Stability Theory*, Nauka, Moscow, 1987. (Russian)
2. Aminov, A.B. and Sirazetdinov, T.K., *The method of Liapunov functions in problems of multistability of motion*, Prikl. Math. Mekh. **51** (1987), 553–558. (Russian)
3. ______, *Lyapunov functions for studying the stability in the large of nonlinear systems*, Prikl. Math. Mekh. **49** (1985), 883–893. (Russian)
4. Antosiewicz, H.A., *A survey of Liapunov's second method.*, in Contribution to the Theory of Nonlinear Oscillations (S.Lefschetz, ed.), vol. 4, Princeton University Press, 1958, pp. 141–166.
5. Arnold, L., *Stochastic Differential Equations: Theory and Applications*, Wiley, New York, 1974.
6. Azimov, R.K., *Aggregation form of stochastic system in terms of matrix Liapunov functions*, Dokl. Akad. Nauk Ukraine, Ser. A (1991), no. 11, 5–8. (Russian)
7. ______, *Stability Analysis of Stochastic Systems by Matrix Liapunov Functions.* Thesis of Diss. paper, Inst. of Mech. Ac. of Sci. of Ukr., Kiev, 1993. (Russian)
8. Azimov, R.K. and Martynyuk, A.A., *Stability Analysis of Stochastic System by the Method of Matrix Liapunov Functions*, Preprint Inst. of Math. AN UkrSSR, Kiev, 1990. (Russian)
9. Bailey, F.N., *The application of Lyapunov's second method to interconnected systems*, J. Soc. Ind. Appl. Math. Ser. A **3** (1965), 443–462.
10. Barbashin, Ye.A., *The Liapunov Functions*, Nauka, Moscow, 1970. (Russian)
11. ______, *Introduction to the Stability Theory*, Nauka, Moscow, 1967. (Russian)
12. Barbashin, Ye.A. and Krasovskii, N.N., *On the stability of motion in the large*, Dokl. Akad. Nauk SSSR **86** (1952), 453–456. (Russian)
13. ______, *On the existence of Liapunov functions in the case of asymptotic stability in the whole*, Prikl. Math. Mekh. **18** (1954), 345–350. (Russian)
14. Barnett, S. and Storey, C., *Matrix Methods in Stability Theory*, Nelson, London, 1970.
15. Bellman, R., *Stability Theory of Differential Equatioins*, Academic Press, New York, 1953.
16. ______, *Vector Lyapunov Functions*, SIAM Journ. Contr. Ser. A **1** (1962), 32–34.
17. Bhatia, N.P. and Szego, G.P., *Dynamical Systems: Stability Theory and Applications*, Springer-Verlag, Berlin, 1967.
18. Byushgens, G.S. and Studnev, R.V., *Dynamics of the Space Motion of Aircraft*, Mashinostroenie, Moscow, 1967. (Russian)

19. Chetaev, N.G., *Stability of Motion*, Nauka, Moscow, 1990. (Russian)

20. Coddington, E.A. and Levinson, N., *Theory of Ordinary Diffferential Equations*, McGraw Hill, New York, 1955.

21. Corduneanu, C., *Applications of differential inequalities to stability theory*, An. Sti. Univ. "Al i Cuza", Iasi Sect. 1 a Mat. **6** (1960), 47–58. (Russian)

22. ———, *Sur la stabilité partielle*, Revue Roumaine de Math. Pures et Appl. **9** (1964), 229–236.

23. Coppel, W.A., *Stability and Asymptotic Behaviour of Differential Equations*, Heath and Co., Boston, 1965.

24. Demidovich, B.P., *Lectures on the Mathematical Theory of Stability*, Nauka, Moscow, 1967. (Russian)

25. Deo, S.G., *On vector Lyapunov functions*, Proc. Amer. Math. Soc. **29** (1971), 575–580.

26. Di Caprio, U. and Saccomano, F., *Application of the Lyapunov direct method to analysis of multimachine power system stability*, 3rd Power System Computation Conf., Rome, Italy, 1969.

27. Djordjević M.Z., *Stability analysis of interconnected systems with possibly unstable subsystems*, Systems and Control Letters **3** (1983), 165–169.

28. ———, *Zur Stabilitat Nichtlinearer Gekoppelter Systeme mit der Matrix-Ljapunov-Methode*, Diss. ETH **7690** (1984), Zurich.

29. ———, *Stability analysis of nonlinear systems by the matrix Lyapunov method*, Large Scale Systems **5** (1983), 252–262.

30. ———, *Stability analysis of nonlinear systems by the matrix Lyapunov method*, Proc. IMACS-IFAC Symposium, June 3-6, IDN (1986), Villeneuve d'Ascq, France, 209–212.

31. Doob, J.L., *Stochastic Processes*, Wiley, New York, 1953.

32. Duboshin, G.N., *Basis of Stability Motion Theory*, Moscow University Publishers, Moscow, 1952. (Russian)

33. Duhem, M.P., *Sur les conditions necessaires pour la stabilite' de l'e'quilibre d'an sisteme visqueux*, Comptes Rendus **126** (1902), 939–941.

34. Dynkin, Ye., *Markov Processes*, vol. 1, Springer-Verlag, Berlin, 1965.

35. El-Abiad, A.H. and Nagappan, K., *Transient stability regions of multimachine power systems*, IEEE Trans. PAS-85 **2** (1966), 169–178.

36. Freedman, H.I., *Deterministic Mathematical Models in Population Ecology*, HIFR Consulting Ltd, Edmonton, 1987.

37. Freedman, H.I. and Martynyuk, A.A., *Stability analysis with respect to two measures for population growth models of Kolmogorov type*, Nonlin. Analysis **25** (1995), 1221–1230.

38. Gantmacher, F.R., *The Theory of Matrices*, Nauka, Moscow, 1966.

39. Gelig, A. Kh., Leonov, G.A. and Yakubovich, V. A., *Stability of Nonlinear Systems with a Nonunique Equilibrium State*, Nauka, Moscow, 1978. (Russian)

40. Geiss, G.R., Cohen, V.D., D'heedene, R., Rotschild, D. and Chomas, A., *An Algorithm for Lyapunov Stability Analysis of Complex Nonlinear System with Application to the Orbiting Astronomical Observatory*, National Aeronautics and Administration Space: NASA CR-1979, Washington D.C., 1971.

41. Geraschenko, E.I. and Geraschenko, S.M., *Method of Motion Decomposition and Optimization of Non-Linear Systems*, Nauka, Moscow, 1975. (Russian)

42. Gikhman, I.I. and Skorokhod, A.V., *Stochastic Differential Equations*, Springer-Verlag, Berlin, 1972.

43. Gradshtein, U.S., *Application of Liapunov's stability theory to the theory of differential equations with small multipliers in the derivative terms*, Dokl. AN SSSR **81** (1951), 985–986. (Russian)

44. Grujić, Lj.T., *Uniform asymptotic stability of dyscrete large scale systems*, IEEE Trans. on System, Man and Cybernetics, November (1973), 636–643.

45. ______, *Novel development of Liapunov stability of motion*, Int. J. Control **22** (1975), 525–549.

46. ______, *General stability analysis of large-scale systems*, Proc. IFAC Symp. on Large-Scale Systems Theory and Applications, 1976, pp. 203–213.

47. ______, *On large-scale systems stability*, Proc. 12th World Congress IMACS, vol. 1, 1988, pp. 224–229.

48. ______, *Vector Liapunov functions and singularly perturbed large-scale systems*, Proc. JACC (1976), 408–416.

49. ______, *Stability and instability of product sets*, Systems Science **3** (1977), 14–31.

50. ______, *Singular perturbations, uniform asymptotic stability and large-scale systems*, Proc. JACC (1978), 339–347.

51. ______, *Absolute stability of nonstationary systems: resolutions and applications*, Proc. JACC (1978), 327–337.

52. ______, *Singular perturbations and large-scale systems*, Int. J. Control **29** (1979), 159–169.

53. ______, *Necessary and sufficient Lyapunov-type conditions for absolute stability and the Aizerman conjecture*, Math. Fiz. **28** (1980), 7–19.

54. ______, *Uniform asymptotic stability of non-linear singularly perturbed general and large-scale systems*, Int. J. Control **33** (1981), 481–504.

55. ______, *Stabilnost Velikih Sistema*, Masinski Fakultet Univerzitet u Beogradu, Beograd, 1974.

56. Grujić, Lj.T. and Burgat, C., *Stability Analysis of Large Scale Generalized Lotka-Volterra systems*, Ricerche di Automatica **9** (1980), 161–170.

57. Grujić, Lj.T., Martynyuk, A.A. and Ribbens-Pavella, M., *Large Scale Systems Stability under Structural and Singular Perturbations*, Springer Verlag, Berlin, 1987.

58. Grujić, Lj.T. and Ribbens-Pavella, M., *New approach to stability domain estimate of large-scale power systems*, Revue E **VIII** (1977), no. 10, 241–249.

59. ______, *Relaxed large-scale systems stability analysis applied to power systems*, 7th World IFAC Congress, 1978, pp. 27–34.

60. Grujić, Lj.T., Ribbens-Pavella, M. and Bouffioux, A., *Asymptotic stability of large-scale systems with application to power systems. Part 1: Domain estimation*, Electrical Power and Energy Systems **1** (1979), 151–157.

61. Grujić, Lj.T. and Shaaban, H., *On transient Stability Analysis of Large-Scale Systems*, Proc. 12th World Congress IMACS, vol. 1, 1988, pp. 283–288.

62. Grujić, Lj.T. and Šiljak D.D., *Stability of large-scale systems with stable and unstable subsystems*, Proc. 1972 JACC Paper 17-3, Stanford University, 1972, pp. 550–555.

63. ______, *Asymptotic stability and instability of large-scale systems*, IEEE Trans. on Aut. Control **AC-18** (1973), no. 6, 636–645.

64. Habets, P., *Stabilite Asymptotique pour des Problemes de Perturbations Singulieres*, Centro Internazionalle Mathematico Estivo, Ed. Cremonese, Roma, 1974.

65. ______, *A consistency theory of singular perturbations of differential equations*, SIAM J. Appl. Math. **26** (1974), 136–152.

66. Hahn, W., *Stability of Motion*, Springer-Verlag, Berlin, 1967.

67. Halanay, A., *Differential Equations*, Academic Press, New York, 1966.

68. Harris, C.J. and Miles, J.F., *Stability of Linear Systems*, Academic Press 1980, New York.

69. Hartman, P., *Ordinary Differential Equations*, John Wiley and Sons, New York, 1964.

70. Hatvani, L., *On the asymptotic stability by nondecrescent Ljapunov function*, Nonlinear Analysis **8** (1984), 66–77.

71. ______, *On application of differential inequalities in the theory of stability*, Vestnik Moscow Univ. Ser 1 Math. Mekh. **3** (1975), 83–88. (Russian)

72. He, J.X. and Wang, M.S., *Remarks on exponential stability by comparison functions of the same order of magnitude*, Ann. of Diff. Eqs. **7(4)** (1991), 409–414.

73. Hoppensteadt, F., *Singular perturbations on the infinite interval*, Trans. Amer. Math. Soc. **123** (1966), 521–535.

74. ______, *Stability in systems with parameter*, J. Math. Anal. Appl. **18** (1967), 129–134.

75. ______, *Asymptotic stability in singular perturbation problems*, J. Diff. Eq. **4** (1968), 350–358.

76. ______, *Asymptotic stability in singular perturbation problems. II: Problems having matched asymptotic expansion solutions*, J. Diff. Eq. **15** (1974), 510–521.

77. ______, *Analysis and Simulation of Chaotic Systems*, Springer-Verlag, Berlin, 1993.

78. Itô, K., *On stochastic differential equations*, Mem. Amer. Math. Soc. **4** (1951), 87–97.

79. Jocic, Lj.B., Ribbens-Pavella, M. and Šiljak, D.D., *On transient stability of multimachine power systems*, Proc. JACC (1977), 627–632.

80. Kalman, R.E. and Bertram, J.E., *Control system analysis and design via the "second method" of Liapunov*, I. Trans. of ASME: J. Basic Enf. **82** (1960), 371–393.

81. Kamke, E., *Zur Theorie der Systeme gewonlicher differentialgleichungen. II*, Acta mathematica **58** (1930), 57–85.

82. Katz, I.Ya. and Krasovskii, N.N., *On the stability of systems with random parameters*, Prikl. Math. Mekh. **24** (1960), 809–823. (Russian)

83. Khas'minskii, R.Z., *Stochastic Stability of Differential Equations*, Sijthoff and Nordhoff, Rockville, Md., 1980.

84. Klimushev, A.I. and Krasovskii, N.N., *Uniform asymptotic stability of systems of differential equations with a small parameter at the highest derivatives*, Prikl. Math. Mekh. **25** (1961), 680–690. (Russian)

85. ———, *Uniform asymptotic stability of systems of differential equations with a small parameter in the derivative terms*, J. Appl. Math. and Mech. **25** (1962), 1011–1025.

86. Kokotovic, P.V., O'Malley Jr. and Sannuti, P., *Singular Perturbation and Order Reduction in Control Theory - An Overview*, Prepr. 6th IFAC World Congress IC, vol. 51.3, 1975.

87. ———, *Singular perturbation and order reduction in control theory - An overview*, Automatica **12** (1976), 123–132.

88. Koksal, S., *Nonuniform stability properties and perturbing Liapunov functions*, Florida Tech. (1990), 1–8. (manuscript)

89. Krasovskii, N.N., *Certain Problems of the Theory of Stability of Motion*, Fizmatgiz, Moscow, 1959. (Russian)

90. Kushner, H.J., *Stochastic Stability and Control*, Academic Press, New York, 1967.

91. Ladde, G.S. and Lakshmikantham, V., *Random Differential Inequalities*, Academic Press, New York etc., 1980.

92. Lakshmikantham, V. and Martynyuk, A.A., *Some directions of the developments of Lyapunov direct method in the stability theory*, Prikl. Mekh. **28** (1992), no. 3, 3–13. (Russian)

93. Lakshmikantham, V. and Salvadori, L., *On Massera type converse theorem in terms of two different measures*, Bull. U.M. **13** (1976), 293–301.

94. Lakshmikantham, V., Leela, S. and Martynyuk, A.A., *Stability Analysis of Nonlinear Systems*, Marcel Dekker, Inc., New York, 1989.

95. Lakshmikantham, V., Leela, S. and Rama Mohana Rao, M., *New directions in the method of vector Lyapunov functions*, Nonlinear Analysis **16** (1991), 252–262.

96. Lakshmikantham V., Matrosov V.M. and Sivasundaram S., *Vector Lyapunov Function and Stability Analysis of Nonlinear Systems*, Kluver Academic Publishers, Dordrecht, etc., 1991.

97. LaSalle, J.P., *The Stability of Dynamical Systems*, SIAM, Philadelphia, Pennsylvania, 1976.

98. LaSalle, J.P. and Lefschetz, S., *Stability by Liapunov's Direct Method*, Academic Press, New York, 1961.

99. Letov, A.M., *Mathematical Theory of Control Processes*, Nauka, Moscow, 1981. (Russian)

100. Lefschetz, S., *Differential Equations: Geometric Theory*, Princeton University and The National University of Mexico, Interscience Publishers, Inc., New York, 1957.

101. Liapunov, A.M., *General Problem of Stability of Motion*, Math. Soc., Harkov, 1892, (Published in Collected Papers, 2, Ac. Sci. USSR. Moscow-Leningrad, 1956, 5–263). (Russian)

102. ———, *Investigation a special case of the problem of stability of motion*, Math. Sbor. **17** (1893), 253–333. (Russian)

103. Likhtarnikov, A.L. and Yakubovich, V.A., *Absolute stability of nonlinear systems*, Appendix, in Rezvan V., The Absolute Stability of Automatic Systems with Delay, Nauka, Moscow, 1983, pp. 287–355. (Russian)

104. Lur'e, A.I., *Einige Nichtlineare Problems aus der Theorie der Automatischen Regelung*, Mir, Moscow, 1951.

105. Lur'e, A.I. and Postnikov, V.N., *On the stability theory of control systems*, Prikl. Math. Mekh. **VIII** (1944), 246–248. (Russian)

106. Malkin, I.G., *On the question of the reciprocal Liapunov's theorem on asymptotic stability of control systems*, Avtom. i Telemeh. (1954), 188–191. (Russian)

107. ______, *Theory of Motion Stability*, Nauka, Moscow, 1966. (Russian)

108. Marachkov, V., *On a theorem on stability*, Bull. Soc. Phys.-Math. **12** (1940), Kazan, 171–174. (Russian)

109. Martynyuk, A.A., *The Lyapunov matrix-function*, Nonlin. Anal. **8** (1984), 1223–1226.

110. ______, *Extension of the state space of dynamical systems and the problem of stability*, Colloquia Mathematica Societaties Janos Bolyai 47. Differential Equations: Qualitative Theory, Szeged (Hungary), 1984, pp. 711–749.

111. ______, *On matrix Liapunov function and stability of motion*, Doklady AN USSR **249** (1985), no. 5, 59–63. (Russian)

112. ______, *On application of the Lyapunov matrix-functions in the theory of stability*, Nonlin. Analysis **9** (1985), 1495–1501.

113. ______, *Liapunov matrix-function and stability of theory*, Proc. IMACS-IFAC Symp., June 3–6, IND, Villeneuve d'Ascq, France, 1986, pp. 261–265.

114. ______, *Uniform asymptotic stability of a singularly perturbed system via the Liapunov matrix-function*, Nonlin. Analysis **11** (1987), 1–4.

115. ______, *A stochastic matrix-valued Liapunov function and its application*, Dokl. Akad. Nauk SSSR **299** (1988), 46–49. (Russian)

116. ______, *On the application of matrix Lyapunov functions in motion analysis of lamped and distributed parameter systems*, Teorijska i Primenjena Mehanika **14** (1988), 73–83.

117. ______, *A theorem on multistability*, Dokl. Akad. Nauk SSSR **318** (1991), 803–811. (Russian)

118. ______, *A new direction in the method of matrix Lyapunov functions*, Dokl. Akad. Nauk SSSR **319** (1991), 554–557. (Russian)

119. ______, *On multistability of motion with respect to some of the variables*, Russian Dokl. Akad. Nauk **324** (1992), 39–41. (Russian)

120. ______, *On matrix method of comparison in the theory of motion stability*, Prikl. Mekh. **29** (1993), 116–122.

121. ______, *Qualitative Analysis of Nonlinear Systems by the Method of Matrix Lyapunov Functions*, Rocky Mountain Journal of Mathematics **25** (1995), 397–415.

122. Martynyuk, V.A., *On the polystability of dynamical systems*, J. Math. Anal. Apll. **191** (1995), 466–472.

123. Martynyuk, A.A. and Gutovskii, R., *Integral Inequalities and Stability of Motion*, Naukova Dumka, Kiev, 1979. (Russian)

124. Martynyuk, A.A. and Krapivny, Yu.N., *On a way application of matrix Liapunov function in the theory stability of motion*, Prikl. Mekh. (1988), no. 6, 98–107. (Russian)

125. Martynyuk, A.A. and Miladzhanov, V.G., *Stability Analysis in Whole of Dynamical System via Matrix Lyapunov Function*, Preprint 87.62, Institute of Mathematics, Kiev, 1987. (Russian)

126. ———, *Absolute stability of a singularly perturbed Lur'e system and Lyapunov's matrix function*, Prikl. Mekh. **23** (1987), no. 9, 103–110. (Russian)

127. ———, *To the stability theory of singularly perturbed Lur'e systems*, Teorijska i Primenjena Mehanika **15** (1989), 75–82. (Russian)

128. ———, *Stability investigation of autonomous singularly perturbed systems on the basis of matrix Lyapunov function*, Diff. Uravn. **24** (1988), 416–424. (Russian)

129. Martynyuk, A.A. and Obolenskii, A.Yu., *Stability of Autonomous Wazewski systems*, Diff. Uravn. **16** (1980), 1392–1407. (Russian)

130. Massera, J.L., *On Liapunov's conditions of stability*, Ann. of Math. **50** (1949), 705–721.

131. ———, *Contribution to stability theory*, Ann. of Math. **64** (1956), 182–206.

132. Matrosov, V.M., *On stability of motion*, Prikl. Math. Mekh. **26** (1962), 885–895. (Russian)

133. ———, *To theory of stability of motion*, Prikl. Math. Mekh. **26** (1962), 992–1002. (Russian)

134. ———, *To theory of stability of motion, II*, Works of Kazan Aviation Institute (KAI), Math. Mekh. **80** (1963), 22–33. (Russian)

135. ———, *To theory of stability of motion, III*, Proceedings of Meshvuz. Conf. on Appl. Theory of stability and analytical mechanics, KAI, pp. 103–109. (Russian)

136. ———, *Vector Lyapunov functions in the analysis of interconnected systems*, Simposia Math. (Roma 1970), 1971, pp. 209–242.

137. Matrosov, V.M., Anapolsky, L.Yu. and Vasiljev, S.N., *Comparison Method in Mathematical System Theory*, Nauka, Novosibirsk, 1980. (Russian)

138. McShane, E.L., *Integration*, Princeton University Press, Princeton, 1944.

139. Melnikov, G.I., *Nonlinear stability of motion of the ship on course*, Vestn. Lengosuniver. **13** (1962), no. 13, 90–98. (Russian)

140. Michel, A.N., *Stability analysis of interconnected systems*, Berichte der Mathematisch-Statistischen Section (1973), no. 4, Graz, 1–107.

141. ———, *Stability analysis of interconnected systems*, J. SIAM Control **12** (1974), 554–579.

142. Michel, A.N., Fouad, A.A. and Vittal, V., *Power system transient stability using individual machine energy functions*, IEEE Trans. on Circuits and Systems **CAS-3O** (1983), no. 5, 266–276.

143. Michel, A.N. and Miller, R.K., *Qualitative Analysis of Large Scale Dynamical Systems*, Acad. Press., New York, 1977.

144. Michel, A.N., Nam, B.H. and Vittal, V., *Computer generated Lyapunov functions for interconnected systems: Impruved results with application to power systems*, IEEE Trans on Circuits and Systems **CAS-31** (1984), no. 2, 189–199.

145. Miladzhanov, V.G., *Application of Matrix Lyapunov Function in the Investigation of Motion Stability of Systems with Slowly and Fast of Variables*, Author's Abstract of Kandidat Dissertation in Physicomathematical Sciences, Kiev, 1988. (Russian)

146. Moiseyev, N.D., *Essays on Development of the Theory of Stability*, GITTL, Moscow–Leningrad, 1949. (Russian)

147. Movchan, A.A., *Stability of processes with respect to two measures*, Prikl. Math. Mekh. **24** (1960), 988–1001. (Russian)

148. Narendra, K.S. and Taylor, J.H., *Frequancy Domain Criteria for Absolute Stability*, Academic Press, New York, 1973.

149. Novozhilov, I.V., *On Correctness of Rows Limiting Models in Mechanics*, Preprint 253, Institute of problems of mechanics AN USSR, Moscow, 1985. (Russian)

150. Olech, C. and Opial, Z., *Sur une inégalité différentielle*, Ann. Polon. Math. **7** (1960), 241–254.

151. Pai, M.A. and Narayana, C.I., *Stability of large scale power systems*, Proc. Sixth IFAC World Congres, Boston, Mass., 1975, pp. 1–10.

152. Persidskii, K.P., *On stability of motion at first approximation*, Math. Sbor. **40** (1933), 284–293. (Russian)

153. Poincaré, H., *Sur ler courbes definies par une equation differentielle*, J. de Math., serie 3 **7** (1881), 375–422; **8** (1882), 251–296.

154. Pontryagin, L.S., *Ordinary Differential Equations*, Nauka, Moscow, 1970. (Russian)

155. Popov, V.M., *On the absolute stability of nonlinear systems of automatic control*, Automat. Telemekh. **22** (1961), 961–979. (Russian)

156. Porter, B., *Singular perturbation methods in the design of stabilizing feedback controllers for multivariable linear systems*, Int. J. Control **20** (1974), 689–692.

157. ______, *Design of stabilizing feedback controllers for a class of multivariable linear systems with slow and fast modes*, Int. J. Control **23** (1976), 49–54.

158. ______, *Singular perturbation methods in the design of stabilizing state-feedback controllers for multivariable linear systems*, Int. J. Control **26** (1977), 583–587.

159. Rouche, N., Habets, P. and Laloy, M., *Stability Theory by Liapunov's Direct Method*, Springer-Verlag, New-York, 1977.

160. Rumiantsev, V.V., *On the stability of a motion in a part of variables*, Vestn. Moskov. Univ. Ser. 1 Math. Mekh. **4** (1957), no. 4, 9–16. (Russian)

161. Rumiantsev, V.V. and Oziraner, A.S., *Stability and Stabilization of Motion with Respect a Part of Variables*, Nauka, Moscow, 1987. (Russian)

162. Salvadori, L., *Some contributions to asymptotic stability theory*, Ann. Soc. Sci. **88** (1974), Bruxelles, 183–194.

163. Santalo, L.A., *Encyclopedia of Mathematics and its Applications, 1*, Addison–Wesley, Reading, Mass, 1976.

164. Shaaban, H. and Grujić, Lj.T., *Transient stability analysis of large-scale power systems with speed governor via vector Lyapunov functions*, IEE Proceedings **132** (1985), no. 2, 45–52.

165. ______, *The decomposition-aggregation method applied to a multimachine power system*, Large Scale Systems **10** (1986), no. 4, 115–132.

166. Šiljak, D.D., *Nonlinear Systems*, John Wiley and Sons, New York, 1969.

167. ______, *Large-Scale Dynamic Systems: Stability and Structure*, North Holland, New York, 1978.

168. ______, *Singular perturbation of absolute stability*, IEEE Trans. AC (1972), 720.

169. Tikhonov, A.N., *Systems of differential equations containing small parameters in the derivatives*, Math. Sbor. **73** (1952), no. 31, 576–586. (Russian)

170. Vasilyeva, A.B. and Butuzov, V.F., *Asymptotic Expansion of the Singularly Perturbed Equations Solutions*, Nauka, Moscow, 1973. (Russian)

171. Wazewski, T., *Systemes des équations et des inéqualitiés differentielles ordinaires aux deuxièmes membres monotones et leurs applicationes*, Annales de la societe Polonaise de mathématiques **23** (1950), 112–166.

172. Wilde, R.R. and Kokotovic, P., *Stability of singularly perturbed systems and networks with parasitics*, IEEE Trans. **AC-17** (1972), 245–246.

173. Yakubovich, V.A., *Absolute Stability of Nonlinear Regulating Systems in Critical Cases. 1*, Automatika i Telemechanika **24** (1963), 293–303. (Russian)

174. Yoshizawa, T., *Stability Theory by Liapunov's Second Method*, The Math. Soc. of Japan, 1966.

175. ———, *Stability Theory and the Existence of Periodic Solutions and Almost Periodic Solutions*, Springer-Verlag, New York–Heidelberg–Berlin, 1975.

176. Zien, L., *An upper bound for the singular parameter in a stable, singularly perturbed system*, J. Franklin Inst. **295** (1973), 373–381.

177. Zubov, V.I., *Methods of Liapunov and Their Applications*, P. Noordhoff Ltd., Groningen, 1964.

178. ———, *Mathematical Investigation of Automatic Control Systems*, Mashinostroeniye, Leningrad, 1979. (Russian)

179. ———, *Theory of Oscillations*, Vyschaya Sckola, Moscow, 1979. (Russian)

AUTHOR INDEX

SUBJECT INDEX